Corrosion and Mechanical Behavior of Metal Materials

Corrosion and Mechanical Behavior of Metal Materials

Editor

Ming Liu

MDPI • Basel • Beijing • Wuhan • Barcelona • Belgrade • Manchester • Tokyo • Cluj • Tianjin

Editor
Ming Liu
State Key Laboratory for
Mechanical Behavior of
Materials
Xi'an Jiaotong University
Xi'an
China

Editorial Office
MDPI
St. Alban-Anlage 66
4052 Basel, Switzerland

This is a reprint of articles from the Special Issue published online in the open access journal *Materials* (ISSN 1996-1944) (available at: www.mdpi.com/journal/materials/special_issues/ Corrosion_Mechanical_Behavior_Metal_Materials).

For citation purposes, cite each article independently as indicated on the article page online and as indicated below:

LastName, A.A.; LastName, B.B.; LastName, C.C. Article Title. *Journal Name* **Year**, *Volume Number*, Page Range.

ISBN 978-3-0365-7367-0 (Hbk)
ISBN 978-3-0365-7366-3 (PDF)

© 2023 by the authors. Articles in this book are Open Access and distributed under the Creative Commons Attribution (CC BY) license, which allows users to download, copy and build upon published articles, as long as the author and publisher are properly credited, which ensures maximum dissemination and a wider impact of our publications.
The book as a whole is distributed by MDPI under the terms and conditions of the Creative Commons license CC BY-NC-ND.

Contents

About the Editor vii

Lina Qiu, Dandan Zhao, Shujia Zheng, Aijun Gong, Zhipeng Liu and Yiran Su et al.
Inhibition Effect of *Pseudomonas stutzeri* on the Corrosion of X70 Pipeline Steel Caused by Sulfate-Reducing Bacteria
Reprinted from: *Materials* **2023**, *16*, 2896, doi:10.3390/ma16072896 1

Iva Betova, Martin Bojinov and Vasil Karastoyanov
Long-Term Oxidation of Zirconium Alloy in Simulated Nuclear Reactor Primary Coolant—Experiments and Modeling
Reprinted from: *Materials* **2023**, *16*, 2577, doi:10.3390/ma16072577 19

Danping Li, Wenwen Song, Junping Zhang, Chengxian Yin, Mifeng Zhao and Hongzhou Chao et al.
Corrosion Inhibition Mechanism of Ultra-High-Temperature Acidizing Corrosion Inhibitor for 2205 Duplex Stainless Steel
Reprinted from: *Materials* **2023**, *16*, 2358, doi:10.3390/ma16062358 37

Ruoqi Dang and Wenshan Yu
Standard Deviation Effect of Average Structure Descriptor on Grain Boundary Energy Prediction
Reprinted from: *Materials* **2023**, *16*, 1197, doi:10.3390/ma16031197 57

Ming Liu
Corrosion and Mechanical Behavior of Metal Materials
Reprinted from: *Materials* **2023**, *16*, 973, doi:10.3390/ma16030973 71

Yue Zhao, Botong Su, Xiaobo Fan, Yangguang Yuan and Yiyun Zhu
Corrosion Fatigue Degradation Characteristics of Galvanized and Galfan High-Strength Steel Wire
Reprinted from: *Materials* **2023**, *16*, 708, doi:10.3390/ma16020708 75

Yue Zhao, Xuelian Guo, Botong Su, Yamin Sun and Xiaolong Li
Evaluation of Flexible Central Buckles on Short Suspenders' Corrosion Fatigue Degradation on a Suspension Bridge under Traffic Load
Reprinted from: *Materials* **2022**, *16*, 290, doi:10.3390/ma16010290 99

Xin Gao and Ming Liu
Corrosion Behavior of High-Strength C71500 Copper-Nickel Alloy in Simulated Seawater with High Concentration of Sulfide
Reprinted from: *Materials* **2022**, *15*, 8513, doi:10.3390/ma15238513 117

Chenglong Li, Katharina Freiberg, Yuntong Tang, Stephanie Lippmann and Yongfu Zhu
Formation of Nanoscale Al_2O_3 Protective Layer by Preheating Treatment for Improving Corrosion Resistance of Dilute Fe-Al Alloys
Reprinted from: *Materials* **2022**, *15*, 7978, doi:10.3390/ma15227978 139

Evgeniy Merson, Vitaliy Poluyanov, Pavel Myagkikh, Dmitri Merson and Alexei Vinogradov
Effect of Air Storage on Stress Corrosion Cracking of ZK60 Alloy Induced by Preliminary Immersion in NaCl-Based Corrosion Solution
Reprinted from: *Materials* **2022**, *15*, 7862, doi:10.3390/ma15217862 149

Guoqiang Ma, Man Zhao, Song Xiang, Wanquan Zhu, Guilin Wu and Xinping Mao
Effect of the Severe Plastic Deformation on the Corrosion Resistance of a Tantalum–Tungsten Alloy
Reprinted from: *Materials* **2022**, *15*, 7806, doi:10.3390/ma15217806 . . . . **171**

Xiaoxiao Liu, Wenbin Zhang, Peng Sun and Ming Liu
Time-Dependent Seismic Fragility of Typical Concrete Girder Bridges under Chloride-Induced Corrosion
Reprinted from: *Materials* **2022**, *15*, 5020, doi:10.3390/ma15145020 . . . . **181**

About the Editor

Ming Liu

Dr. Ming Liu is now an associate professor at MSE of XJTU. He received his PhD at USTB under the supervision of Prof. Xiaogang Li in January 2017. He was a postdoctoral fellow in TU Delft from October 2019 to October 2021 under the supervision of Prof. J.M.C. Mol.

Currently, he is mainly engaged in the research of "Corrosion and mechanical behavior of metal materials". He is responsible for many grants from NSFC, International Postdoctoral Exchange Program, Shaanxi Natural Science Foundation, Postdoctoral Project and enterprise projects. He has published many high level papers and has been cited more than 1000 times.

Article

Inhibition Effect of *Pseudomonas stutzeri* on the Corrosion of X70 Pipeline Steel Caused by Sulfate-Reducing Bacteria

Lina Qiu [1,2], Dandan Zhao [1], Shujia Zheng [1], Aijun Gong [1,2,*], Zhipeng Liu [1], Yiran Su [1] and Ziyi Liu [1]

1. School of Chemistry and Biological Engineering, University of Science and Technology Beijing, Beijing 100083, China
2. Beijing Key Laboratory for Science and Application of Functional Molecular and Crystalline Materials, University of Science and Technology Beijing, Beijing 100083, China
* Correspondence: gongaijun5661@ustb.edu.cn

Abstract: Microbiologically influenced corrosion (MIC) is a common phenomenon in water treatment, shipping, construction, marine and other industries. Sulfate-reducing bacteria (SRB) often lead to MIC. In this paper, a strain of *Pseudomonas stutzeri* (*P. stutzeri*) with the ability to inhibit SRB corrosion is isolated from the soil through enrichment culture. *P. stutzeri* is a short, rod-shaped, white and transparent colony with denitrification ability. Our 16SrDNA sequencing results verify the properties of *P. stutzeri* strains. The growth conditions of *P. stutzeri* bacteria and SRB are similar, and the optimal culture conditions are about 30 °C, pH 7, and the stable stage is reached in about seven days. The bacteria can coexist in the same growth environment. Using the weight loss method, electrochemical experiments and composition analysis techniques we found that *P. stutzeri* can inhibit the corrosion of X70 steel by SRB at 20~40 °C, pH 6~8. Furthermore, long-term tests at 3, 6 and 9 months reveal that *P. stutzeri* can effectively inhibit the corrosion of X70 steel caused by SRB.

Keywords: microbiologically influenced corrosion inhibition; sulfate-reducing bacteria; *Pseudomonas stutzeri*; X70 steel

1. Introduction

Metal corrosion is a spontaneous chemical, electrochemical or biological process that destroys metal and its physical and chemical properties [1]. It affects global security in areas as diverse as oil and gas transportation, offshore engineering equipment and water treatment [2–4]. Among all kinds of metal corrosion, microbially induced corrosion (MIC) accounts for about 20% of the total corrosion [5], and can cause economic losses of tens of billions of dollars every year [6]. Many large accidents have been directly or indirectly caused by MIC, such as the fire on the offshore drilling platform of Mexican oil giant PEMEX in 2015 [7], the Prudhoe Bay oil spill on the North Slope of Alaska [8] and a large explosion caused by a natural gas pipeline leak in Carlsbad, New Mexico [9]. Many types of microorganisms are involved in MIC, which can be divided into aerobic, anaerobic and facultative, according to different requirements for oxygen. These microorganisms generally form biofilms on the surface of metal materials simultaneously. Aerobic and facultative bacteria can consume O_2 in the system during their growth and provide local anaerobic growth conditions for anaerobic bacteria while growing [10]. Common corrosive microorganisms include bacteria, archaea and fungi, such as sulfate-reducing bacteria (SRB) [11], sulfate-reducing archaea (SRA) [12], nitrate-reducing bacteria (NRB) [13], methanogenic bacteria [14], acid-producing bacteria (APB) [15], iron-reducing bacteria (IRB) [16], iron-oxidizing bacteria (IOB) [17], sulfur-oxidizing bacteria (SOB) [18], manganese-oxidizing bacteria (MOB) [19], a variety fungi [20] etc. Because sulfate is widely distributed in many systems, such as seawater, brackish water and agricultural runoff water [21], SRB have been extensively studied for decades as major pathogenic microorganisms in microbial systems. Traditional anti-corrosion measures such as coating, surface treatment and corrosion inhibitors can control most corrosion problems, but inevitably bring environmental problems because of the accumulated release of toxic substances [6,22,23].

Microbiologically influenced corrosion inhibition (MICI) is caused by microorganisms' direct or indirect actions [6]. Compared with traditional anti-corrosion methods, MICI is characterized by low cost, environmental friendliness as well as moderate application and maintenance requirements. Therefore, it has ecological and economic significance and will become the development direction of the next generation of preservative technology. However, due to the diversity of microorganisms and the complexity of their metabolic processes influenced by environmental factors, MICI is still facing the challenge of practical application. Therefore, the research on MICI in China and abroad is still in the initial stage. In addition, some problems remain to be solved, such as single species and a lack of systematic understanding of the preservative mechanism.

Pseudomonas stutzeri is a Gram-negative facultative anaerobic bacterium with strong denitrification activity and high denitrification capacity. It is widely used in biological denitrification [24–27]. Liu et al. showed that the inhibition of corrosion of deposit-covered X80 pipeline steel was due to *P. stutzeri* in seawater containing CO_2 [28]. Fu et al. studied the effects of *P. stutzeri* on the biocorrosion of X80 pipeline steel for different nitrate and nitrite concentrations [29]. In this work, a strain of *P. stutzeri* with SRB corrosion resistance was isolated and purified from the soil by enrichment culture, and the anti-corrosion behavior of the bacterium against X70 steel in the presence of SRB was investigated. This provides a theoretical basis for the further development of new green, low-cost and long-term novel SRB corrosion inhibition technologies.

2. Materials and Methods

2.1. Isolation and Identification of P. stutzeri

The liquid culture medium used for *P. stutzeri* was as follows (g/L) [25,30]: K_2HPO_4 (2.0), NH_4Cl (0.5), $Na_2S_2O_3 \cdot 5H_2O$ (5.0), $MgSO_4 \cdot 7H_2O$ (0.6), $FeSO_4 \cdot 7H_2O$ (0.01), KNO_3 (2.0), $NaHCO_3$ (1.0), $NaKC_4H_4O_6$ (20.0) and pH 1.0~7.2. Before inoculation, the liquid medium was sterilized by autoclaving at 120 °C for 20 min. Then, 1.5–2% agar was added to the solid medium.

The soil sample was weighed to 10 g, inoculated into a conical flask containing 100 mL liquid medium, and shaken for 10 min so that the soil sample was well-mixed and incubated at a constant temperature of 30 °C. After the upper layer of liquid was turbid, 5 mL of the upper layer of clear liquid was inoculated into 50 mL of the liquid medium in the sterile table, mixed well and continued to incubate at a constant temperature. When the bacterial fluid became turbid again, 0.2, 0.5, 1.0, 1.5 and 2.0 mL of bacterial fluid were taken to the sterile table, quickly added to sterilized Petri dishes and poured into sterilized warm solid media (the temperature was below 60 °C). The media and the bacterial solutions were mixed well by gentle shaking. After cooling and solidifying, the plates were inverted into the constant-temperature incubator at 30 °C for observation. After 7 days of incubation, scattered single colonies appeared on the surface and inside of the plates, and single colonies were separately picked and inoculated onto solid plates to continue the culture of the purified strains.

The single purified colonies were inoculated into sterilized test tubes with 10 mL of liquid medium and incubated at 30 °C. Once the bacterial solutions were turbid, the strains were preserved and set aside.

The identification of the strains was determined by microscopic observation, Giltay medium denitrification ability identification and 16S rDNA sequencing.

2.2. Growth Characteristics Experiment

The growth of bacteria has a greater impact on corrosion and anti-corrosion, and the corrosion effect of the strain is more evident at higher concentrations. For this experiment, the optical density at 600 nm (OD_{600}) values of the experimental strains under different conditions of time, temperature and pH were studied by using a spectrophotometer to determine the growth characteristics of the strains.

1. Effect of temperature on the growth of SRB and *P. stutzeri*

SRB and *P. stutzeri* strains were inoculated in a liquid medium and incubated anaerobically at 10 °C, 20 °C, 30 °C, 40 °C and 50 °C for 7 days. The OD_{600} values at each temperature were measured to determine the optimal growth temperature of the strains.

2. Effect of pH on the growth of SRB and *P. stutzeri*

The SRB and *P. stutzeri* strains were inoculated equally in a liquid medium at pH 1.0, 3.0, 5.0, 7.0, 9.0, 11.0 and 13.0 for 7 days. The OD_{600} values were measured at each pH to determine the optimal growth pH of the strains.

3. Graphical growth curves of *P. stutzeri*, SRB and co-culture of *P. stutzeri* + SRB

The media of *P. stutzeri* and SRB were prepared separately, as well as the mixed medium mixed at a ratio of 1:1. *P. stutzeri*, SRB, as well as the mixture of SRB and *P. stutzeri* were inoculated in a 1/100 ratio separately and incubated anaerobically in blue-capped culture flasks. Then the OD_{600} values were measured by sampling every 24 h.

All experiments were implemented in three replicates and results were based on the mean after three replicates with a standard deviation range of 0.5–5.0%.

2.3. Exposure Experiments

2.3.1. EDS and Weight Loss Measurement

Before soaking the specimens, holes were punched at the top middle, and each side was sanded to mirror-smooth using SiC sandpaper from 400 to 800, 1000 and 2000 mesh. Then, the specimens were washed with deionized water and anhydrous ethanol, blown dry and weighed and irradiated with UV light for at least 30 min. The steel sheets were suspended vertically inside the frosted jars with waterproof wires. The surface of the liquid was covered with liquid paraffin. After nitrogen was flushed into the bottle, the gap between the stopper and the bottle body was sealed with the sealing film to isolate the air. The experiments were divided into blank, SRB and SRB + *P. stutzeri*, with different control of incubation time, temperature, pH etc. All experiments were done three times in parallel. When the immersion time was reached, the steel sheets were removed and the surface was descaled according to the Chinese Standard "Corrosion of metals and alloys-Removal of corrosion products from corrosion test specimens" GB/T16545-2015.

The rust remover comprised 500 mL hydrochloric acid (ρ = 1.19 g/mL), 500 mL deionized water and 3.5 g of hexamethylene tetramine, well-mixed. After rust removal, samples were rinsed with deionized water and anhydrous ethanol, then dried and weighed. The inhibition effect of *P. stutzeri* bacteria on the corrosion caused by SRB under each condition was analyzed by the weight loss of steel sheets. The additional specimens were removed after soaking for 7, 14 and 21 days, before scanning X-ray energy-dispersive spectrometer (EDS, from Oxford Instruments, Abingdon, UK). For analysis of corrosion products, the specimens were pretreated by soaking in a phosphate-buffered saline solution containing 2.5% (v/v) glutaraldehyde for 8 h to immobilize cells. The specimens were then dehydrated using a serial dilution of ethanol (30%, 50%, 70%, 90% v/v), each for 15 min, except the final step for 30 min [31].

2.3.2. Electrochemical Measurements

The steel sheets were polished to smooth grade by grade, soldered with copper conductor, then put into the mold and sealed with epoxy resin, with ethylenediamine as the curing agent. Then, the samples were reserved with a working surface of 10 mm × 10 mm, polished with sandpaper in steps to 2000 mesh before use, and then polished with w 2.5 and w 0.5 diamond grinding paste in turn, rinsed with deionized water, cleaned with anhydrous ethanol, sterilized under UV for 30 min and set aside. A three-electrode system was used for the electrochemical tests, with X70 steel specimen as the working electrode, platinum sheet as the auxiliary electrode, and Saturated Calomel Electrode (SCE) as the reference electrode, using a Chi760e electrochemical workstation. Specimens were divided into two experimental groups using the co-culture medium—one group was inoculated with SRB only and the other group was inoculated with SRB + *P. stutzeri*. For every 100 mL of medium, 10 mL of bacterial solution was inoculated. The test scan potential range was ±0.5 V vs. OCP, and the potential scan rate was 2 mV·s^{-1}. The polarization curve (Tafel curve) of the steel sheet was measured. It is worth noting that potential scan rate has an essential role in minimizing the effects of distortion in Tafel slopes and corrosion current density analyses, as previously reported [32–34]. However, based on these reports, the adopted 2 mV/s has no deleterious effects on those Tafel extrapolations [32,33] to determine the corrosion current densities (i_{corr}) of the examined samples.

2.4. Long-Term Corrosion Experiments

Initially, nine plastic reagent bottles (500 mL) were placed in 500 mL co-culture medium and autoclaved for 30 min. The coupons of X70 pipeline steel (18.7 mm × 7.8 mm × 1 mm) were provided by School of Materials Science and Engineering, USTB. Three experimental groups were set up, and the inocula are shown in Table 1. The samples were sealed and placed at room temperature for 3, 6 and 9 months to observe the corrosion and corrosion inhibition over a long period of time.

Table 1. The inocula volumes of the three experimental groups.

	Control (mL)	SRB (mL)	*P. stutzeri* + SRB (mL)
Sterile water	5	----	----
SRB	----	5	5
P. stutzeri	----	----	5

3. Results and Discussion

3.1. Identification of P. stutzeri Strain

3.1.1. Observation of Colonies

After the collected soil samples were cultured in liquid and enriched several times with strains, the samples were isolated and purified by the spread plate method. The growth rate of colonies in the autotrophic medium was slow, and small beige or white colonies appeared on the plates after about 5 days of incubation. After 7 days of incubation, there was a significant difference between each soil sample plate's surface and internal colony morphology. The small yellowish translucent needle-tip colonies on the surface of the plates were *P. stutzeri*.

The single colonies of *P. stutzeri* were inoculated in the sterilized medium for more than three generations of enrichment. The acceleration voltage of SEM test was set to 3 kV and the emission current was set to 100 µA. Then, the purified strains were subjected to Gram staining and SEM observation, as shown in Figure 1. *P. stutzeri* stained dark red, indicating that they were Gram-negative bacteria, and the cells were short rods with a length between 0.5 and 1 µm. The Gram staining experiment also showed that the purified strains were relatively pure, and no other miscellaneous bacteria were mixed in.

3.1.2. Identification of the Denitrification Capacity of P. stutzeri

After incubation with the Giltay medium, *P. stutzeri* could be observed for its discoloration reaction. *P. stutzeri* can turn the color of Giltay liquid medium blue-green as they are autotrophic denitrifying bacteria. Under anaerobic conditions, *P. stutzeri* can produce nitrogen, so air bubbles (N_2) can be trapped in the Duchenne tubules. Without the presence of the denitrifying bacteria *P. stutzeri*, the Giltay medium was still dark green, but no air bubbles appeared in the Duchenne tubules. The results of the experiment are shown in Figure 2. *P. stutzeri* bacteria discolored the Giltay medium, and thus had a denitrification ability that could be used for subsequent experiments.

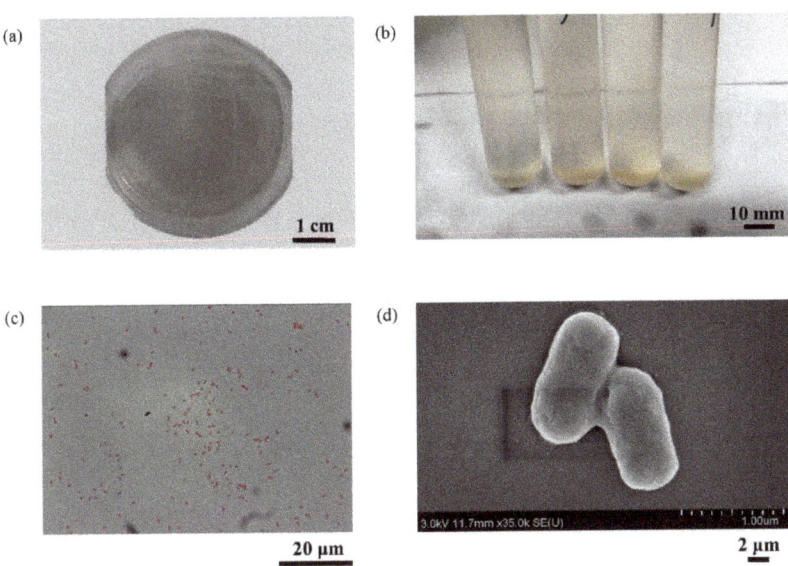

Figure 1. *P. stutzeri* colony (**a**); *P. stutzeri* liquid culture (**b**); *P. stutzeri* Gram staining (**c**); *P. stutzeri* SEM (**d**).

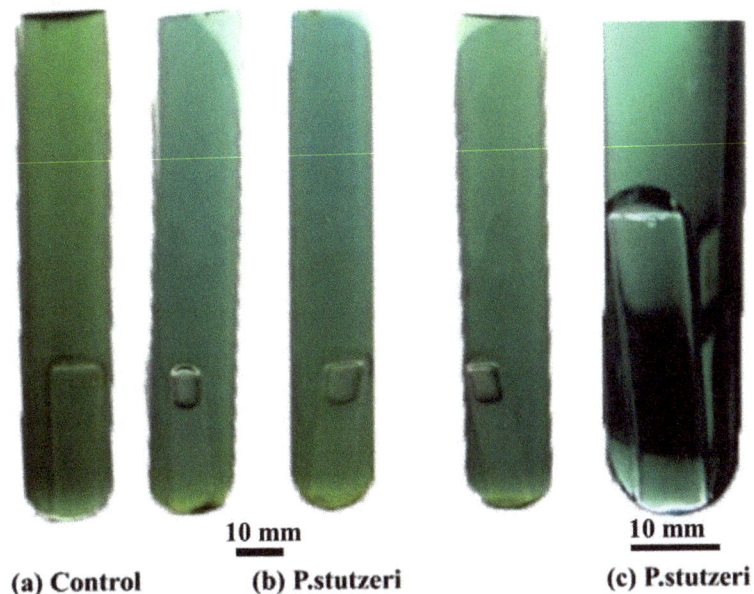

(a) Control   (b) P.stutzeri   (c) P.stutzeri

Figure 2. Identification of denitrifying bacteria using Giltay medium. (**a**) The blank Giltay medium; (**b**) medium for inoculation with *P. stutzeri*; (**c**) local amplification of (**b**).

3.1.3. Identification by 16S rDNA Sequencing

The sequencing of the strains was conducted by Sangon Biotech (Shanghai) Co., Ltd. (Shanghai, China). Then, the sequencing results were spliced with ContigExpress and the faulty parts at both ends were removed. Next, the spliced sequences were compared in the NCBI database (blast.ncbi.nlm.nih.gov) with the standard strains' rRNA type strains/16S_ribosomal_RNA database. After that, the species with the highest homology was selected and an evolutionary tree was constructed to confirm that the strain was *P. stutzeri*.

3.2. Growth Characteristics of P. stutzeri and SRB

3.2.1. Effect of Temperature on the Growth of the Strain

Activated SRB and *P. stutzeri* were inoculated into liquid medium at pH 7 and incubated at a temperature ranging from 10 to 50 °C. After 7 days, the OD_{600} values of the bacterial broths were measured and the results are shown in Figure 3.

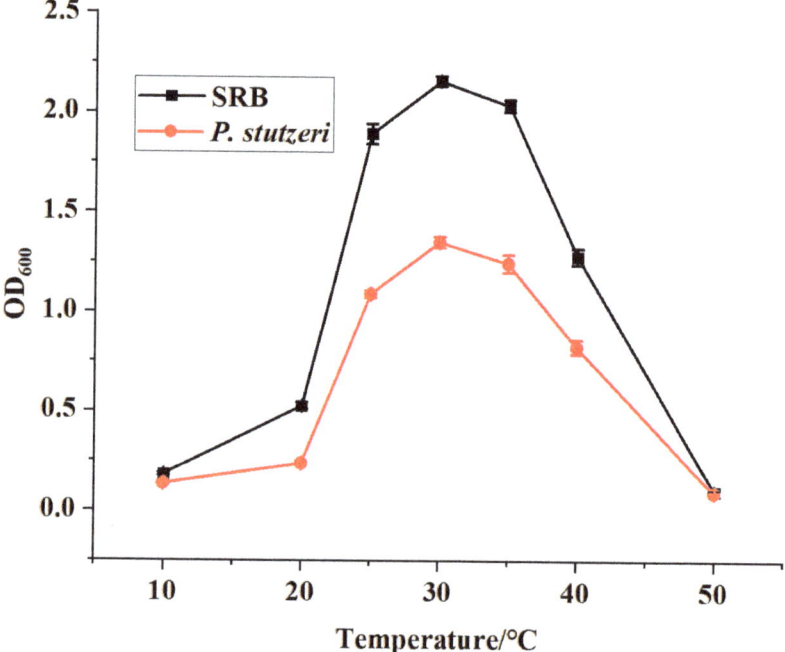

Figure 3. Effect of temperature on strain growth.

From Figure 3, it can be seen that the isolated and purified strains of SRB and *P. stutzeri* maintained high bacterial concentrations in the range of 25–40 °C. The absorbance of both SRB and *P. stutzeri* strains reached the maximum at around 30 °C. In this experiment, the optimum growth temperature of both SRB and *P. stutzeri* was around 30 °C. Thus, temperature affected the growth of SRB and *P. stutzeri* to a similar extent.

3.2.2. Effect of pH on the Growth of the Strain

After the activation, SRB and *P. stutzeri* were respectively inoculated into a liquid medium (the incubation temperature was 30 °C) with pH ranging from 1 to 13 to determine the effect of pH on the growth of SRB and *P. stutzeri*. OD_{600} values of the bacterial solution were measured after 7 days, and the results are shown in Figure 4.

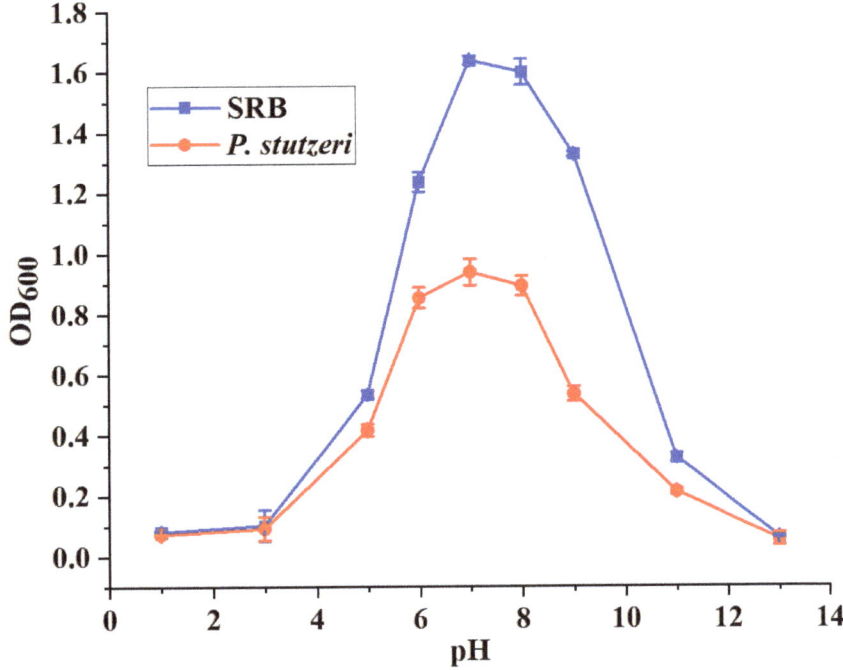

Figure 4. Effect of pH on strain growth.

As shown in Figure 4, the maximum concentration of both SRB and *P. stutzeri* was reached at pH 7. The experiment showed that both SRB and *P. stutzeri* were able to grow in a wide pH range from 5 to 8, and the optimum growth pH was about 7. Thus, the effect of environmental pH on the growth of SRB and *P. stutzeri* was similar. The above experiments provide a feasible basis for the coexistence of SRB and *P. stutzeri* in the same growth environment.

3.2.3. Graphical Growth Curves of *P. stutzeri*, SRB and Co-Culture of *P. stutzeri* + SRB

First, it was necessary to separately prepare the media of *P. stutzeri*, SRB, and mixed medium (mixed in the ratio of 1:1). Then, *P. stutzeri*, SRB and SRB + *P. stutzeri* mixture were inoculated into the above media at a 1/100 ratio and incubated in blue-capped culture flasks under anaerobic conditions at 30 °C constant temperature. After two days, samples were taken every 24 h and analyzed for OD_{600} values. Based on the results, the growth curves of SRB, *P. stutzeri* and co-culture of *P. stutzeri* + SRB were plotted, and the results are shown in Figure 5.

From Figure 5, it can be seen that SRB reached the logarithmic growth phase on the third day and entered the stable phase after 8 days. However, after 12 days, SRB gradually began to enter the decay phase due to nutrient depletion, which changed the optimal growth environment of SRB. In contrast, *P. stutzeri* entered the stable phase after 5 days and slowly started to enter the decay phase after 11 days. As a result, it can be concluded that the growth cycles of SRB and *P. stutzeri* are consistent.

When SRB was co-cultured with *P. stutzeri*, the growth of SRB was not inhibited by *P. stutzeri* in the first 2 days; the concentrations of the mixed bacteria on the third and fourth days were greater than those of the two bacteria in separate cultures; after the fifth day, the concentrations of the two bacteria in co-culture were higher than those of *P. stutzeri*, but lower than those of SRB in separate cultures. It can be tentatively assumed that the growth of SRB was inhibited by *P. stutzeri* when the two bacteria were co-cultured together.

This is likely because when the two bacteria were co-cultured in the same medium, they grew competitively as the medium substrate was consumed over a longer incubation time. Namely, possible factors inhibiting the growth of SRB by *P. stutzeri* include their substrate utilization or their secretions.

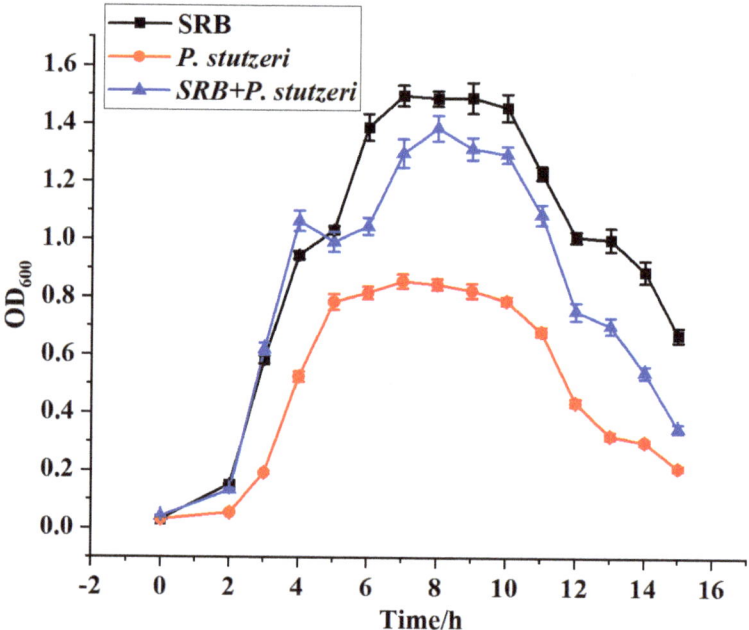

Figure 5. Growth curves of SRB, *P. stutzeri* and co-culture of *P. stutzeri* and SRB.

3.3. *Corrosion Inhibition of X70 Pipeline Steel*

3.3.1. Effect of Temperature on the Inhibition of SRB Corrosion by *P. stutzeri*

With the temperature change, there was a fairly obvious change in the weight loss of the samples, and the results are shown in Figure 6.

The weight loss was relatively low when the temperature was between 10 °C and 20 °C. With the increase in temperature, the weight loss increased significantly. The corrosion of the steel sheets was more severe in the system in which only SRB was inoculated at 30 °C, while at 50 °C severe corrosion was noted in all three systems. From the analysis, the main factors were identified: most of the microorganisms had a suitable growth temperature of 20–35 °C—the higher the temperature, the stronger the microbial activity. If the external temperature were reduced or increased, the growth of the microorganisms would be affected to a certain extent. At temperatures lower than 20 °C, the metabolic activity of microorganisms (e.g., related enzymes) would be inhibited; at temperatures as low as 10 °C, the growth rate of microorganisms would be significantly reduced and they would basically be in a dormant state, making the corrosion of X70 steel relatively weak. The corrosion under these conditions would mainly be based on electrochemical corrosion, and since the temperature is low, the electrochemical corrosion would also be relatively low. With the increase in temperature, the activity of bacteria increased. Since the weight loss of steel sheets in the medium only inoculated with SRB was more apparent between 30 and 40 °C, it can be concluded that SRB grew faster and were more active.

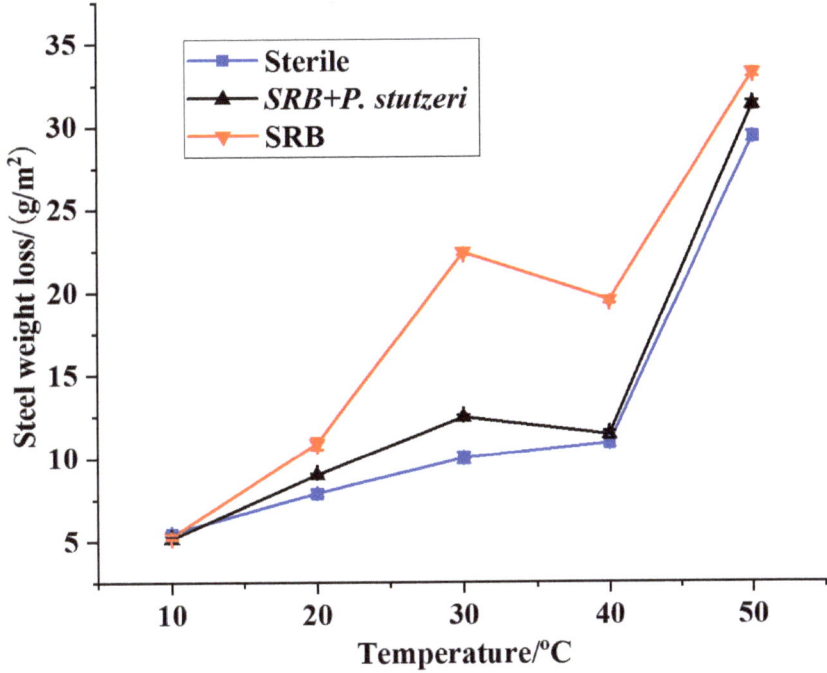

Figure 6. Weight loss curves (7 days) of X70 steel under different temperature conditions.

Meanwhile, there was a slight trend of increased weight loss of steel sheets in the blank medium, mainly due to electrochemical corrosion. In the system inoculated with *P. stutzeri* + SRB, the weight loss of steel sheets was greatly reduced compared with the system inoculated with SRB only, indicating that *P. stutzeri* reduced the corrosion of X70 steel by SRB. When the temperature reached 40 °C, the growth of bacteria and the activity of enzymes would be somewhat inhibited by the high temperature. Hence, the corrosion induced by SRB was weakened, and the electrochemical corrosion was enhanced. Moreover, when the temperature reached 50 °C, as the temperature was far too high, the activity of bacteria was reduced by the influence of temperature. However, this high temperature promoted electrochemical corrosion, which caused more severe corrosion and a greater loss of weight. Thus, it can be concluded that *P. stutzeri* may have a good inhibitory effect on the corrosion of steel sheets caused by SRB in a wide temperature range.

3.3.2. Effect of Time on the Inhibition of SRB Corrosion by *P. stutzeri*

There was a gradual weight loss per area of the steel sheets as the incubation time increased (the incubation temperature was 30 °C), as shown in Figure 7.

The corrosion loss of the steel sheets inoculated with SRB was much greater than that observed in the other two tested systems. The 30-day loss was 8.57 g/m^2, measured from the steel sheets in the blank group, which was mainly due to electrochemical corrosion caused by certain salts in the incubation medium. The corrosion rate of X70 steel in the test system inoculated with SRB alone was greater, which reached 15.67 g/m^2 at 10 days, 19.26 g/m^2 at 15 days, 21.87 g/m^2 at 20 days and 22.57 g/m^2 at 30 days. The growth curve of SRB bacteria showed four typical phases: retardation phase (0–2 days), log phase (3–6 days), stabilization phase (7–11 days) and decay phase (>12 days). In the range of 0–5 days, SRB were in the activation period after inoculation, and the number of active bacteria was small. Therefore, the corrosion was mainly caused by chemical corrosion; in days 5–10, the number of SRB increased drastically, and the corrosion enhanced. Notably, after 7 days, the number of strains reached the maximum, the number of active bacteria was high, both enzyme activation and metabolism were at the peak, and the corrosion also intensified. Although the number of bacteria remained at a high state in the period of 10–15 days, the weight loss was lower than that in the 5–10 days period, probably because the growth of high quantities of bacteria led to the formation of biofilms on the surface of the steel sheets, which slowed the rate of increase of corrosion. Furthermore, at 20–30 days, corrosion was significantly delayed due to the large consumption of nutrients at this stage, while the toxic metabolites accumulated in large quantities, leading to an increase in SRB mortality and a decrease in viable bacteria. At the same time, the chemical corrosion also slowed because inorganic substances were consumed during their growth. In that case, it can also play a role in mitigating the corrosion on the surface layer as the corrosion product film became thicker and its combination with the microbial film was denser, which led to the corrosion into a slower stage after 15 days. The weight loss of steel sheets in the medium inoculated with SRB + *P. stutzeri* was 11.52 g/m^2 after 30 d immersion experiment. During the 0–30 days period, the weight loss in the mixed bacteria culture was significantly lower than that of SRB alone; between 15 and 30 days, the weight loss was basically at a stable stage. It was assumed that in the co-culture of *P. stutzeri* + SRB, the presence of *P. stutzeri* influenced the growth of SRB and slowed the corrosion of steel sheets by SRB, while on the other hand, the bacteria, extracellular polymers and corrosion products were mixed to generate a dense biofilm, which played a protective role on the steel sheets. Thus, it can be noted that *P. stutzeri* has a better protective effect on the corrosion of steel sheets caused by SRB with the extension of time.

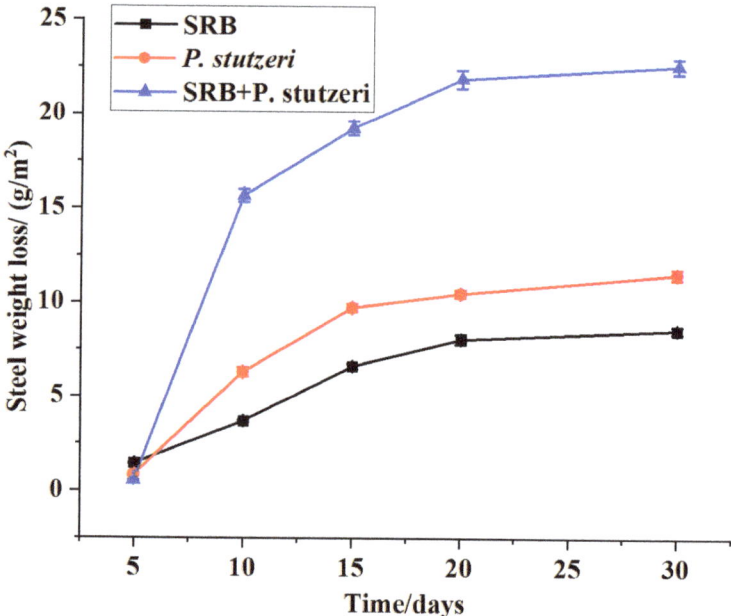

Figure 7. Weight loss curves of X70 steel under different test times.

3.3.3. Effect of Initial pH on the Inhibition of SRB Corrosion by *P. stutzeri*

The rules of electrochemical corrosion of steel affected by pH were as follows: when pH < 4, the corrosion of carbon steel was severe, mainly caused by hydrogen evolution corrosion; at pH 5–13, the corrosion was slower, mainly caused by oxygen absorption corrosion. From Figure 8, it can be seen that when pH 3, steel corrosion was more serious. At this stage, the role of microorganisms was not prominent; the bacteria in the acidic medium grew slowly in general, so it the main corrosion mechanism was chemical hydrogen precipitation corrosion due to acidic conditions, and the rate of X70 corrosion was higher. With increased pH, electrochemical corrosion weakened, and microbial corrosion increased. At pH 6–8, conditions were more suitable for microbial growth, and SRB and chemistry caused the corrosion. At pH 7, the corrosion caused by SRB was more severe, and the weight loss reached 39.31 g/m^2, but the weight loss in the combined system of *P. stutzeri* + SRB was weakened to 12.33 g/m^2; meanwhile, in the blank medium system, the weight loss was 9.89 g/m^2, which was mainly due to chemical oxygen absorption corrosion. Discounting the electrochemical corrosion, the addition of *P. stutzeri* resulted in an 83% reduction in weight loss. Therefore, *P. stutzeri* provided better protection against steel sheet corrosion at pH 7. At pH > 9, the corrosion was relatively weaker because both chemical and microbial corrosion were slowed down under alkaline conditions. Overall, the growth of microorganisms was somewhat restricted in alkaline environments, enzyme activity was reduced, and there was a relative decrease in both corrosion and corrosion resistance. Li et al. also showed that in an alkaline environment, a thick oxide film was generated on the surface layer of pipeline steel, which acted as a passivation layer and slowed the corrosion rate [35]. Therefore, the protection of *P. stutzeri* against SRB corrosion was better in the pH 6–8 environment.

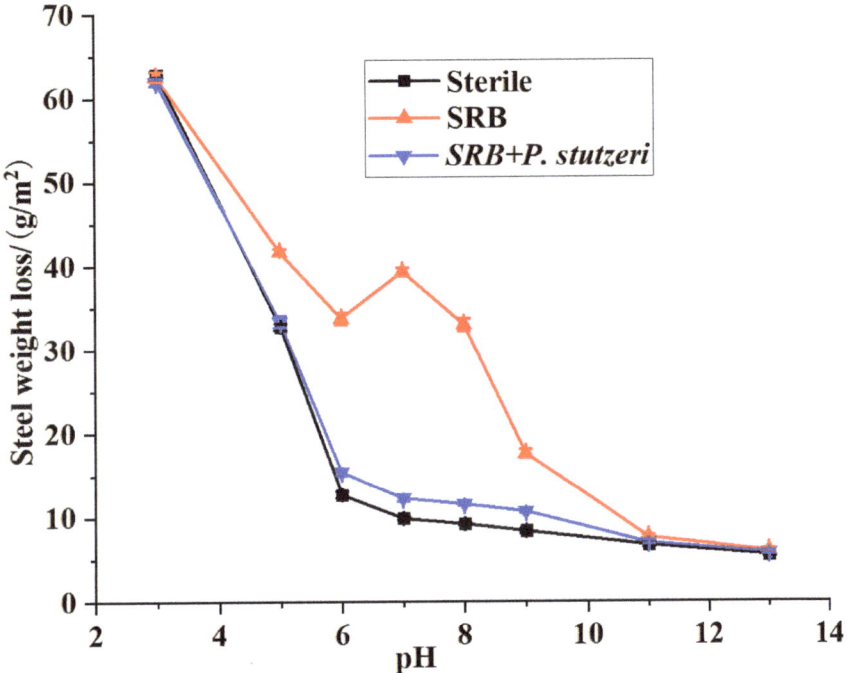

Figure 8. Weight loss curves of X70 steel under different pH.

3.3.4. Analysis of Elements in Corrosion Products

From the analysis of corrosion products (Figure 9), it can be seen that the corrosion products in the sterile environment were mainly with inorganic compounds, such as iron oxides, and the P elements were mainly derived from K$_2$HPO$_4$ in the medium. The content of Fe and S in the corrosion products inoculated with SRB was significantly higher than that in the sterile environment and in the environment with *P. stutzeri*; presumably, the corrosion products mainly contained sulfides and iron oxides. In the environment with mixed bacteria, the corrosion products mainly contained C, O, Fe and S, but the content of S was significantly lower than that of corrosion products inoculated with SRB. The oxidation of sulfatide induced by *P. stutzeri* hindered the accumulation of corrosive sulfide, and the SRB-involved corrosion can be weakened thereby.

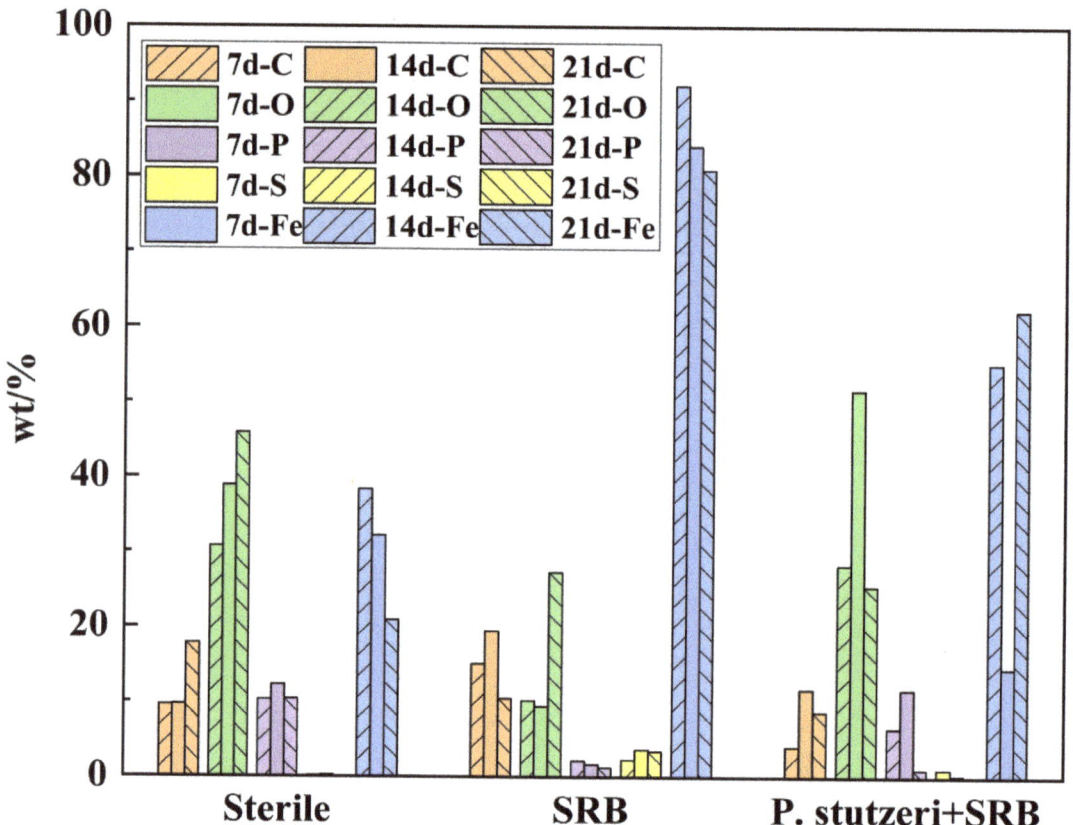

Figure 9. Element content of corrosion products on steel coupon surface.

3.3.5. Tafel Curve

The polarization curve test is damaging, and hence it is also known as a disposable test, where irreversible damage is caused to the sample surface during the measurement. Because the Tafel polarization can damage the sample surface, which can effectively exclude interference, the relationship between the polarization current and electrode potential of the respective anodic and cathodic reactions can be observed separately, rendering the polarization curves capable of revealing the mechanism of electrochemical reactions and their kinetic characteristics in depth.

The polarization curves of X70 steel samples in SRB and SRB + *P. stutzeri* solutions from 0 to 14 days of inoculation were measured with activated strains. In addition, the corrosion current density (i_{corr}) was obtained after fitting using extrapolation software (Origin 2018), and the variation trend is shown in Figure 10.

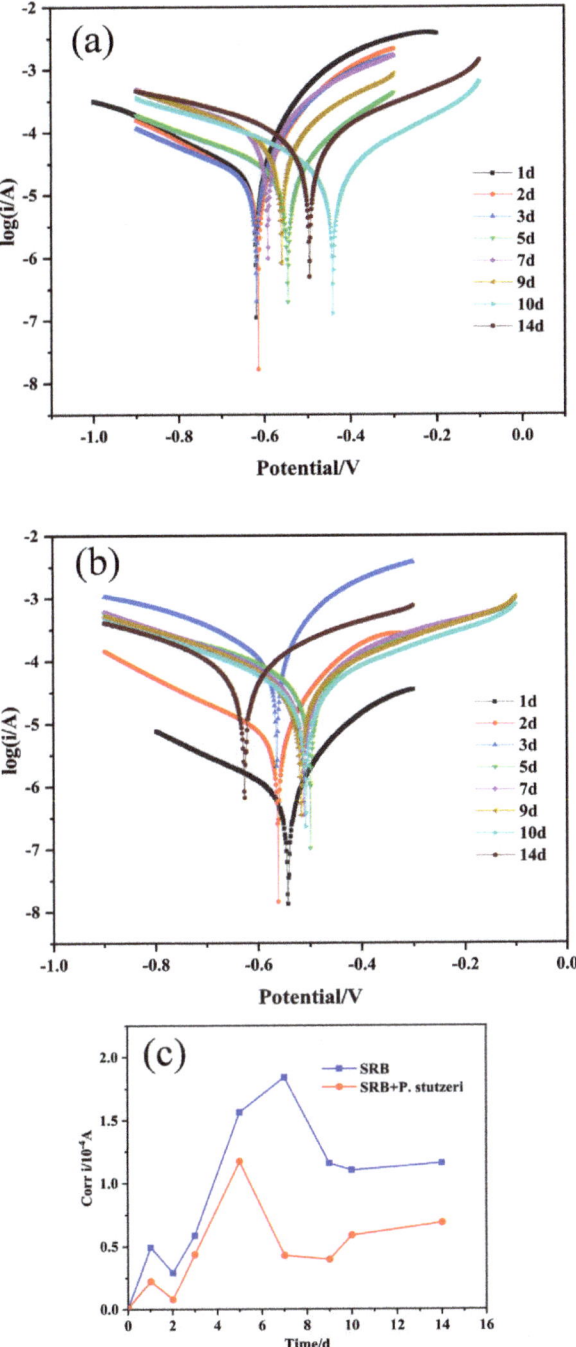

Figure 10. (**a**) The potentiodynamic polarization curves of SRB; (**b**) the potentiodynamic polarization curves of SRB + *P. stutzeri*; (**c**) the corrosion current density.

Figure 10a,b are the corrosion polarization curves for SRB and SRB + *P. stutzeri* samples, respectively, and Figure 10c shows the corrosion current density in both media. As can be seen from Figure 10, the corrosion current density in both media was the same at the beginning of the experiment, and with the extension of the incubation time, the corrosion current density of X70 steel in the media inoculated with SRB only reached the maximum at 7 days. At this stage, the growth rate of SRB was faster and the number of strains increased, while corrosion intensified. In the next 7 days, there was a decrease in corrosion current density compared to the seventh day owing to the formation of corrosion products on the surface, which had a protective effect on the specimens. As can be seen from the test results of 14 days of inoculation, after 2 days, the corrosion current density of X70 steel in the mixed medium of SRB + *P. stutzeri* was always lower than that of the medium inoculated with SRB alone, which indicated that the addition of *P. stutzeri* had a certain inhibitory effect on the corrosion caused by SRB. In brief, the corrosion current density generally showed a trend of first increasing and then decreasing, which means the corrosion was initially enhanced and then weakened.

3.3.6. Long-Term Exposure Weight Loss Test

The weight losses of X70 steel sheets exposed to media for 3, 6 and 9 months are shown in Figure 11.

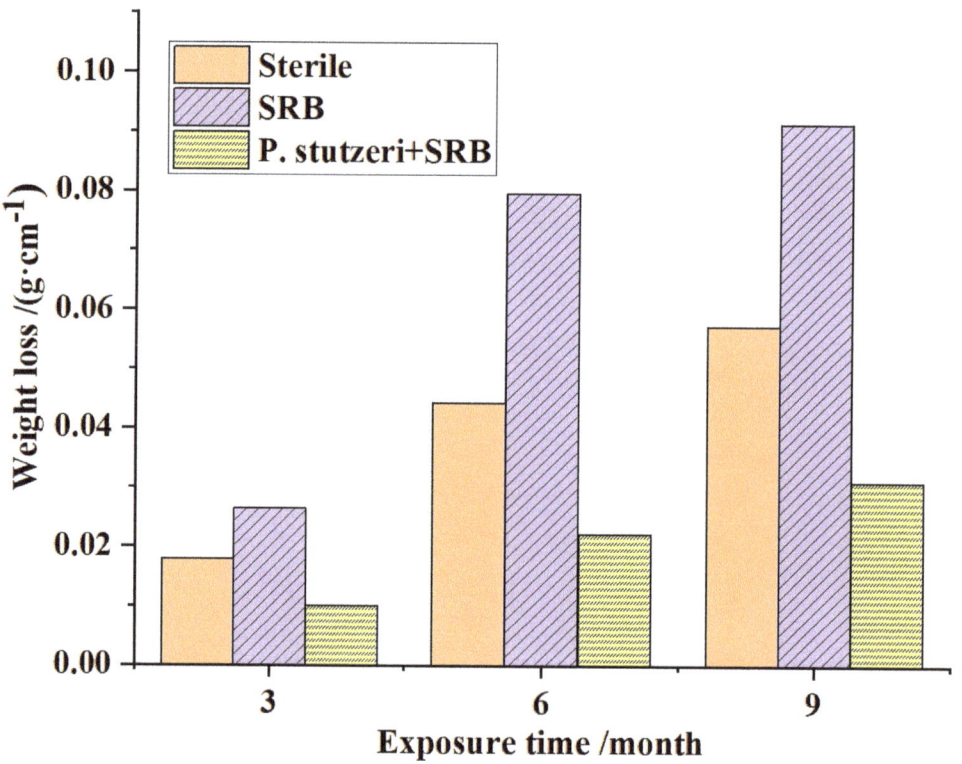

Figure 11. Weight losses of steel coupons from long-term exposure (3, 6, 9 months).

The weight loss of the steel sheets in all systems gradually increased with time, which indicates that the degree of corrosion varied in different environments. Still, as shown in Figure 11, the media inoculated with SRB alone showed more weight loss and more severe corrosion. The weight of specimens in the blank group lost slightly more than those inoculated with SRB + *P. stutzeri* because the blank medium had a higher electrolyte concentration than the solution inoculated with SRB + *P. stutzeri*, where bacteria grew and consumed part of the electrolytes. This is consistent with the results of the short-time corrosion weight loss experiments described above. In addition, the weight loss of steel sheets in the media inoculated with SRB + *P. stutzeri* was significantly lower than that in the system inoculated with SRB, which suggests that *P. stutzeri* had a better effect on preventing and controlling SRB-induced corrosion on steel sheets.

4. Conclusions

(1) A strain of *P. stutzeri* was obtained by isolation and purification. The strain is a Gram-negative bacterium with short rod-shaped cells, about 0.5–1 μm in length, and white transparent colonies, which could discolor Giltay medium and produce gas, showing that it has denitrification ability.

(2) The sequencing results of 16SrDNA also verified the properties of the *P. stutzeri* strain: its growth conditions were similar to those of SRB, with an optimum culture temperature of about 30 °C and an optimum growth pH of about 7. It entered the stabilization phase after 5 days and started to enter the decay phase slowly after 11 days. The growth cycle was the same as that of SRB, and the two bacteria can co-exist in the common growth environment.

(3) The results of the weight loss tests showed that *P. stutzeri* could inhibit the corrosion of X70 steel caused by SRB at 20–40 °C and pH 6–8. The electrochemical results showed that SRB promoted the corrosion of X70 steel, and the corrosion of X70 steel was inhibited in the *P. stutzeri* + SRB media. In addition, the corrosion current density of X70 steel in the media containing mixed bacteria was less than that of the environment inoculated with SRB alone, and the EDS results showed that the elemental S content was significantly lower than that of the corrosion products inoculated with SRB. The results of the weight loss tests also showed that *P. stutzeri* had a better inhibitory effect on the corrosion of X70 steel caused by SRB in the medium at 3, 6 and 9 months.

Author Contributions: Conceptualization, L.Q. and A.G.; methodology, L.Q. and D.Z.; software, Z.L. (Zhipeng Liu) and S.Z.; validation, Y.S., Z.L. (Ziyi Liu) and S.Z.; data curation, L.Q.; supervision, A.G.; project administration, A.G.; funding acquisition, A.G. and L.Q. All authors have read and agreed to the published version of the manuscript.

Funding: This research was funded by the Chinese National Natural Science Foundation (No. 51701016).

Institutional Review Board Statement: Not applicable.

Informed Consent Statement: Not applicable.

Data Availability Statement: Not applicable.

Acknowledgments: This work was financially supported by the Chinese National Natural Science Foundation (No. 51701016).

Conflicts of Interest: The authors declare no conflict of interest.

References

1. Enning, D.; Garrelfs, J. Corrosion of iron by sufate-reducing bacteria, new views of an old problem. *Appl. Environ. Microbiol.* **2014**, *80*, 1226–1236. [CrossRef] [PubMed]
2. Jogdeo, P.; Chai, R.; Shuyang, S.; Saballus, M.; Constancias, F.; Wijesinghe, S.L.; Thierry, D.; Blackwood, D.J.; McDougald, D.; Rice, S.A.; et al. Onset of microbial influenced corrosion (MIC) in stainless steel exposed to mixed species biofilms from equatorial seawater. *Electrochem. Soc.* **2017**, *164*, C532–C538. [CrossRef]
3. Liu, H.; Frank Cheng, Y. Mechanism of microbiologically influenced corrosion of X52 pipeline steel in a wet soil containing sulfate-reduced bacteria. *Electrochim. Acta* **2017**, *253*, 368–378. [CrossRef]
4. Sheng, X.; Ting, Y.-P. The influence of sulphate-reducing bacteria biofilm on the corrosion of stainless steel aisi 316. *Corros. Sci.* **2007**, *49*, 2159–2176. [CrossRef]

5. Fatah, M.C.; Ismail, M.C. Effects of sulphide ion on corrosion behaviour of X52 steel in simulated solution containing metabolic products species, a study pertaining to microbiologically influenced corrosion (MIC). *Corros. Eng. Sci. Technol.* **2013**, *48*, 211–220. [CrossRef]
6. Lou, Y.; Chang, W. Microbiologically influenced corrosion inhibition mechanisms in corrosion protection, A review. *Bioelectrochemistry* **2021**, *141*, 107883. [CrossRef]
7. Kannan, P.; Su, S.S. A review of characterization and quantification tools for microbiologically influenced corrosion in the oil and gas industry, current and future trends. *Ind. Eng. Chem. Res.* **2018**, *57*, 13895–13922. [CrossRef]
8. Jacobson, G.A. Corrosion at Prudhoe Bay, a lesson on the line. *Mater. Perform.* **2007**, *46*, 26–35.
9. Parthipan, P.; AlSalhi, M.S.; Devanesan, S.; Rajasekar, A. Evaluation of Syzygium aromaticum aqueous extract as an ecofriendly inhibitor for microbiologically influenced corrosion of carbon steel in oil reservoir environment. *Bioproc. Biosyst. Eng.* **2021**, *44*, 1441–1452. [CrossRef]
10. Li, Y.C.; Xu, D.K. Anaerobic microbiologically influenced corrosion mechanisms interpreted using bioenergetics and bioelectrochemistry, A review. *J. Mater. Sci. Technol.* **2018**, *34*, 1713–1718. [CrossRef]
11. Chan, K.Y.; Xu, L.C. Anaerobic electrochemical corrosion of mild steel in the presence of extracellular polymeric substances produced by a culture enriched in sulfate-reducing bacteria. *Environ. Sci. Technol.* **2002**, *36*, 1720–1727. [CrossRef]
12. San, N.O.; Nazir, H. Evaluation of microbiologically influenced corrosion inhibition on Ni-Co alloy coatings by Aeromonas salmonicida and Clavibacter michiganensis. *Corros. Sci.* **2012**, *65*, 113–118. [CrossRef]
13. Xu, D.; Li, Y. Laboratory investigation of microbiologically influenced corrosion of C1018 carbon steel by nitrate reducing bacterium Bacillus licheniformis. *Corros. Sci.* **2013**, *77*, 385–390. [CrossRef]
14. Tan, J.L.; Goh, P.C. Influence of H_2S-producing chemical species in culture medium and energy source starvation on carbon steel corrosion caused by methanogens. *Corros. Sci.* **2017**, *119*, 102–111. [CrossRef]
15. Xu, D.; Li, Y. Mechanistic modeling of biocorrosion caused by biofilms of sulfate reducing bacteria and acid producing bacteria. *Bioelectrochemistry* **2016**, *110*, 52–58. [CrossRef] [PubMed]
16. Duan, J.; Wu, S. Corrosion of carbon steel influenced by anaerobic biofilm in natural seawater. *Electrochim. Acta* **2008**, *54*, 22–28. [CrossRef]
17. Chandrasatheesh, C.; Jayapriya, J. Detection and analysis of microbiologically influenced corrosion of 316 L stainless steel with electrochemical noise technique. *Eng. Fail. Anal.* **2014**, *42*, 133–142. [CrossRef]
18. Okabe, S.; Odagiri, M. Succession of sulfur-oxidizing bacteria in the microbial community on corroding concrete in sewer systems. *Appl. Environ. Microbiol.* **2007**, *73*, 971–980. [CrossRef]
19. Ashassi-Sorkhabi, H.; Moradi-Haghighi, M. The effect of Pseudoxanthomonas sp. as manganese oxidizing bacterium on the corrosion behavior of carbon steel. *MSE* **2012**, *32*, 303–309. [CrossRef]
20. Ching, T.H.; Yoza, B.A. Biodegradation of biodiesel and microbiologically induced corrosion of 1018 steel by Moniliella wahieum Y12. *Int. Biodeterior. Biodegrad.* **2016**, *108*, 122–126. [CrossRef]
21. Jia, R.; Yang, D. Mitigation of the Desulfovibrio vulgaris biofilm using alkyldimethylbenzylammonium chloride enhanced by D-amino acids. *Int. Biodeterior. Biodegrad.* **2017**, *117*, 97–104. [CrossRef]
22. Zhang, F.; Ju, P. Self-healing mechanisms smart protective coatings, A review. *Corros. Sci.* **2018**, *144*, 74–88. [CrossRef]
23. Hao, X.; Chen, S. Antifouling and antibacterial behaviors of capsaicin-based pH responsive smart coatings in marine environments. *Mater. Sci. Eng. C* **2020**, *108*, 110361. [CrossRef] [PubMed]
24. Gunasekaran, G.; Chongdar, S.; Gaonkar, S.N.; Kumar, P. Influence of bacteria on fifilm formation inhibiting corrosion. *Corros. Sci.* **2004**, *46*, 1953–1967. [CrossRef]
25. Rajasekar, A.; Ting, Y.P. Role of Inorganic and Organic Medium in the Corrosion Behavior of Bacillus megaterium and Pseudomonas sp. in, Stainless Steel SS 304. *Ind. Eng. Chem. Res.* **2011**, *50*, 12534–12541. [CrossRef]
26. Jia, R.; Yang, D. Microbiologically influenced corrosion of C1018 carbon steel by nitrate reducing Pseudomonas aeruginosa biofilm under organic carbon starvation. *Corros. Sci.* **2017**, *127*, 19–46. [CrossRef]
27. Liu, H.; Chen, W. Characterizations of the biomineralization film caused by marine Pseudomonas stutzeri and its mechanistic effects on X80 pipeline steel corrosion. *J. Mater. Sci. Technol.* **2022**, *125*, 15–28. [CrossRef]
28. Liu, H.; Jin, Z.; Wang, Z.; Liu, H.; Meng, G.; Liu, H. Corrosion inhibition of deposit-covered X80 pipeline steel in seawater containing Pseudomonas stutzeri. *Bioelectrochemistry* **2023**, *149*, 108279. [CrossRef]
29. Fu, Q.; Xu, J.; Wei, B.; Qin, Q.; Bai, Y.; Yu, C.; Sun, C. Mechanistic diversity between nitrate and nitrite on biocorrosion of X80 pipeline steel caused by Desulfovibrio desulfurican and Pseudomonas stutzeri. *Corros. Sci.* **2022**, *207*, 110573. [CrossRef]
30. Yang, X.; Wang, S.; Zhou, L. Effect of Carbon Source, C/N Ratio, Nitrate and dissolved oxygen concentration on nitrite and ammonium production from denitrification process by *Pseudomonas stutzeri* D6. *Bioresour. Tchnol.* **2012**, *104*, 65–72. [CrossRef]
31. Lv, M.; Du, M.; Li, X.; Yue, Y.; Chen, X. Mechanism of microbiologically influenced corrosion of X65 steel in seawater containing sulfate-reducing bacteria and iron-oxidizing bacteria. *J. Mater. Res. Technol.* **2019**, *8*, 4066–4078. [CrossRef]
32. Duarte, T.; Meyer, Y.A.; Osório, W.R. The Holes of Zn Phosphate and Hot Dip Galvanizing on Electrochemical Behaviors of Multicoatings on Steel Substrates. *Metals* **2022**, *12*, 863. [CrossRef]
33. Osorio, W.R.; Cheung, N.; Peixoto, L.C.; Garcia, A. Corrsion resistance and mechanical properties of an Al 9wt%Si alloy treated by laser surface remelting. *Int. J. Electrochem. Sci.* **2009**, *4*, 820–831.

34. Zhang, X.L.; Jiang, Z.H.; Yao, Z.P.; Song, Y.; Wu, Z.D. Effects of scan rate on the potentiodynamic polarization curve obtained to determine the Tafel slopes and corrosion current density. *Corros. Sci.* **2009**, *51*, 581–587. [CrossRef]
35. Li, Y.; Qin, H. Corrosion Mechanism Research of X70 Pipeline Steel in Simulated Solution of Guangxi Soil. *J. Qinzhou Univ.* **2017**, *32*, 23–26.

Disclaimer/Publisher's Note: The statements, opinions and data contained in all publications are solely those of the individual author(s) and contributor(s) and not of MDPI and/or the editor(s). MDPI and/or the editor(s) disclaim responsibility for any injury to people or property resulting from any ideas, methods, instructions or products referred to in the content.

Article

Long-Term Oxidation of Zirconium Alloy in Simulated Nuclear Reactor Primary Coolant—Experiments and Modeling

Iva Betova [1], Martin Bojinov [2,*] and Vasil Karastoyanov [2]

[1] Institute of Electrochemistry and Energy Systems, Bulgarian Academy of Sciences, 1113 Sofia, Bulgaria
[2] Department of Physical Chemistry, University of Chemical Technology and Metallurgy, 1756 Sofia, Bulgaria
* Correspondence: martin@uctm.edu

Abstract: Oxidation of Zr-1%Nb fuel cladding alloy in simulated primary coolant of a pressurized water nuclear reactor is followed by in-situ electrochemical impedance spectroscopy. In-depth composition and thickness of the oxide are estimated by ex-situ analytical techniques. A kinetic model of the oxidation process featuring interfacial reactions of metal oxidation and water reduction, as well as electron and ion transport through the oxide governed by diffusion-migration, is parameterized by quantitative comparison to impedance data. The effects of compressive stress on diffusion and ionic space charge on migration of ionic point defects are introduced to rationalize the dependence of transport parameters on thickness (or oxidation time). The influence of ex-situ and in-situ hydrogen charging on kinetic and transport parameters is also studied.

Keywords: zirconium alloy; nuclear reactor primary coolant; electrochemical impedance spectroscopy; oxidation model; compressive stress; ionic space charge

Citation: Betova, I.; Bojinov, M.; Karastoyanov, V. Long-Term Oxidation of Zirconium Alloy in Simulated Nuclear Reactor Primary Coolant—Experiments and Modeling. *Materials* **2023**, *16*, 2577. https://doi.org/10.3390/ma16072577

Academic Editor: Ming Liu

Received: 17 February 2023
Revised: 22 March 2023
Accepted: 22 March 2023
Published: 24 March 2023

Copyright: © 2023 by the authors. Licensee MDPI, Basel, Switzerland. This article is an open access article distributed under the terms and conditions of the Creative Commons Attribution (CC BY) license (https:// creativecommons.org/licenses/by/ 4.0/).

1. Introduction

A typical pressurized water reactor (PWR) contains around 200 fuel assemblies, with each fuel assembly featuring hundreds of fuel rods. A fuel rod is composed of fuel pellets stacked inside a cladding tube [1,2], The mechanical integrity of the cladding is essential to nuclear safety because it provides the first barrier for fission products. The technical specifications used to choose PWR cladding material have long been: (1) high neutron transparency, i.e., low thermal neutron cross sections, (2) good thermal conductivity, (3) low creep rate, (4) good ultimate elongation, (5) sound mechanical properties, and (6) good corrosion resistance [1].

The trend towards more severe operating conditions for this type of structural material, induced by the need to extend the lifetime and increase fuel burnup, leads to a need for a detailed characterization of correlations between composition, microstructure, and morphology of such alloys, growth kinetics of oxide layers on them, and their susceptibility to localized corrosion [1]. Even if the qualitative picture of the oxidation process of Zr alloys in nuclear power plant coolants is well established, the intimate mechanism of the transfer of matter and charge through the oxide remains largely unclear. It is generally believed that the growth of the oxide proceeds according to the so-called coupled-currents mechanism [3]. According to that mechanism, transport of oxygen by a vacancy mechanism along grain boundaries of the already formed zirconium oxide is the rate-determining step of the overall reaction. It is assumed that water reduction at the oxide/coolant interface consumes electrons generated by metal oxidation, electronic conduction in the oxide proceeding along preferred pathways associated with secondary phase particles incorporated into the zirconium oxide matrix [1,4,5]. The extent of the coupling of the electron and ion fluxes is not quantitatively assessed.

Increasing the concentration of lithium hydroxide in PWRs with higher fuel burnup and fuel cycles longer than 12 months is necessary to maintain the primary coolant pH within an acceptable range. Higher lithium concentrations associated with supercooled condensate boiling can lead to crud formation on fuel cladding and reactor internals. In turn, crud formation induces axial power anomaly (AOA) and localized corrosion [6–9]. The mechanism of AOA is still unclear, but a number of studies have indicated a correlation between concentration of alkali, boric acid, zirconium alloy composition, and occurrence of this type of anomaly. The use of KOH in water cooled—water moderated energy reactor (WWER) coolants have considerable advantages due to its higher solubility, and accordingly, a weaker tendency towards crud formation on fuel cladding and internals. All this added complexity calls for the use of advanced in-situ methods for the characterization of oxidation processes of Zr alloys.

A promising non-destructive characterization technique is electrochemical impedance spectroscopy (EIS), which has been used to study zirconium alloys in WWER and PWR coolants for several decades now [10–18]. Despite numerous experimental data found in the literature, EIS has been used almost exclusively to assess oxide conductivity as depending on environmental conditions. Only a few deterministic models that describe quantitatively the whole impedance spectrum were advanced on the basis of the Point Defect Model (PDM) [13–15] and the Mixed-Conduction Model (MCM) [19,20].

The aim of the present paper is to further develop and test a deterministic model of growth of a protective oxide layer on a Zr-1%Nb alloy. The model features electrochemical processes at the alloy/oxide and oxide/coolant interfaces coupled via diffusion-migration of point defects in the oxide at the atomic and mesoscopic level. The model is based on the generalized quantitative model of growth of passive oxide films in high temperature electrolytes, the MCM, which has recently been adapted to interpret initial stages of corrosion of zirconium alloys in PWR and WWER primary coolants [19–21]. First, experimental data on the oxidation of the alloy in WWER coolant with or without LiOH are presented. Measurements on samples that were ex-situ and in-situ electrochemically charged with hydrogen are also discussed. The thickness and in-depth elemental composition of oxides is estimated by Glow Discharge Optical Emission Spectrometry (GDOES) and scanning electron microscopy (SEM) of oxide cross sections. Further, a new version of the model that considers the influence of compressive stress generated during growth on oxygen diffusion and space charge on the electric field strength in the oxide is described. This extension allows rationalization of the thickness (i.e., oxidation time) dependences of diffusion and migration parameters. Finally, the relative importance of rate constants, diffusion coefficients, and field strength for the kinetics of oxide growth and hydrogen pick-up is discussed.

2. Materials and Methods

Working electrodes were cut from Zr-1%Nb fuel cladding tubes. The chemical composition of the alloy is presented in Table 1. Pretreatment of electrodes consisted of mechanical polishing with emery paper grade 1200, chemical polishing in a mixture of 30% HNO_3 (70%) + 30% H_2SO_4 (96%) + 9% HF (50%) + 31% H_2O, and drying with hot air. Experiments were carried out using a three-electrode system with the alloy as working electrode, a Pt (99.9%, Goodfellow) sheet symmetrically arranged around it as a counter electrode, and a Pd (99.9%, Goodfellow) pseudo-reference electrode mounted in close proximity to the tube (at a distance of 2 mm). Pd was continuously polarized with a current of -30 µA against an additional Pt electrode in order to approximate the reversible hydrogen electrode (RHE). To simulate defective microstructure resulting from proton irradiation, samples were subjected either to preliminary cathodic polarization with a current density of -10 mA cm^{-2} in a 0.1 M KOH solution for 24 h at room temperature, galvanostatically (current density -1 mA cm^{-2}) or potentiostatically (at -2.0 V vs. RHE) in a beginning-of-cycle WWER primary coolant without LiOH (0.283 mmol kg^{-1} K as KOH and 0.13 mol kg^{-1} B as H_3BO_3) at 80 °C. As a result of this type of treatment, hydrogen atoms formed during water reduction are introduced into the alloy structure.

Table 1. Chemical composition of the studied material.

Element	Zr	Nb	Sn	Fe	N	C	O
Content/wt.%	Base	1.05	0.04	0.05	<0.01	0.02	0.10

Experiments were conducted in a nominal beginning-of-cycle WWER coolant with a composition of 0.283 mmol kg^{-1} (16 ppm) K as KOH, 0.13 mol kg^{-1} (1400 ppm) B as H_3BO_3, with and without the addition of 0.042 mmol kg^{-1} (1 ppm) Li (as LiOH). For its preparation, p.a. H_3BO_3, LiOH and KOH (Sigma Aldrich, St. Louis, MO, USA) were used. No NH_3 was added to the solutions in order to minimize the effect of dissolved H_2 (formed by decomposition of ammonia at high temperature) on oxidation kinetics. The conductivity of the coolant at room temperature was 37 ± 0.5 µS cm^{-1} (without LiOH) and 42 ± 0.5 µS cm^{-1} (with LiOH), whereas its pH was 6.1 ± 0.1 (without LiOH) and 6.2 ± 0.1 (with LiOH).

All experiments were performed in an autoclave made of 316 L stainless steel (Parr, volume 3.75 dm^3) connected to a laboratory made re-circulation loop. The respective electrolyte from a 20 dm^3 reservoir was continuously pumped through the autoclave at a flow rate of 5 dm^3 h^{-1}, resulting in a full re-circulation every 1.5 h. Prior to its use for this type of measurement, the autoclave and all tube connections were pre-oxidized for 168 h at 280 °C in a coolant containing 0.21 mmol kg^{-1} (12 ppm) K as KOH, in order to form a stable protective layer on all surfaces.

In a typical experiment, after mounting the electrodes and filling the loop with coolant, the system was heated to 80 °C and purged with N_2 (99.999%) for 16 h. The residual dissolved oxygen concentration after this procedure was below 0.31 µmol kg^{-1}. After reaching this value, the temperature was gradually increased, and the target value of 300 ± 1 °C at a pressure of 8.8 ± 0.01 MPa was reached in 2–2.5 h. Electrochemical impedance measurements were started at that temperature and were carried out for exposures up to 720 h. A 10030 Compactstat (Ivium, Eindhoven, the Netherlands) operating in floating mode and driven by IviumSoft 4.9 software was used. The frequency range of the measurements was from 30 kHz to 1 mHz at an ac amplitude of 50 mV (rms). The linearity of the spectra was checked by measurements using amplitudes between 10 and 50 mV, whereas causality was ensured using a Kramers-Kronig compatibility test using the procedure of Boukamp [22]. During the first 2–3 days of an experiment, the spectra were recorded every 2–3 h, during the next 3–4 days—every 4 h, and subsequently—every 12 or 24 h, depending on the variation of spectra with time (a typical measurement of an impedance spectrum took about 2–2.5 h). All the experiments were at least triplicated to ensure reproducibility. In order to discriminate more clearly between processes by their time constants in the high-frequency domain, 90% of the ohmic resistance of the coolant between the working and reference electrodes was subtracted from the real part of each spectrum. Complex non-linear fitting of impedance data to the transfer function of the proposed model was performed by a custom routine on an Origin Pro platform (Originlab, Northampton, MA, USA). A Levenberg-Marquardt algorithm with statistical weighting of the respective datasets was used.

In-depth elemental profiles of the samples after exposure were obtained by GDOES over an area of 5 mm^2 with a GDA750 instrument (Spectruma Analytik, Hof, Germany) equipped with a polychromator (focal length 750 mm and grating of 2400 channels/mm). Typical operating parameters were: primary voltage 950 V, current 9 mA and pressure 3 hPa. Calibration was based on certified reference materials chosen to cover the elements present in a wide range of nickel alloys in the relevant concentration ranges. Cross-sectional scanning electron microscopic images of the samples were also taken after exposure to estimate oxide thickness.

3. Results

3.1. Electrochemical Impedance Spectroscopy

Electrochemical impedance spectra in LiOH-free coolant as depending on exposure time are presented in Figure 1 in Bode coordinates (magnitude and phase shift of impedance vs. frequency). Analogous dependences of the impedance on oxidation time in WWER coolant with the addition of 0.042 mmol kg^{-1} Li (as LiOH) are presented in Figure 2.

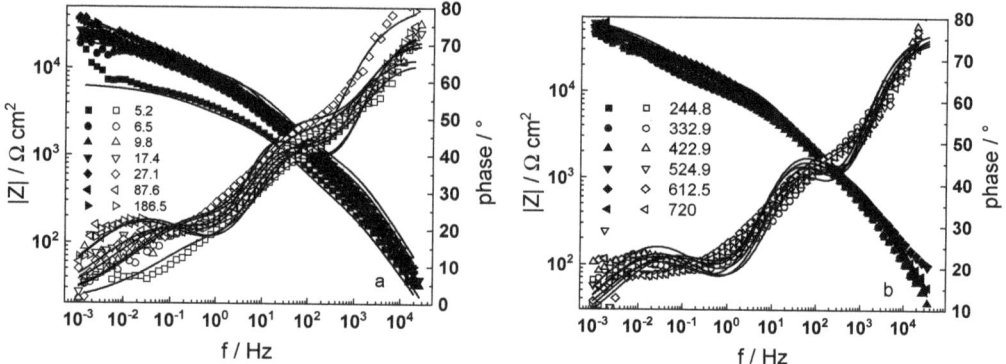

Figure 1. Electrochemical impedance spectra of zirconium alloy in WWER coolant without LiOH at 300 °C depending on the oxidation time. Left ordinate-impedance magnitude (full symbols) and phase shift (open symbols) vs. frequency. Points-experimental data, solid lines-best-fit calculation. The legend gives oxidation time in h.

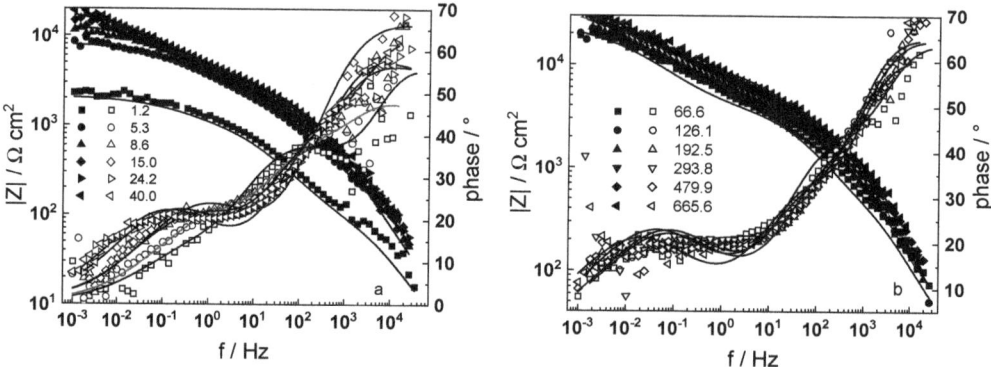

Figure 2. Electrochemical impedance spectra of zirconium alloy in WWER coolant with 0.042 mmol kg^{-1} Li (as LiOH) at 300 °C depending on the oxidation time. Left ordinate-impedance magnitude (full symbols) and phase shift (open symbols) vs. frequency. Points-experimental data, solid lines-best-fit calculation. The legend gives oxidation time in h.

Impedance spectra of a sample that was pre-hydrogenated for 24 h by polarization with −10 mA cm^{-2} at room temperature in 0.1 M KOH are presented in Figure 3. The respective spectra measured after 24 h cathodic polarization at 80 °C in a WWER coolant without LiOH in galvanostatic (−1 mA cm^{-2}) or potentiostatic (−2.0 V vs. RHE) modes are summarized in Figures 4 and 5.

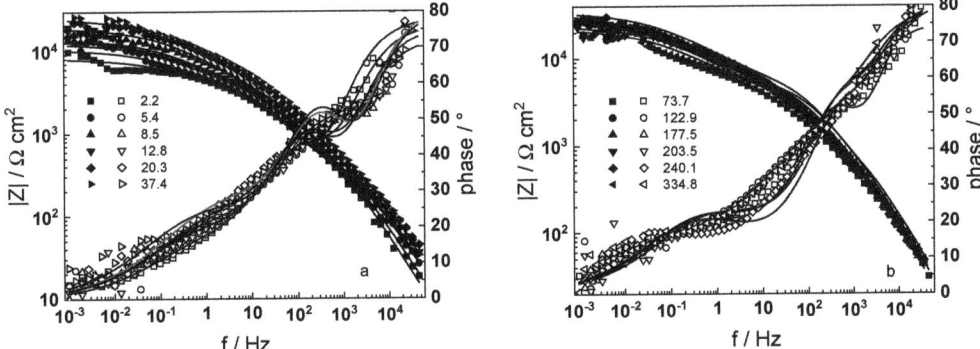

Figure 3. Electrochemical impedance spectra of ex-situ pre-hydrogenated sample in WWER coolant without LiOH at 300 °C depending on the oxidation time. Left ordinate-impedance magnitude (full symbols) and phase shift (open symbols) vs. frequency. Points-experimental data, solid lines-best-fit calculation. The legend gives oxidation time in h.

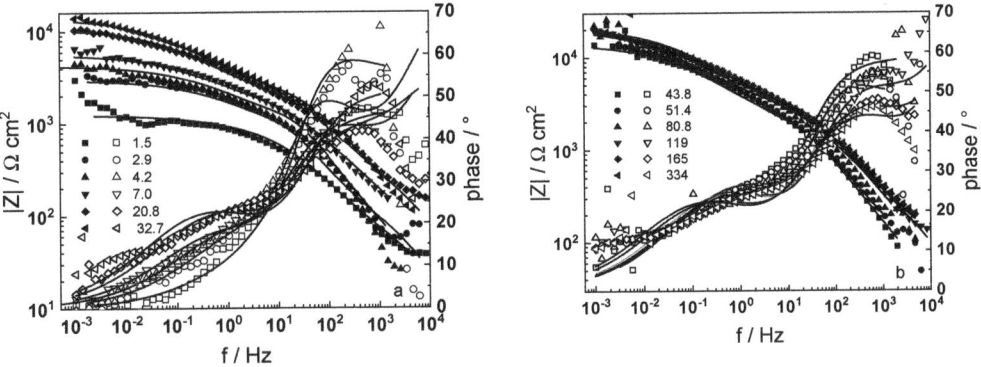

Figure 4. Electrochemical impedance spectra of in-situ pre-hydrogenated sample (galvanostatic mode) in WWER coolant without LiOH at 300 °C depending on the oxidation time. Left ordinate-impedance magnitude (full symbols) and phase shift (open symbols) vs. frequency. Points-experimental data, solid lines-best-fit calculation. The legend gives oxidation time in h.

A common feature of all spectra is the increase of the impedance magnitude at low frequencies (e.g., 1 mHz) with oxidation time, indicating a decrease in the rate of the-limiting step of corrosion. The extent of this decrease depends on oxidation conditions. The larger scatter in the low-frequency data for in-situ hydrogen charged samples is probably due to a dynamic steady state achieved at a larger local variation of transport rates in the oxide. In the phase shift vs. frequency curves, three-time constants are observed, corresponding to the electric properties of the oxide layer, the charge transfer process at the oxide/coolant interface and the ion conduction through the growing film, as discussed also in our previous studies [19,20].

3.2. Chemical Composition and Thickness

GDOES depth profiles of samples after exposure to the experimental conditions described above are presented in Figures 6–8. Nb is slightly impoverished in the oxides in comparison to bulk content, and the estimates of the position of alloy/oxide interface using sigmoidal fitting of Zr and O profiles coincide within ±5%.

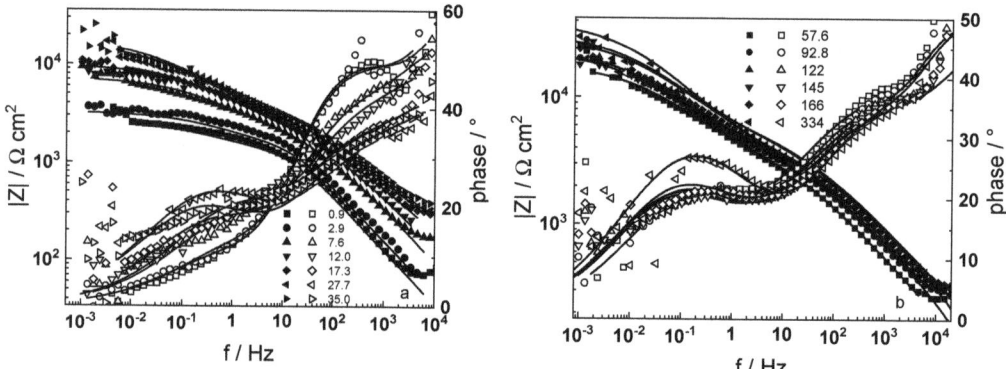

Figure 5. Electrochemical impedance spectra of in-situ pre-hydrogenated sample (potentiostatic mode) in WWER coolant without LiOH at 300 °C depending on the oxidation time. Left ordinate-impedance magnitude (full symbols) and phase shift (open symbols) vs. frequency. Points-experimental data, solid lines-best-fit calculation. The legend gives oxidation time in h.

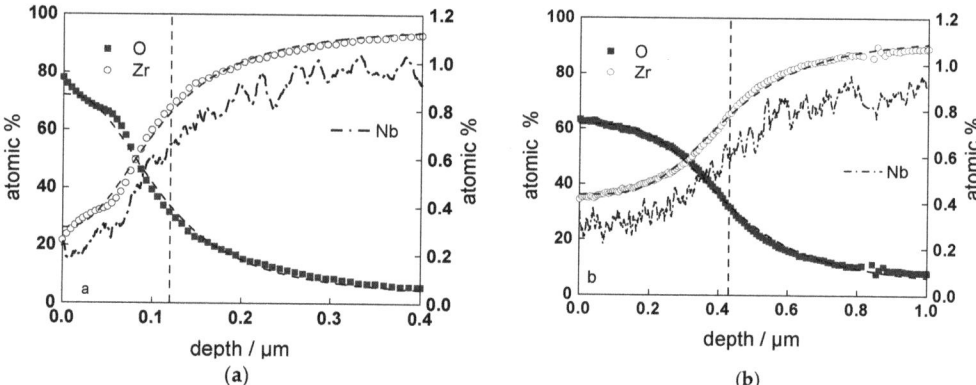

Figure 6. Elemental composition profile of the oxide formed for 24 (**a**) and 720 h (**b**) in WWER coolant without LiOH at 300 °C. Dashed lines show sigmoidal regression of O and Zr profiles to estimate the position of the alloy/oxide interface (shown by vertical line).

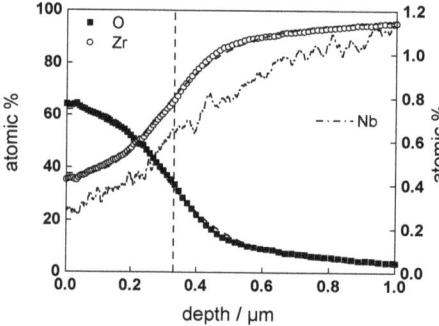

Figure 7. Elemental composition profile of the oxide on Zr-1%Nb alloy formed for 720 h in WWER coolant with LiOH at 300 °C. Dashed lines show sigmoidal regression of O and Zr profiles to estimate the position of the alloy/oxide interface (shown by vertical line).

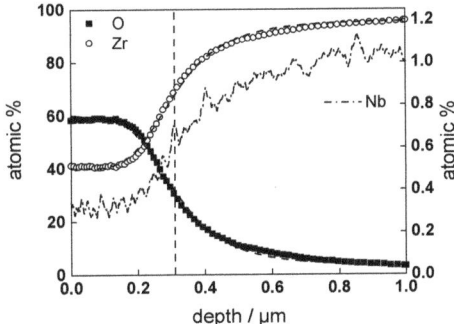

Figure 8. Elemental composition profile of the oxide on formed for 360 h in WWER coolant without LiOH at 300 °C on ex-situ pre-hydrogenated material. Dashed lines show sigmoidal regression of O and Zr profiles to estimate the position of the alloy/oxide interface (shown by vertical line).

Typical scanning electron micrographs of the cross-sections of samples exposed to WWER coolant without LiOH at 300 °C for 500 and 720 h are shown in Figure 9. It can be concluded that the oxides are uniform and the variation of thickness is less than 5%. No micro-cracks are observed at the magnification used.

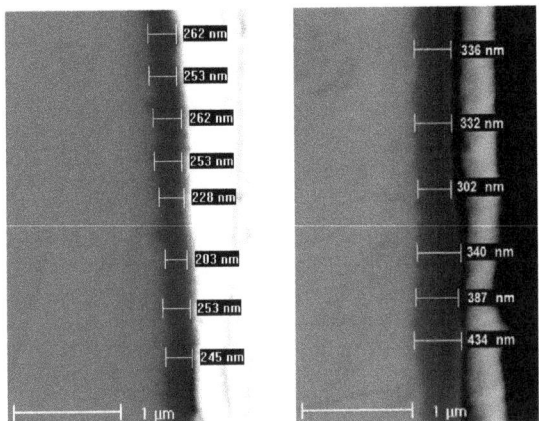

Figure 9. Scanning electron micrographs of the cross-section of oxides formed in WWER coolant without LiOH for 500 (**left**) and 720 h (**right**).

A summary of oxide thicknesses estimated by GDOES and SEM is collected in Table 2 and indicates a good agreement between the two methods. A larger difference was observed for the pre-hydrogenated sample, which could be due to the significant porosity of the outer layer on it. This layer is most likely a result of oxidation of the hydrogen-enriched layer of the alloy formed during cathodic charging.

Table 2. Comparison of estimates of oxide thickness obtained by GDOES and SEM.

Experiment	Oxide Thickness (GDOES) μm/2 Points	Oxide Thickness (SEM) μm/22 Points
w/o LiOH, 720 h	0.44 ± 0.01	0.40 ± 0.04
with LiOH, 720 h	0.31 ± 0.01	0.27 ± 0.04
w/o LiOH, ex-situ charged, 360 h	0.32 ± 0.01	0.33 ± 0.04

4. Discussion

4.1. Physical Model and Basic Equations

The qualitative picture of the oxidation of zirconium alloys in high-temperature electrolytes suggests the following generalized reaction of the protective layer growth:

Alloy/oxide interface: $Zr_m \xrightarrow{k_1} Zr_{Zr} + 2V_O^{\bullet\bullet} + 4e'$ (metal oxidation)

Oxide/coolant interface: $2V_O^{\bullet\bullet} + 2H_2O + 4e' \xrightarrow{k_2} 2O_O + 4H_{oxide}$ (reduction of water and entry of oxygen ions and hydrogen atoms into the oxide),

where Zr_m is a zirconium atom in the alloy, Zr_{Zr} and O_O are zirconium and oxygen positions in the crystal lattice of ZrO_2, $V_O^{\bullet\bullet}$ is an oxygen vacancy in this lattice, and k_1 and k_2 are the rate constants of the interfacial reactions. The above processes require transport of oxygen and electrons through the oxide. Oxygen is transferred by a vacancy diffusion/migration mechanism along grain boundaries of already formed oxide, while electrons—by a polaron hopping mechanism via secondary phases and/or hetero-valent substituted ions (e.g., $Nb^{II}{}_{Zr}{}''$, $Nb^{III}{}_{Zr}{}'$, $Nb^V{}_{Zr}{}^{\bullet}$). At the barrier film/coolant interface, restructuring of the barrier film and formation of a secondary outer oxide layer proceeds. This restructuring is assumed to proceed by a dissolution/deposition mechanism, the rate of which depends on the concentration of alkali:

$$ZrO_2 + OH^- \xrightarrow{k_d} HZrO_3^- \rightarrow Zr(OH)_4$$

As mentioned above, the flux of oxygen by vacancy mechanism $J_O(x,t)$ is determined by diffusion and migration:

$$J_O(x,t) = -D_O \frac{\partial c_O(x,t)}{\partial x} - 2\frac{FE}{RT} D_O c_O(x,t) \tag{1}$$

In this equation, $c_O(x,t)$ is the concentration of oxygen vacancies as a function of time t and distance x in the oxide, E is the electric field strength, and D_O the diffusion coefficient of oxygen by a vacancy mechanism; F, R, and T have their usual meanings. The steady-state solution of the above equation subject to the boundary conditions $J_O(L) = k_1$, $J_O(0) = k_2 c_O(0)$ in a coordinate system where $x = 0$ is at the oxide/electrolyte interface and $x = L_b$ at the alloy/oxide interface, L_b being the barrier oxide thickness, leads to the following expression for the concentration profile of oxygen vacancies:

$$c_O(x) = 2k_1 \left[\frac{e^{-2\frac{FE}{RT}x}}{k_2} + \frac{RT}{2FED_O} \right] \tag{2}$$

The total impedance of the coolant/outer layer/barrier layer/alloy layer/coolant system can be written as

$$Z = R_{el} + Z_{out} + Z_b = R_{el} + Z_{out} + \left(Z_e^{-1} + Z_{ion}^{-1} \right)^{-1} \tag{3}$$

where R_{el} is the uncompensated electrolyte resistance and Z_{out} is the impedance function describing the dielectric properties of the outer oxide film. Following previous treatments [19,20], the impedance of the outer layer, Z_{out}, is described with the so-called Havriliak-Negami impedance [23]

$$Z_{out} = \frac{R_{out}}{\left[1 + (j\omega R_{out} C_{out})^u\right]^n} \tag{4}$$

where R_{out} is the electrical resistance of the outer layer, C_{out}—its capacity, and u and n are constants with values between 0 and 1. Equation (4) describes charge transport through a porous layer by migration along conductive linear defects with no concentration gradient at difference to the diffusion-migration mechanism in the barrier layer described by Equation (1).

In Equation (3), Z_e is the impedance function describing the dielectric properties of the barrier oxide layer on the surface of the zirconium alloy, which is related to the variation of the steady-state concentration of oxygen vacancies, playing the role of donors, with distance in the oxide. Using Equation (2) as a starting point, the following expression is obtained for Z_e within the frames of the MCM [21]

$$Z_e = \frac{RT\varepsilon\varepsilon_0}{2j\omega FE} \ln\left[\frac{1 + j\omega\frac{RT}{F^2 D_e}\frac{k_2}{k_1}\varepsilon\varepsilon_0 e^{2\frac{FE}{RT}L}}{1 + j\omega\frac{RT}{F^2 D_e}\frac{k_2}{k_1}\varepsilon\varepsilon_0}\right] \tag{5}$$

where ε is the dielectric constant of the oxide, assumed to be equal to 22 [20], ε_0 is the dielectric permittivity of vacuum and D_e is the diffusion coefficient of electronic carriers.

In turn, the impedance of ion transport, Z_{ion}, is obtained by solving Equation (1) in the frequency domain for a low-amplitude sinusoidal perturbation [21]

$$Z_{ion} = R_t + \frac{(RT)^2}{4F^3 E D_O c_O(L_b)(1-\alpha)\left[1 + \sqrt{1 + \frac{j\omega(RT)^2}{F^2 E^2 D_O}}\right]} \tag{6}$$

where R_t is charge transfer resistance inversely proportional to the exchange current of the zirconium oxidation reaction, $c_O(L)$ is the concentration of oxygen vacancies at the alloy/oxide interface and α the part of the potential consumed at the film/coolant interface compared to film bulk. In the simplest case, it can be assumed that $\alpha = 0$, i.e., the entire potential drop in the alloy/oxide/electrolyte system is located in the barrier layer. On the other hand, based on Equation (2), the concentration $c_O(L)$ can be approximated as

$$c_O(L_b) = 2k_1\left[\frac{e^{-2KL}}{k_2} + \frac{RT}{2FED_O}\right] \approx \frac{RTk_1}{FED_O} \tag{7}$$

Finally, inserting the assumed value of α and the approximate expression for $c_O(L_b)$ in Equation (6), an equation for the impedance of ion transport across the oxide barrier sublayer is obtained.

$$Z_{ion} = R_t + \frac{RT}{4F^2 k_1\left[1 + \sqrt{1 + \frac{j\omega(RT)^2}{F^2 E^2 D_O}}\right]} \tag{8}$$

The total impedance transfer function of the system, then, can be expressed by substituting Equations (4), (5) and (8) in Equation (3):

$$Z = R_{el} + \frac{R_{out}}{[1+(j\omega R_{out}C_{out})^n]} + \left[\left(\frac{RT\varepsilon\varepsilon_0}{2j\omega FE}\ln\left[\frac{1+j\omega\frac{RT}{F^2 D_e}\frac{k_2}{k_1}\varepsilon\varepsilon_0 e^{2\frac{FE}{RT}L}}{1+j\omega\frac{RT}{F^2 D_e}\frac{k_2}{k_1}\varepsilon\varepsilon_0}\right]\right)^{-1} + \left(R_t + \frac{RT}{4F^2 k_1\left[1+\sqrt{1+\frac{j\omega(RT)^2}{F^2 E^2 D_O}}\right]}\right)^{-1}\right]^{-1} \tag{9}$$

4.2. Parameter Estimation

The experimental data were fitted to the transfer function expressed by Equations (3)–(8) using the procedure described in the Experimental section. The comparison of experimental and best-fit calculated impedance spectra (Figures 1–5) demonstrates the ability of the proposed model to reproduce quantitatively both the magnitude and the frequency distribution of the impedance function. Therefore, estimates of the kinetic parameters can be considered viable and suited to reproduce oxidation rates of the material in a simulated WWER coolant. The dependences of the main parameters of the processes at the alloy/oxide and oxide/coolant interfaces (rate constants k_1 and k_2, and resistance of the outer oxide layer R_{out} and C_{out}), as well as those characterizing defect transport (electric field intensity E, diffusion coefficients of oxygen anions and electrons) are presented in Figures 10 and 11 as depending on oxidation time. The estimated values of the metal oxidation rate constant at the alloy/oxide interface, the electric field strength, and the diffusion coefficients in the oxide are in line with our previous work [20].

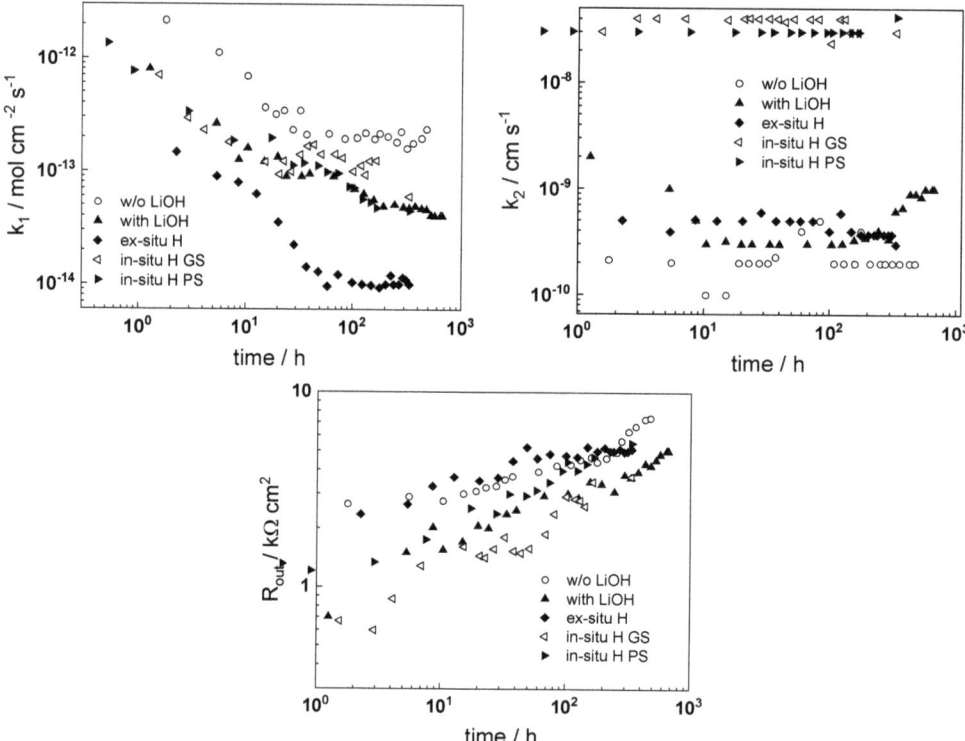

Figure 10. Parameters of interfacial reactions as a function of time and oxidation conditions.

Based on the dependence of the kinetic and transport parameters with oxidation time, the following conclusions on their relative significance in the overall oxidation process can be drawn:

- The rate constant of metal oxidation at the alloy/oxide interface decreases significantly with time indicating that the corrosion rate decreases with increasing film thickness. Nearly constant values are reached after ca. 100–200 h of oxidation, i.e., a quasi-steady state is achieved. On the other hand, the rate of incorporation of oxygen at the oxide/coolant interface is almost independent on time, suggesting that this process is of secondary significance when compared to metal oxidation at the inner interface;
- The charge transfer resistance at the oxide/coolant interface increases with oxidation time, which means that the respective rate of water reduction decreases. This is in line with the decrease of metal oxidation rate leading to a smaller electron supply rate for the coupled cathodic reaction, taking into account the fact that the diffusion coefficient of electrons does not exhibit any dependence on oxidation time.

A rationalization of the decrease of the oxygen diffusion coefficient and field strength in the oxide with oxidation time (or equivalently, film thickness) is attempted in the next section.

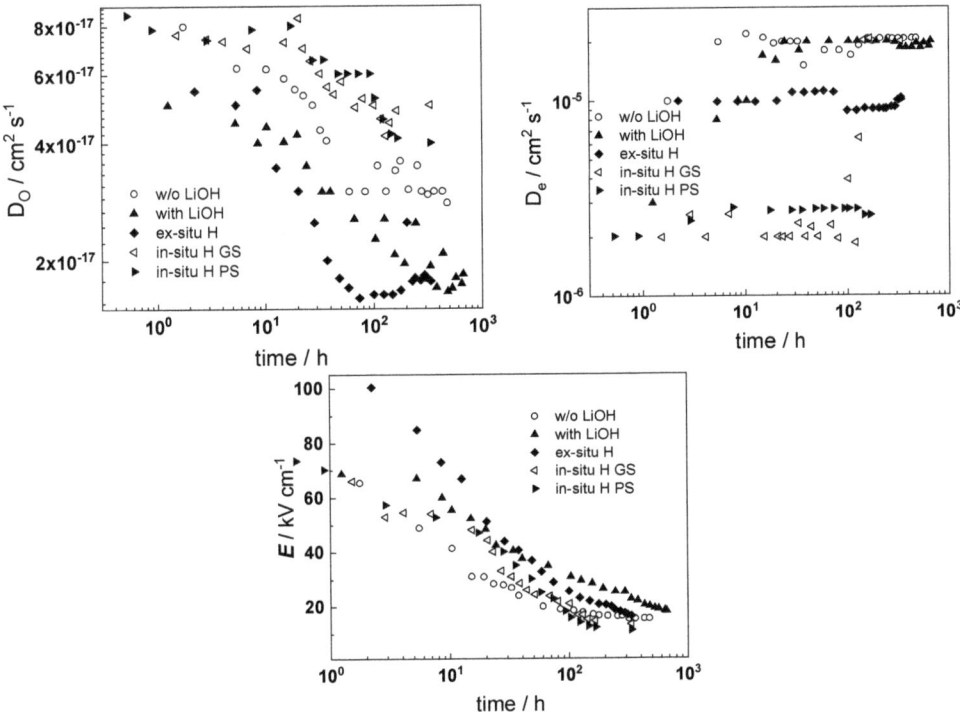

Figure 11. Defect transport parameters as a function of time and oxidation conditions.

4.3. Influence of Internal Stresses on Oxygen Transport by Vacancy Mechanism

Atomistic simulations by density functional theory (DFT) were reported to estimate the effect of internal stresses on the energy of formation and migration of oxygen vacancies during oxide film growth on zirconium alloys [24]. Table 3 shows the variations in the formation energies of oxygen vacancies in the m-and t-ZrO$_2$, as well as migration energies t-ZrO$_2$, depending on stress direction with a value of 1 GPa. This value corresponds to the limiting stress in oxides formed on Zircaloy-4 as estimated by synchrotron X-ray diffraction [25].

Table 3. Energy of oxygen vacancy formation (E_f) and migration (E_v) for different ZrO$_2$ modifications depending on the direction of a compressive stress of 1 GPa [24].

Oxide Structure	Isotropic Compression	Compression in Direction a	Compression in Direction b	Compression in Direction c
E_f m-ZrO$_2$(O1)/eV	0.010	0.012	0.011	0.032
E_f m-ZrO$_2$(O2)/eV	0.012	0.016	0.010	0.022
E_f t-ZrO$_2$/eV	0.015	0.011		0.006
E_V t-ZrO$_2$/eV	0.034	0.070	0.008	0.030

Based on these calculations and the expressions for the diffusion coefficient of oxygen via vacancies in the presence and absence of stress:

$$D_{O,\sigma} = \tfrac{1}{6}zva^2 \exp\left(-\tfrac{E_f+\Delta E_f}{RT}\right)\exp\left(-\tfrac{E_m+\Delta E_m}{RT}\right)$$
$$D_{O,\sigma=0} = \tfrac{1}{6}zva^2 \exp\left(-\tfrac{E_f}{RT}\right)\exp\left(-\tfrac{E_m}{RT}\right) \quad (10)$$

ratios of diffusion coefficients in the presence and absence of compressive stress were estimated:

$$\frac{D_{O,\sigma}}{D_{O,\sigma=0}} = \exp\left(-\frac{\Delta E_f + \Delta E_m}{RT}\right) = \exp\left(-\frac{\Delta E}{RT}\right) \quad (11)$$

The values of this ratio at a stress of 1 GPa are presented in Table 4.

Table 4. Ratio of diffusion coefficients of oxygen vacancies in the presence and absence of 1 GPa compressive stresses with different orientation.

Ratio of Diffusion Coefficients	Isotropic Compression	Compression in Direction a	Compression in Direction b	Compression in Direction c
$\frac{D_{O,\sigma}}{D_{O,\sigma=0}}$	0.37	0.19	0.68	0.48

The evolution of compressive stress with oxide thickness (i.e., with oxidation time) was adopted from synchrotron X-ray diffraction data of zirconium alloys oxidized at 350 °C in a simulated primary coolant [25]. In that paper. internal stresses were reported to increase quasi-linearly with thickness and reach limiting values close to 1 GPa, i.e., the corresponding diffusion coefficients for oxides of different thicknesses was corrected using data presented in Table 4 and the corresponding compressive stress-time dependence.

4.4. Influence of Space Charge of Substitutional Ions on the Field Strength in the Oxide

According to the scheme of the oxidation process, transport of oxygen proceeds from the outer to the inner interface by a vacancy mechanism, and electrons transfer in the opposite direction (they are consumed by the reaction of reduction of water at the oxide/coolant interface) [5]. Since transport of electrons is significantly faster than that of oxygen ions, a space charge builds up in the oxide and creates an additional electric field, influencing the transport of current carriers. In addition, hetero-valent impurities, such as niobium ions in different oxidation states (from two to four, according to synchrotron X-ray absorption fine structure data [26,27]), that are present in the oxide partly compensate the space charge. The generalized expression for the field strength has the form:

$$E(L) = E_0 + \frac{e}{\varepsilon\varepsilon_0}\int_0^L \sum_i z_i c_i(x)dx \quad (12)$$

Assuming that hydrogen is transferred to the oxide in atomic form (i.e., has no charge), the sum under the integral is written as

$$\sum_i z_i c_i(x) = 2c_o(x) - c_{e'}(x) - z_{Nb}c_{Nb}(x) \quad (13)$$

Introducing a homogeneous variation of defect concentrations, we obtain a simplified expression:

$$E(L) = E_0 + L\frac{F\Delta c(0)}{\varepsilon\varepsilon_0\left(1+\frac{L}{x_0}\right)} \quad (14)$$

in which $x_0 = \frac{\varepsilon\varepsilon_0 RT}{(2F)^2 ac_o(0)}$ is the so-called field shielding parameter from space charge [28].

4.5. Kinetics of Barrier Oxide Growth

The thickness of the barrier oxide film increases with time according to a logarithmic law derived based on the MCM.

$$L_b(t) = L_{b,t=0} + \frac{1}{b}\ln\left[1 + \Omega k_1 b e^{-bL_{b,t=0}} t\right], \quad b = \frac{2\alpha_1 FE}{RT} \quad (15)$$

α_1 being the transfer coefficient of Zr oxidation reaction at the alloy/barrier layer interface, $L_{b,t=0}$—the initial barrier layer thickness and Ω the molar volume of the barrier oxide.

4.6. Model Validation Based on Computational Results

Figure 12 shows the dependences of the barrier and outer layer thicknesses on oxidation time under different experimental conditions, evaluated by quantitative interpretation of the impedance spectra. The lines represent a non-linear regression of this type of data to Equation (15), with a very good fit, i.e., the proposed equation adequately describes thickness of the protective layer—time dependences. Thickness estimates obtained from analysis of impedance spectra are in good agreement with average values obtained from interpretation of GDOES depth profiles and cross-sectional microscopic observations. The outer layer thickness is smaller than that of the barrier layer, with the notable exception of the sample that was pre-hydrogenated ex-situ in 0.1 M KOH. A tentative explanation is that in this case, the outer layer is formed at the expense of a hydrogen-rich layer in the underlying alloy.

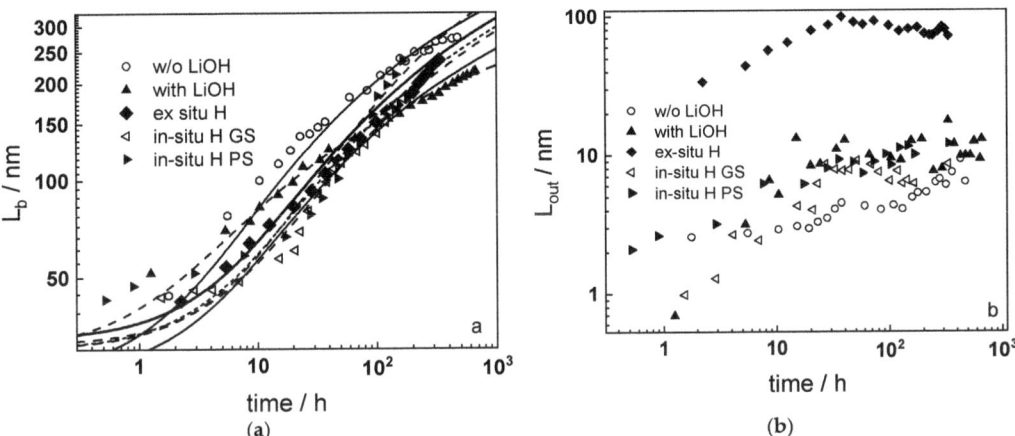

Figure 12. Thicknesses of the barrier oxide layer (**a**) and the outer layer (**b**) estimated from EIS depending on the oxidation time. Lines in (**a**) are best-fit calculations to Equation (15).

Dependences of the oxygen diffusion coefficient on barrier layer thickness in different experimental conditions are presented in Figure 13. The diffusion coefficient decreases linearly with thickness, which is due to a linear increase of compressive stress in the oxide with thickness, causing an increase in the energy of the formation/migration of oxygen vacancies. A significant deviation from the predicted dependence was observed only for the samples pre-hydrogenated for 24 h in 0.1 M KOH, which could be explained by the formation of a hydrogen-rich layer during charging. The oxidation of such a layer can be assumed to proceed by a mechanism that is different from the thickening of the native oxide on untreated samples.

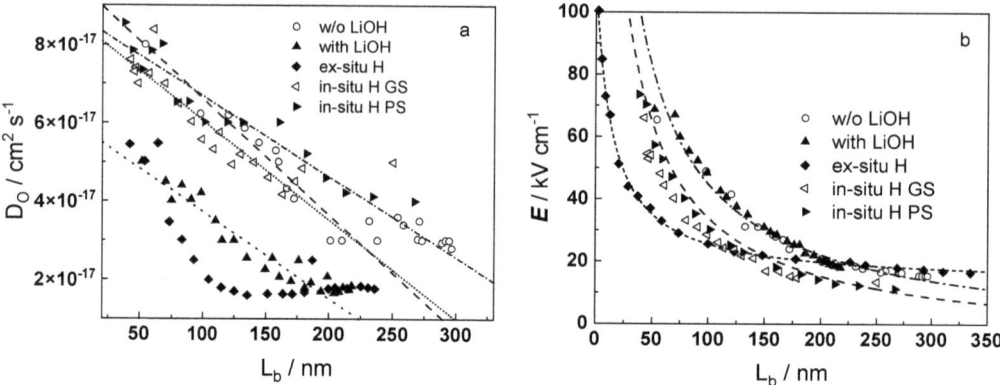

Figure 13. (a) Dependence of oxygen diffusion coefficient on oxide thickness under different conditions. Points-calculated values based on nonlinear regression of impedance spectra; lines-regression based on the assumption for linear increase of compressive stresses with oxide layer thickness; (b) Dependence of the field strength on the thickness of the oxide obtained under different conditions. Points-calculated values based on nonlinear regression of impedance spectra, lines-nonlinear regression by Equation (14).

Figure 13b illustrates the thickness dependences of the field strength in the oxides. The solid lines represent the results of non-linear regression of these dependences according to Equation (14). The computational results show agreement with the estimates determined from impedance spectra, which also validates the hypothesis of the proposed approach on the influence of the space charge of mobile point defects and immobile aliovalent impurities on field strength in the growing oxide.

A comparison of the model predictions with oxide thickness vs. time data for Zr-1%Nb alloys in WWER and PWR coolants at different temperatures [18,19,25,27,29,30] is presented in Figure 14a. The quality of prediction is satisfactory, considering that calculations on the basis of parameterization of impedance spectra give a measure of the instantaneous oxide thickness, whereas literature data based on weight gain measurements and microscopic observations estimate the cumulative thickness of the growing layer. Notably, the model predicts the growth of the barrier film, whereas the formation of an external layer with pores and cracks that does not limit further oxidation of the alloy is not quantified.

As a next step of model verification, its predictions are compared with data on oxide thickness distribution by fuel rod height after three and six years of service [31] in a WWER-1000 reactor (Figure 14b), using the appropriate temperature distribution [32]. The quality of prediction is quite good for three years of service and becomes somewhat worse after six years. This could be due to the fact that similar temperature-height distributions were used in both cases.

In general, it can be concluded that in-situ measurements using EIS and their quantitative interpretation by the MCM represent an important step towards the elucidation of the mechanism of compact protective layer growth on zirconium alloy as fuel cladding and internals in nuclear reactors. The quantification of hydrogen pick-up and hydride formation, as well as the prediction of their effects on service life of zirconium alloys, is underway and will be reported in the near future.

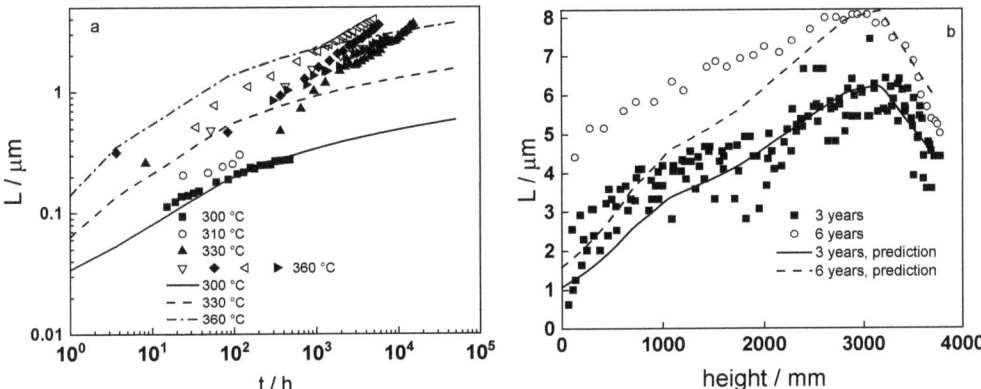

Figure 14. (a) Thickness of oxide layers on Zr-1%Nb as depending on temperature and time of oxidation; (b) Distribution of oxide thickness on Zr-1%Nb fuel cladding by fuel rod height after 3 and 6 years of service in a WWER-1000 reactor: points-experimental data from various sources, lines-model predictions.

5. Conclusions

In the present paper, oxidation of Zr-1%Nb alloy in simulated primary coolants of WWER reactor is studied with a combination of in-situ electrochemical measurements and ex-situ analytical methods of thin film characterization. A new version of a kinetic model of oxide growth including the effects of compressive stress and ionic space charge on defect transport is proposed to interpret the results obtained. The following main conclusions can be drawn from the experimental results and model calculations:

- Electrochemical impedance measurements allow to discernment of the contribution of barrier and outer layer conductivities, kinetics of interfacial reactions, and ionic defect transport in the overall oxidation process;
- Layer thicknesses estimated from EIS are in good agreement with those determined by ex-situ analysis techniques (GDOES and SEM). The thickness of the barrier protective layer is much larger than that of the outer layer except for oxidation of ex-situ pre-hydrogenated samples;
- The proposed kinetic model is able to reproduce quantitatively the impedance spectra as depending on oxidation time in a variety of experimental conditions;
- Taking into account the influence of compressive stress and space charge in the oxide allows rationalization of the dependences of the diffusion coefficient of oxygen via vacancy mechanism and the field strength in the oxide on film thickness, or equivalent oxidation time. Barrier thickness vs. time data are successfully interpreted with the same set of kinetic parameters furnishing further credibility to the model.

Author Contributions: Conceptualization, M.B. and I.B.; methodology, M.B.; validation, I.B. and V.K.; formal analysis, I.B.; investigation, V.K.; resources, M.B.; data curation, V.K.; writing—original draft preparation, I.B.; writing—review and editing, M.B.; visualization, I.B. All authors have read and agreed to the published version of the manuscript.

Funding: This research was funded by The National Scientific Fund of Bulgaria, grant number КП-06-Н59/4-2021 "Deterministic modeling of degradation of structural materials for energy systems in high-temperature electrolytes".

Data Availability Statement: The data presented in this study are available on request from the corresponding author.

Conflicts of Interest: The authors declare no conflict of interest.

References

1. Tupin, M. Understanding the Corrosion Processes of Fuel Cladding in Pressurized Water Reactors. In *Nuclear Corrosion, European Federation of Corrosion*; Elsevier: Amsterdam, The Netherlands, 2020; pp. 252–299.
2. Yagnik, S.; Garde, A. Zirconium Alloys for LWR Fuel Cladding and Core Internals. In *Structural Alloys for Nuclear Energy Applications*; Elsevier: Amsterdam, The Netherlands, 2019; pp. 247–291.
3. Fromhold, A.T. *Theory of Metal Oxidation-Vol. I: Fundamentals*; North-Holland: Amsterdam, The Netherlands, 1976.
4. Cox, B. Some thoughts on the mechanisms of in-reactor corrosion of zirconium alloys. *J. Nucl. Mater.* 2005, *336*, 331–368. [CrossRef]
5. Motta, A.T. Mechanistic Understanding of Zirconium Alloy Fuel Cladding Performance. In *Zirconium in the Nuclear Industry, Proceedings of the 18th International Symposium*; Comstock, R.J., Motta, A.T., Eds.; ASTM International: West Conshohocken, PA, USA, 2018; pp. 19–51.
6. Actis-Dato, L.O.; Aldave de Las Heras, L.; Betti, M.; Toscano, E.H.; Miserque, F.; Gouder, T. Investigation of mechanisms of corrosion due to diffusion of impurities by direct current glow discharge mass spectrometry depth profiling. *J. Anal. Atom. Spectrom.* 2000, *15*, 1479–1484. [CrossRef]
7. Moya, J.S.; Diaz, M.; Bartolomeé, J.F.; Roman, E.; Sacedon, J.L.; Izquierdo, J. Zirconium oxide film formation on zircaloy by water corrosion. *Acta Mater.* 2000, *48*, 4749–4754. [CrossRef]
8. Sawicki, J. Evidence of Ni_2FeBO_5 and m-ZrO_2 precipitates in fuel rod deposits in AOA-affected high boiling duty PWR core. *J. Nucl. Mater.* 2008, *374*, 248–269. [CrossRef]
9. Henshaw, J.; McGurk, J.C.; Sims, H.E.; Tuson, A.; Dickinson, S.; Deshon, J. A model of chemistry and thermal hydraulics in PWR fuel crud deposits. *J. Nucl. Mater.* 2006, *353*, 1–11. [CrossRef]
10. Nagy, G.; Kerner, Z.; Pajkossy, T. In situ electrochemical impedance spectroscopy of Zr–1%Nb under VVER primary circuit conditions. *J. Nucl. Mater.* 2002, *300*, 230–236. [CrossRef]
11. Vermoyal, J.J.; Hammou, A.; Dessemond, L.; Frichet, A. Electrical characterization of waterside corrosion films formed on ZrNb(1%)O(0.13%). *Electrochim. Acta* 2002, *47*, 2679–2695. [CrossRef]
12. Schefold, J.; Lincot, D.; Ambard, A.; Kerrec, O. The Cyclic Nature of Corrosion of Zr and Zr-Sn in High-Temperature Water at 633 K. A Long-Term In Situ Impedance Spectroscopic Study. *J. Electrochem. Soc.* 2003, *150*, B451–B461. [CrossRef]
13. Chen, Y.; Urquidi-Macdonald, M.; Macdonald, D.D. The electrochemistry of zirconium in aqueous solutions at elevated temperatures and pressures. *J. Nucl. Mater.* 2006, *348*, 133–147. [CrossRef]
14. Ai, J.Y.; Chen, M.; Urquidi-Macdonald, M.; Macdonald, D.D. Electrochemical Impedance Spectroscopic Study of Passive Zirconium. I. High-Temperature, Deaerated Aqueous Solutions. *J. Electrochem. Soc.* 2007, *154*, C43. [CrossRef]
15. Ai, J.Y.; Chen, M.; Urquidi-Macdonald, M.; Macdonald, D.D. Electrochemical Impedance Spectroscopic Study of Passive Zirconium. II. High-Temperature, Hydrogenated Aqueous Solutions. *J. Electrochem. Soc.* 2007, *154*, C52. [CrossRef]
16. Couet, A.; Motta, A.T.; Ambard, A.; Livigni, D. In-situ electrochemical impedance spectroscopy measurements of zirconium alloy oxide conductivity: Relationship to hydrogen pickup. *Corros. Sci.* 2017, *119*, 1–13. [CrossRef]
17. Kim, T.; Couet, A.; Kim, S.; Lee, Y.; Yoo, S.C.; Kim, J.H. In-situ electrochemical study of zirconium alloy in high temperature hydrogenated water conditions. *Corros. Sci.* 2020, *173*, 108745. [CrossRef]
18. Renciukova, V.; Macak, J.; Sajdl, P.; Novotný, R.; Krausova, A. Corrosion of zirconium alloys demonstrated by using impedance spectroscopy. *J. Nucl. Mater.* 2018, *510*, 312–321. [CrossRef]
19. Bojinov, M.; Cai, W.; Kinnunen, P.; Saario, T. Kinetic parameters of the oxidation of zirconium alloys in simulated WWER water—Effect of KOH content. *J. Nucl. Mater.* 2008, *378*, 45–54. [CrossRef]
20. Bojinov, M.; Karastoyanov, V.; Kinnunen, P.; Saario, T. Influence of water chemistry on the corrosion mechanism of a zirconium-niobium alloy in simulated light water reactor coolant conditions. *Corros. Sci.* 2010, *52*, 54–67. [CrossRef]
21. Beverskog, B.; Bojinov, M.; Kinnunen, P.; Laitinen, T.; Makela, K.; Saario, T. A mixed-conduction model for oxide films on. Fe, Cr and Fe–Cr alloys in high-temperature aqueous electrolytes—II. Adaptation and justification of the model. *Corros. Sci.* 2002, *44*, 1923–1940. [CrossRef]
22. Boukamp, B. Practical application of the Kramers-Kronig transformation on impedance measurements in solid state electrochemistry. *Solid State Ion.* 1993, *62*, 131–141. [CrossRef]
23. Havriliak, S.; Negami, S. A complex plane analysis of α-dispersions in some polymer systems. *J. Polym. Sci. C* 1966, *14*, 99–117.
24. Yamamoto, Y.; Morishita, K.; Iwakiri, H.; Kaneta, Y. Stress dependence of oxygen diffusion in ZrO_2 film. *Nucl. Instrum. Methods Phys. Res.* 2013, *B303*, 42–45. [CrossRef]
25. Swan, H.; Blackmur, M.S.; Hyde, J.M.; Laferrere, A.; Ortner, S.R.; Styman, P.D.; Staines, C.; Gass, M.; Hulme, H.; Cole-Baker, A.; et al. The measurement of stress and phase fraction distributions in pre and post-transition Zircaloy oxides using nano-beam synchrotron X-ray diffraction. *J. Nucl. Mater.* 2016, *479*, 559–575. [CrossRef]
26. Moorehead, M.; Yu, Z.; Borrel, L.; Hu, J.; Cai, Z.; Couet, A. Comprehensive investigation of the role of Nb on the oxidation kinetics of Zr-Nb alloys. *Corros. Sci.* 2019, *155*, 173–181. [CrossRef]
27. Couet, A.; Borrel, L.; Liu, J.; Hu, J.; Grovenor, C. An integrated modeling and experimental approach to study hydrogen pickup mechanism in zirconium alloys. *Corros. Sci.* 2019, *159*, 108134. [CrossRef]
28. Fromhold, A., Jr. Single Carrier Steady-State Theory for Formation of Anodic Films Under Conditions of High Space Charge in Very Large Electric Fields. *J. Electrochem. Soc.* 1977, *124*, 538–549. [CrossRef]

29. Moorehead, M.; Hu, J.; Couet, A.; Cai, Z. Progressing Zirconium-Alloy Corrosion Models Using Synchrotron XANES. In Proceedings of the 18th International Conference on Environmental Degradation of Materials in Nuclear Power Systems—Water Reactors, Portland, OR, USA, 13–17 August 2017; Jackson, J., Paraventi, D., Wright, M., Eds.; The Minerals, Metals & Materials Series. Springer: Cham, Switzerland, 2018; pp. 565–576.
30. Hu, J.; Garner, A.; Frankel, P.; Li, M.; Kirk, M.A.; Lozano-Perez, S.; Preuss, M.; Grovenor, C. Effect of neutron and ion irradiation on the metal matrix and oxide corrosion layer on Zr-1.0Nb cladding alloys. *Acta Mater.* **2019**, *173*, 313–326. [CrossRef]
31. Likhanskii, V.V.; Evdokimov, I.A.; Aliev, T.N.; Kon'kov, V.F.; Markelov, V.A.; Novikov, V.V.; Khokhunova, T.N. Corrosion model for zirconium-niobium alloys in pressurized water reactors. *At. Energy* **2014**, *116*, 186–193. [CrossRef]
32. Ramezani, L.; Mansouri, M.; Rahgoshay, M. Modeling the Water Side Corrosion and Hydrogen Pickup of the WWER fuel clad. *Nucl. Technol. Radiat. Prot.* **2018**, *33*, 334–340. [CrossRef]

Disclaimer/Publisher's Note: The statements, opinions and data contained in all publications are solely those of the individual author(s) and contributor(s) and not of MDPI and/or the editor(s). MDPI and/or the editor(s) disclaim responsibility for any injury to people or property resulting from any ideas, methods, instructions or products referred to in the content.

Article

Corrosion Inhibition Mechanism of Ultra-High-Temperature Acidizing Corrosion Inhibitor for 2205 Duplex Stainless Steel

Danping Li [1,2,*], Wenwen Song [3], Junping Zhang [4], Chengxian Yin [1], Mifeng Zhao [3], Hongzhou Chao [3], Juantao Zhang [1], Zigang Lei [1,2], Lei Fan [1], Wan Liu [1,2] and Xiaolong Li [1]

1. Tubular Goods Research Institute of CNPC, Xi'an 710077, China
2. Xi'an Dexincheng Technology Co., Ltd., Xi'an 710075, China
3. PetroChina Tarim Oilfield Company, Korla 841000, China
4. School of Chemistry and Chemical Engineering, Northwestern Polytechnical University, Xi'an 710082, China
* Correspondence: lidanping006@cnpc.com.cn

Abstract: The acidizing corrosion inhibitors reported so far have a poor effect on duplex stainless steel in high-temperature and high-concentration acid systems and cannot effectively inhibit the occurrence of selective corrosion. In this paper, a new acidizing corrosion inhibitor was designed, which was mainly composed of Mannich base and antimony salt. The inorganic substance in the corrosion inhibitor had good stability at high temperatures and could quickly form a complex with the metal matrix to enhance the binding ability. The organic substance can make up for the non-dense part of the inorganic film. The properties of developed corrosion inhibitors were analyzed by quantum chemical calculation, molecular dynamics simulation, and scanning electron microscopy. The results showed that a double-layer membrane structure could be constructed after adding the corrosion inhibitor, which could play a good role in blocking the diffusion of acid solution at high-temperature. The uniform corrosion rate of 2205 duplex stainless steel after adding acidizing corrosion inhibitor immersion in a simulated service condition (9 wt.% HCl + 1.5 wt.% HF + 3 wt.% CH_3COOH + 4~6 wt.%) at 140 °C, 160 °C and 180 °C for a 4 h test is 6.9350 $g \cdot m^{-2} \cdot h^{-1}$, 6.3899 $g \cdot m^{-2} \cdot h^{-1}$ and 12.1881 $g \cdot m^{-2} \cdot h^{-1}$, respectively, which shows excellent corrosion inhibition effect and is far lower than that of the commonly accepted 81 $g \cdot m^{-2} \cdot h^{-1}$ and no selective corrosion could be detected.

Keywords: duplex stainless steel; selective corrosion; high-temperature acidification; corrosion inhibitor; molecular dynamics simulation

1. Introduction

With the increasing demand for energy, the exploration and development of oil and gas fields have gradually developed from conventional working conditions to the harsh working conditions of high-temperature and high-pressure or even ultra-high-temperature and high-pressure. The high-temperature and high-pressure well is defined with wellhead pressure and bottom hole pressure greater than 70 and 105 MPa, respectively, and the bottom hole temperature greater than 150 °C by the International high-temperature and high-pressure Well Association, the wells with wellhead pressure greater than 105 MPa, bottom-hole pressure greater than 140 MPa and bottom-hole temperature greater than 175 °C is defined as ultra-high-temperature and pressure wells [1]. At present, high-temperature and high-pressure oil and gas wells in the world are mainly distributed in the Gulf of Mexico in the United States, the North Sea in the United Kingdom, Southeast Asia, Africa, and the Tarim Basin, the South China Sea and Sichuan in China [2–5]. The acidizing operation temperature of oil wells in the North Sea has exceeded 160 °C [6]. In China, representative ultra-high-temperature and high-pressure wells are mainly distributed in the Tarim Basin of Xinjiang, with deep burial depths and extremely harsh service conditions. The bottom hole temperature usually exceeds 180 °C [7,8]. In addition, the acid with extremely low pH value in the process of reservoir reconstruction and the CO_2 and Cl^- in the fluid produced in the later production process all put forward high requirements for the corrosion resistance of oil pipes. According to the SUMITOMO oil well pipe material selection spectrum [9], duplex stainless steel should be used for H_2S partial pressure $\leq$0.1 bar when the temperature is over 180 °C, while nickel base alloy with higher corrosion resistance should be selected for H_2S partial pressure > 0.1 bar, and the traditional super 13Cr stainless steel could no longer meet the requirement of the working conditions.

Generally, a passive oxide (mainly Cr_2O_3) layer could be formed on the stainless steel to resist various corrosion, and the absence or breakdown of this layer could result in accelerated corrosion rates [10]. Stainless steel is mainly divided into the following categories: ferritic stainless steels, austenitic stainless steels, martensitic stainless steels, duplex stainless steels, precipitation-hardening stainless steels, and Mn-N substituted austenitic stainless steels [11]. Due to the particularity of its composition, duplex stainless steel has the characteristics of ferrite and austenite stainless steel. Duplex stainless steel has higher plastic toughness, and its weldability and intergranular corrosion resistance are significantly stronger than that of ferritic stainless steel. Compared with austenitic stainless steel, the yield strength, intergranular corrosion, and Cl^- corrosion resistance of duplex steel is much better. In addition, duplex stainless steel also has the advantage of better resistance to stress corrosion, pitting, and crevice corrosion [12].

During the whole production cycle of the well, applying a strong acid (such as 10–28 wt.% HCl) to increase the output of the reservoir is inevitable. The passive film of duplex stainless steel could be destroyed under such harsh conditions. Duplex stainless steel has good corrosion resistance in general conditions, but it will exhibit selective corrosion of different phases in specific conditions [13–25]. Due to the different crystal structure and chemical composition of α and γ phases, which shows different electrochemical potential in a specific solution, a micro-electric couple will be formed and cause selective corrosion [26,27]. For example, the anodic polarization curve of 2205 duplex stainless steel sample in acid solution usually shows two anodic dissolution peaks, the lower and the higher peaks are the ferrite phase and austenite phase activation peaks, respectively. The ferrite phase is preferentially dissolved when the applied potential corresponds to the lower peak. Similarly, the austenite phase is preferentially dissolved when the applied potential is close to the higher peak [16]. Yau and Streicher [24] found that ferritic selective corrosion of FeCr–10%Ni duplex stainless steel occurred in reducing acid. Sridhar and Kolts [25] discovered that the austenitic phase selective corrosion of high nitrogen content duplex stainless steel occurred in sulfuric acid and phosphoric acid, while the ferrite phase selective corrosion occurred in hydrochloric acid. Tsai et al. [18] showed that the activity of ferrite was stronger than austenite in 2 mol/L H_2SO_4 + 0.5 mol/L HCl solution; however, austenite was in a more active state and is preferentially corroded as the anode in 1.5 mol/L HNO_3 solution.

Acid corrosion inhibitors must be used to reduce the corrosion damage of steel in service [26]. At present, most of the research on acidizing corrosion inhibitors for duplex stainless steel was mainly focused on the acid solution system under 120 °C [12,27–30] or high-temperature, low-concentration acid [31]. Hirotaka et al. [28] adopted 7 wt.% HCl at 60 °C as the test medium to simulate the acidizing bottom-hole conditions of onshore oil fields in Japan. The corrosion inhibition mechanism of 25 wt.% cinnamaldehyde + 20 wt.% long-chain alkyl imidazoline + 55 wt.% methanol plus $CuI/KI/Cu/CuCl_2/CuSO_4$ on 2205 duplex stainless steel in an acid system was studied, the results showed that the corrosion inhibition rate was only 34.9 wt.%, and the ferrite selective corrosion could be observed, and the corrosion inhibition rate reaches more than 98% after the addition of Cu compound, and no selective corrosion could be seen. Wang et al. [12] showed that the corrosion rate of 2205 duplex stainless steel was 26.8053 $g \cdot m^{-2} \cdot h^{-1}$ after adding corrosion inhibitor in 10 wt.% HCl + 1.5% wt.HF + 3 wt.% HAC solution at 120 °C, but ferrite selective corrosion could be seen on the surface of the sample. Du et al. [31] showed that the corrosion rate of

2205 duplex stainless steel could be controlled at 40.3092 g·m^{-2}·h^{-1} in the 4 wt.% HCl + 3 wt.% CH$_3$COOH + 5 wt.% QASE + 1 wt.% Sb$_2$O$_3$ at 200 °C test environment.

Molecular simulation techniques play an important role in the modern characterization of materials [32], which are beneficial to the study of corrosion behavior by increasing the understanding of chemical and physical processes at the molecular and atomic levels. Cl$^-$ has a great influence on metal pitting corrosion, and the molecular dynamics (MD) simulations method is also used to study the pitting behavior of Cl$^-$ on metals [32–36]. By using molecular dynamics simulation, Sepehr Y et al. [33] studied the pitting corrosion behavior of nano-diamonds. Chen et al. [34], using MD simulation, found that the mixed corrosive anions (Cl$^-$ and HSO$_3^-$) adsorbed much more strongly to the passive film, and their diffusion coefficient was also significantly improved compared to the solution containing only one kind of anion.

The above-mentioned acidizing corrosion inhibitor had a certain corrosion inhibition effect on 2205 duplex stainless steel. However, there have been no reports on corrosion inhibitors suitable for ultra-high temperature and high-concentration acid solution systems. Herein, it is essential to develop a corrosion inhibitor suitable for the ultra-high-temperature and high-concentration acidizing environment and to clarify its corrosion inhibition mechanism, which is critical to prolonging the service life of duplex stainless steel. In this paper, quantum chemical calculation and molecular dynamics simulation were applied to design a new acid corrosion inhibitor, and the suitability was further verified by high-temperature and high-pressure corrosion simulation tests. This research has great engineering application value to broaden the selection of materials for ultra-high-temperature and high-pressure wells and reduce the cost of oil and gas field string.

2. Research Methods

2.1. Test Materials

The special ultra-high-temperature acidizing corrosion inhibitor for duplex stainless steel was developed by China Petroleum Engineering Materials Research Institute Co., Ltd., Xi'an, China. and it is mainly composed of 40 wt.% Mannich base + 6 wt.% antimony trioxide + 15 wt.% hydrochloric acid + 5 wt.% OP + 34 wt.% DMF.

The test 2205 duplex stainless steel was developed and produced by Zhejiang Jiuli Special Material Technology Co., Ltd., Huzhou China. and its chemical composition was 0.019 wt.% C, 0.41 wt.%Si, 0.63 wt.% Mn, 0.023 wt.% P, 0.005 wt.% S, 2.38 wt.% Cr, 52 wt.% Mo, 5.47 wt.% Ni, 0.181 wt.%N and residual Fe. The size of the high-temperature and high-pressure corrosion test sample was cut from a pipe material with the dimension of 50 mm × 10 mm × 3 mm.

Before testing, the samples were ground by using the abrasive paper from 400 # to 800 # step by step, then immersed in anhydrous ethanol for 5 min by ultrasonic cleaning, dried with cold air, placed in a dryer for at least 1 h, weighed (accurate to 0.0001 g), and measured the size (accurate to 0.01 mm). The samples should be mutually insulated and installed on the fixture during the test. After testing, the acid solution on the surface of the sample was removed by deionization, the corrosion inhibitor film on the surface of the sample was removed by N, N–dimethylformamide, and then the sample was ultrasonically cleaned with anhydrous ethanol for 5 min, dried by cold air, placed in a desiccator for at least 1 h and weighed (accurate to 0.0001 g).

2.2. High-Temperature and High-Pressure Corrosion Test

2.2.1. Test Conditions

The typical operating temperature and acidizing operation system of an oil and gas field in western China were selected to evaluate the performance of corrosion inhibitors. The test period was selected based on normal operation time and possible long operation time in the oil field. Table 1 shows the specific test conditions.

Table 1. The specific corrosion inhibitor test conditions of 2205 duplex stainless steel in ultra-high-temperature acidizing.

Medium	Temperature (°C)	Time (h)	Concentration (wt.%)
9 wt.%HCl + 1.5 wt.% HF + 3 wt.% CH$_3$COOH	140	4	4
	160	4	5
	180	4	6
	180	12	6

The high-temperature and high-pressure test was carried out in TFCZ5 25/450 autoclave, which the test temperature ≤ 450 °C and pressure ≤ 25 MPa.

The uniform corrosion rate was calculated by the weight-loss method according to the following equation:

$$V = \frac{10^6 \Delta m}{A \cdot \Delta t} \quad (1)$$

where V is uniform corrosion rate; g·m^{-2}·h^{-1}; Δt is reaction time, h; Δm is weight loss, g; and A is surface area of test sample, mm^2.

2.2.2. Microstructure Analysis

The depth of local corrosion was measured and counted by the Smart Zoom5 (Smart Zoom 5, Zeiss, Oberkochen, Germany), with a magnification of 10–500×. The phase composition of 2205 duplex stainless steel was analyzed by an X-ray diffractometer (SmartLab, Rigaku, Tokyo, Japan). The step length was 0.02°, the time of each step was 5 s, and the angular spacing was between 30° and 90°. Scanning electron microscope (SEM, JSM-IT500LA, JEOL Ltd., Japan) and energy spectrum analysis (EDS, JEOL Ltd., Japan) were used to test the corrosion morphology, inhibitor film thickness, and the composition of the film after high-temperature and high-pressure tests. The magnification was 5–300,000×, the secondary electron was 3.0 nm (30 kV), and the backscattered electron was 4.0 mm (30 kV). When measuring the thickness of the film, the sample was embedded in the epoxy resin, and a 50 mm × 3 mm section was observed. Before observing the inner film of the corrosion inhibitor, soak the sample in N, N-dimethylformamide for 10 s to remove the outer film, and then wash it with clean water, dehydrate it with anhydrous ethanol, and dry it with cold air.

2.3. Molecular Dynamics Simulation

The molecular dynamics simulation software Accelrys MS Modeling 7.2 (Accelrys, San Diego, CA, USA) was used to build the model of the water-Fe corrosion inhibitor system, and the diffusion behavior of the corrosion medium particles H$_2$O, H$_3$O$^+$, and Cl$^-$ in the corrosion inhibitor film was established and calculated.

2.3.1. Construction of Water-Fe-Corrosion Inhibitor System Model

Firstly, construct a cell of metal Fe, cut the cell along the (001) plane into a surface with a thickness of 17.198 Å, and construct it into (11 × 11) Two-dimensional supercell, and then build it as 31.53 Å × 31.53 Å × 15.76 Å three-dimensional supercell, and 31.53 Å × 31.53 Å × 24.06 Å of liquid water three-dimensional amorphous unit is constructed by the Amorphous Cell module. The constructed inhibitor molecules were immersed into the amorphous unit. The molecular positions of water and corrosion inhibitor are determined randomly. Finally, a three-layer supercell structure was constructed by the "vacuum layer-corrosion inhibitor solution layer-metal Fe supercell" in top-down order. In order to study the interaction between the inhibitor molecule and the Fe surface, the inhibitor molecule was manually moved to the appropriate position near the Fe surface, and the position was appropriate to not form a short-range interaction. The simulation process adopts a group-based truncation method (that is, group-based truncation), with a truncation radius of 9.5 Å, to ensure the calculation accuracy and minimize the calculation time. Then the Fe layer of the final structure is fixed. The smart method is used to minimize the energy of the built three-layer supercell structure by 5000 steps to remove the local high potential energy points for dynamic balance and data acquisition.

The NVT ensemble is used for dynamic balance and data acquisition. The Andersen temperature control method was used to conduct dynamic balance of 160 ps at first and then conduct data acquisition of 80 ps. One track file was output every 800 fs, and a total of 100 track files were output. The time step in the simulation process was 0.8 fs. COMPASS force field was used in the simulation process.

2.3.2. Establishment and Calculation of Diffusion Behavior Model of Corrosion Medium Particles H_2O and H_3O^+ in Corrosion Inhibitor Film

Three corrosion particles, H_2O, H_3O^+, and Cl^-, were selected. The simulation system consists of one corrosion medium particle and 100 corrosion inhibitor molecules. Firstly, the Amorphous Cell module was used to build an amorphous structure containing 100 corrosion inhibitor molecules with periodic boundary conditions, and then the NPT ensemble was chosen to simulate the system with molecular dynamics at 298 K at one atmospheric pressure. The time step was 1 fs, and the total simulation time was 300 ps. After balancing, the average density of the system was calculated. Then the Particle number, Volume, Temperature(NVT) ensemble was selected to simulate the dynamics of the system. The temperature was 298 K, the time step was 1fs, and the total simulation time was 200 ps. One frame was output every 2 ps, a total of 100 frames. The COMPASS force field was used to optimize the system in all dynamic simulations, which were completed through the Fortite module. The charge group method was used for the interaction between van der Waals and Coulomb, and the truncation radius was 9.5 nm. The charge of each anion and cation was assigned by using the current method, and the universal force field was applied to define the potential energy [33].

3. Experimental Results and Analysis

3.1. Microstructure

According to ASTM A923-2014, the polished sample of 2205 was etched in 40 g reagent grade NaOH plus 100 g water solution weight at 2 V DC for 20 s to obtain the structure in Figure 1. It can be seen that the microstructure of 2205 duplex stainless steel consists of an elongated austenite phase (γ, light color) inlaid on a ferrite matrix (α, Dark), the two phases are evenly distributed, and the contents of ferrite and austenite phases are 49.13% and 50.87% respectively.

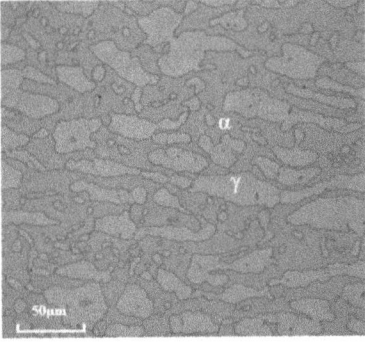

Figure 1. Metallographic morphology of 2205 duplex stainless steel for test.

3.2. Optimization Calculation and Design of Corrosion Inhibitor

The quantum chemical method was used to calculate the geometric optimization of the corrosion inhibitor, and the corresponding frontier orbital charge distribution was obtained. The results are shown in Figure 2. The electronic density of the nucleophilic frontier orbital (HOMO) of the organic matter in the corrosion inhibitor is mainly distributed on the O and N atoms in the molecule (Figure 2a), and the electronic density of the electrophilic frontier orbital (LUMO) is respectively distributed on the benzene ring (Figure 2b, providing electrons or receiving electrons on the 4 s orbital of the Fe atom respectively.

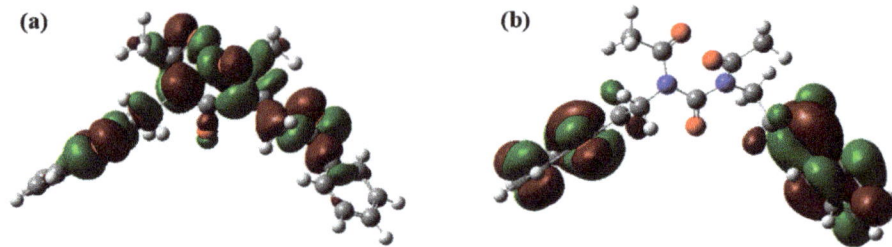

Figure 2. Charge distribution of organic matter frontier orbital in corrosion inhibitor ((**a**)-nucleophilic frontier orbital HOMO, (**b**)-electrophilic frontier orbital LUMO).

The adsorption behavior of the corrosion inhibitor and Fe surface in an aqueous solution was simulated by molecular dynamics, and the kinetic equilibrium configuration of the corresponding system at different temperatures was obtained (Figure 3). Figure 4 shows the adsorption mechanism of organic corrosion inhibitors on the iron surface. The combination of chemical adsorption and physical adsorption could form a protective film to reduce the corrosion of acid solution on the metal surface. When the corrosion inhibitor interacts with the iron surface, the O and N atoms in the benzene ring and molecule are close to the iron surface and tend to adsorb on the iron surface in parallel to form an adsorption layer. The alkyl chain deviates from the iron surface at a certain angle to form a thicker hydrophobic layer (Figures 3 and 4). With the increase in temperature, the planarity of adsorption between the benzene ring and the iron surface becomes weaker, indicating that the adsorption capacity of the molecule and iron surface would decrease with the increase in temperature (Figure 3c).

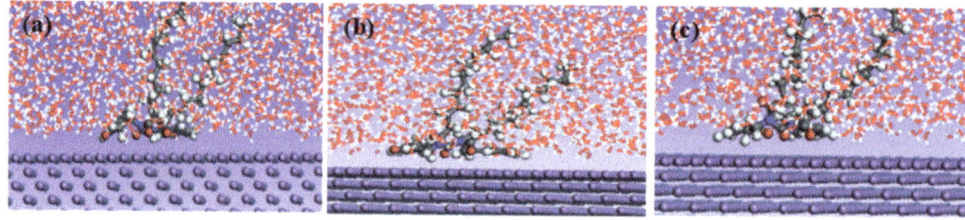

Figure 3. Adsorption behavior of corrosion inhibitor and Fe surface in different temperature systems. ((**a**) 140 °C, (**b**) 160 °C, (**c**) 180 °C).

The interaction energy between the inhibitor molecule and the iron surface at different temperatures was calculated according to the track file collected during the dynamic simulation [37].

$$\Delta E = E_{inhibitor+Fe} - E_{Fe} - E_{inhibitor} \qquad (2)$$

where ΔE is the interaction energy of the corrosion inhibitor and iron surface, $E_{inhibitor+Fe}$ is the total energy of the corrosion inhibitor and iron surface, E_{Fe} is the energy of the iron surface, and $E_{inhibitor}$ is the energy of the inhibitor molecule. The binding energy is defined as the negative value of the interaction energy, i.e., $E_{binding} = -\Delta E$.

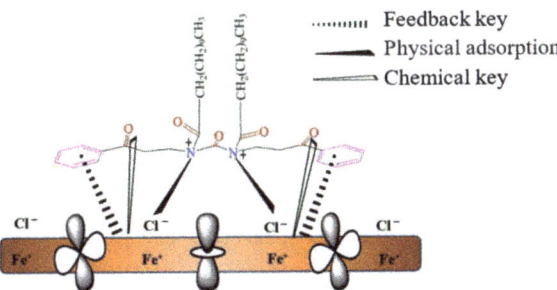

Figure 4. Adsorption mechanism of organic corrosion inhibitor on iron surface.

Table 2 depicts the calculated binding energy of the corrosion inhibitor and iron surface in an aqueous solution at different temperatures. The binding energy of the corrosion inhibitor and iron surface decreases gradually with the increase in temperature. The temperature has a great impact on the adsorption performance of corrosion inhibitors. Desorption is easy to occur with the increase in temperature, thus reducing its adsorption performance.

Table 2. Binding energy of corrosion inhibitor and iron surface in aqueous solution at different temperatures.

Temperatures (°C)	140	160	180
Binding energy (kcal/mol)	188.39	183.26	180.75

Corrosion inhibitor molecules could adsorb on the metal surface to form a protective film exhibiting corrosion inhibition performance, which slows down the anode reaction and the diffusion of metal ions, thus preventing the corrosion of the corrosive medium particles on the metal. The most important criterion of corrosion inhibition performance is whether corrosion inhibitors can adsorb on the metal surface to form a stable, protective film and maintain a high coverage for a long period of time, and can effectively prevent the migration of corrosive medium particles to the metal surface so as to block the reaction path of corrosion, which could be judged by the diffusion of the corrosion medium particles in the corrosion inhibitor film: "the stronger the diffusion ability, the worse the barrier performance of the film and the corrosion inhibitor molecules, on the contrary, the corrosion inhibition is stronger [38].

The diffusion behavior of corrosion medium particles H_2O, H_3O^+, and Cl^- in corrosion inhibitor film was studied by molecular dynamics simulation. The diffusion coefficient of corrosion medium particles in the inhibitor film and the interaction energy between the inhibitor film and two kinds of corrosion medium particles were calculated. The diffusion coefficient is the most direct measure of the diffusion and migration ability of particles in the system. The larger the diffusion coefficient, the stronger the diffusion and migration ability of particles in the system, and the weaker the diffusion and migration ability of particles in the system.

According to Einstein's relation [39], the diffusion coefficient can be expressed as Equations (3) and (4):

$$D = \frac{1}{6} \lim_{t \to \infty} \frac{d}{dt} \sum_{i}^{n} \langle |R_i(t) - R_i(0)| \rangle^2 \qquad (3)$$

$$D = \frac{1}{6} \lim_{t \to \infty} \frac{dMSD}{dt} \qquad (4)$$

By applying the finite difference approximation, the diffusion coefficient can be expressed as Equation (5):

$$D = m/6 \qquad (5)$$

where m is the slope of the MSD curve, and its value can be directly calculated by the software.

Table 3 shows the diffusion coefficients of H_2O and corrosive medium particles in the inhibitor film. It shows that the diffusion coefficient is significantly reduced compared with that in the water system. The corrosion inhibitor film has a good blocking effect on the diffusion of corrosive medium particles, which could avoid its migration to the metal interface as much as possible and effectively inhibit the corrosion of metals.

Table 3. Diffusion coefficient of two corrosion media particles in the inhibitor film, the self-diffusion coefficient of water is 2.34×10^{-9} m^2/s.

System	Corrosion Inhibitor -H_2O	Corrosion Inhibitor -H_3O^+
Diffusion coefficient (10^{-9} m^2/s)	0.0108	0.0099

Table 4 shows the diffusion coefficients of Cl^- in inhibitor film at different temperatures. The diffusion coefficient of Cl^- is relatively large in an aqueous solution, and it is easy to diffuse and migrate, resulting in corrosion pits. After the addition of the corrosion inhibitor, the diffusion coefficient decreases sharply, indicating that the diffusion and migration of Cl^- were effectively prevented and pitting is mitigated. The diffusion coefficient increases with the increase in temperature and the pitting corrosion is accelerated.

Table 4. Diffusion coefficients of Cl^- in corrosion inhibitor films at different temperatures.

System	H_2O [40]	Corrosion Inhibitor		
Temperature		140	160	180
Diffusion coefficient (10^{-9} m^2/s)	0.3220	0.0177	0.0189	0.0226

The interaction will have a greater impact on the diffusion and migration behavior of particles in it. If the interaction energy is positive, the medium has a repulsive effect on the particle, which can accelerate the diffusion of the particle. On the contrary, which means the medium is attractive to hinder the diffusion of particles. The interaction energy between corrosion medium particles and corrosion inhibitor film can be expressed by the following Equation:

$$E = E_{film+particle} - E_{film} - E_{particle} \qquad (6)$$

where E represents the interaction energy of particle and film, $E_{film+particle}$ is the total energy of particle and film, E_{film} is the energy of the corrosion inhibitor film, and $E_{particle}$ is the energy of the corrosion medium particles.

Table 5 shows the interaction energy between corrosion medium particles and corrosion inhibitor film. It can be seen that the interaction energy of water molecules and the inhibitor film is much smaller than that of the interaction energy of H_3O^+ and the inhibitor film, indicating that there may be a stronger interaction between H_3O^+ and the inhibitor film, which has excellent corrosion inhibition effect.

Table 5. Interaction energy of corrosion medium particles and corrosion inhibitor film.

System	Corrosion Inhibitor Film -H_2O	Corrosion Inhibitor Film -H_3O^+
E(kcal/mol)	−99.25	−155.91

Considering the serious desorption of organic corrosion inhibitors at high temperatures, the formulation is to combine organic matter and inorganic salts that can form complexes so as to achieve rapid and stable film formation at high temperatures.

3.3. Corrosion Inhibition Mechanism Analysis at High-Temperature and High-Pressure

3.3.1. Effect of Temperature on Performance of Corrosion Inhibitor

Figures 5–7 show the uniform corrosion rate, inhibitor film thickness, and local corrosion depth of 2205 duplex stainless steel in acid solution at 140 °C, 160 °C, and 180 °C, respectively. It can be seen that the uniform corrosion rate shows no significant difference between 140 °C and 160 °C but rises sharply at 180 °C, which is twice that at 160 °C, which is far lower than the commonly accepted 81 $g \cdot m^{-2} \cdot h^{-1}$ [41] and 26.8053 $g \cdot m^{-2} \cdot h^{-1}$ of the 2205 duplex stainless steel in the reported 120 °C 15 wt.% HCl + 1.5% wt. HF + 3 wt.% HAC + 5.1 wt.% acid corrosion inhibitor [12]. The thickness of both layers showed a decreasing trend with the increase in temperature. When combined with the local corrosion and the morphology of the corrosion inhibitor film, it can be seen that the density of the film decreases with the increase in temperature.

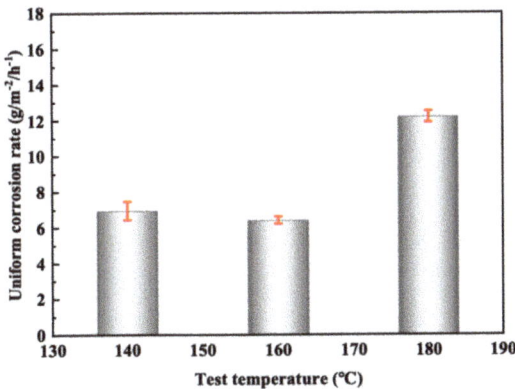

Figure 5. Uniform corrosion rate of 2205 duplex stainless steel at different temperatures.

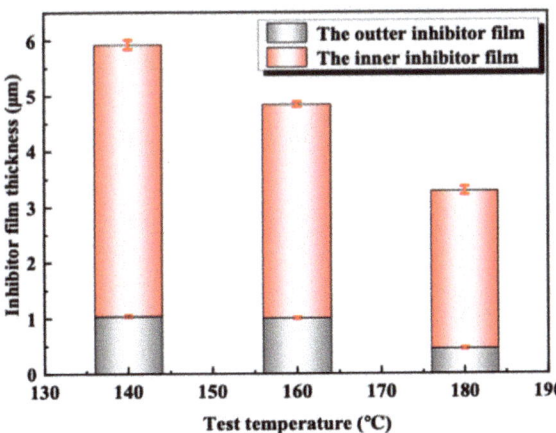

Figure 6. Thickness of corrosion inhibitor film on 2205 duplex stainless steel at different temperatures.

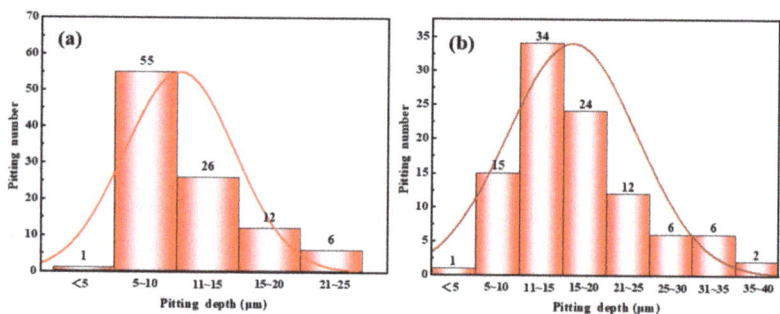

Figure 7. Localized corrosion depth of 2205 duplex stainless steel in acid solution system at different temperatures ((**a**) 160 °C; (**b**) 180 °C).

No obvious localized corrosion was observed on the specimen surface after the 140 °C test. But the sample surface appeared to have localized corrosion at 160 °C and 180 °C tests. The maximum localized corrosion depth of the 160 °C sample is 22 μm. The average localized corrosion depth is 10 μm. The depth of the corrosion pit is mainly concentrated in 5~10 μm. The maximum localized corrosion depth of 180 °C sample surface is 37 μm. The average localized corrosion depth is 16 μm. The depth of the corrosion pit is mainly concentrated at 11~15 μm. It can be seen that the localized corrosion depth increases with the increase in test temperature.

Figure 8 shows the SEM morphology of the corrosion inhibitor film at different temperatures. At 140 °C, the inner and outer films of the corrosion inhibitor on the surface of the sample were relatively intact without obvious damage. With the increase in temperature, the corrosion inhibitor film was damaged, and the inner and outer films on the surface of the 180 °C sample were damaged. Thus, localized corrosion occurred.

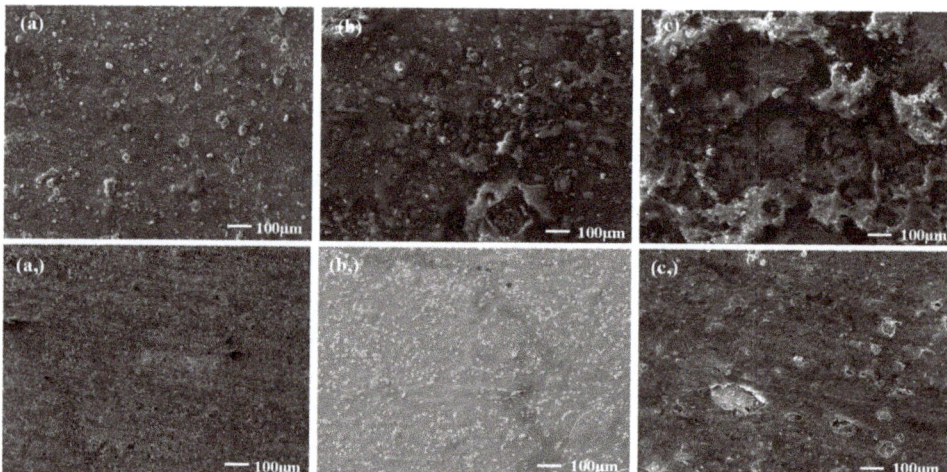

Figure 8. Micromorphology of corrosion inhibitor films on the surface of 2205 duplex stainless steel sample at 100× magnification ((**a**) 140 °C the outer film, (**a,**) 140 °C the inner film, (**b**) 160 °C outer film, (**b,**) 160 °C the inner film, (**c**) 180 °C outer film, (**c,**) 180 °C the inner film).

The EDS results of the main element contents of outer and inner films are displayed in Table 5. The main elements of the film are C, O, Sb, Cl, Fe, S and Cr, and the content of C in the outer film is much higher than that of in the inner film, while the content of Sb in the inner film is much higher than that of in the outer film. The contents of C and Sb indicate that the outer layer is mainly organic film, and the inner layer is mainly inorganic salt. The content of Cl^- in the outer film is higher than that of in the inner film, indicating that the corrosion inhibitor film has an excellent blocking effect in the acid solution, which is consistent with the simulation results (see Table 6). The content of Fe in the inner film increases with the increase in temperature, indicating that the corrosion resistance of the film is reduced.

Table 6. Main element contents EDS analysis result of outer and inner films at different test temperatures (wt.%).

Temperature (°C)	Test Location	C	O	F	S	Cl	Cr	Fe	Ni	Mo	Sb
140	Outer film	40.38	7.39	-	1.27	5.42	0.37	2.22	0.52	-	42.43
	Inner film	4.63	2.94	-	0.48	0.89	1.35	5.3	-	-	78.99
160	Outer film	53.95	7.47	1.84	0.98	7.77	-	1.23	-	-	26.33
	Inner film	3.12	-	-	-	-	1.26	5.30	-	-	89.64
180	Outer film	64.55	7.62	-	1.04	7.81	0.52	1.82	2.36	-	14.28
	Inner film	6.84	7.64	-	0.28	1.84	5.54	15.22	18.32	1.78	42.53

3.3.2. Effect of Time on Performance of Corrosion Inhibitor

The uniform corrosion rate of 2205 duplex stainless steel after 4 h and 12 h tests in acid solution at 180 °C is shown in Figure 9. It can be seen with the extension of test cycles, and the corrosion rate generally shows a decreasing trend. Figure 10 shows the thickness of two corrosion inhibitor film layers at different test cycles. With the extension of the test cycles, the thickness of the two layers displays an increasing trend. According to the local corrosion situation and the micro-morphology of the inhibitor film, although the film thickness increases with the increase in temperature, the compactness decreases, and the localized corrosion density and depth increase.

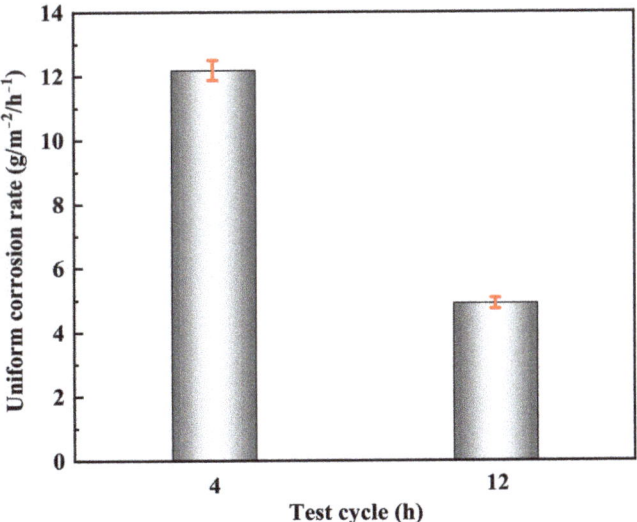

Figure 9. Uniform corrosion rate at different test cycle.

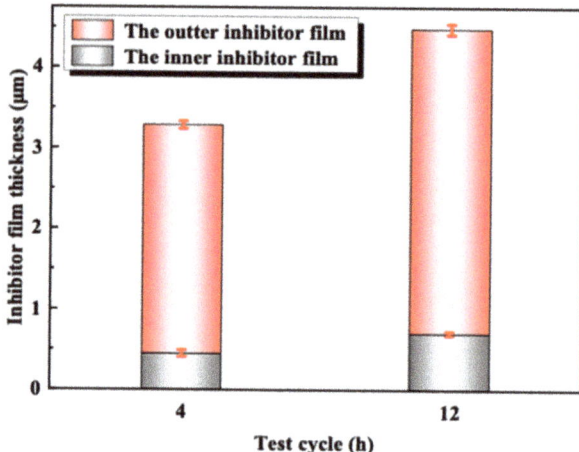

Figure 10. Thickness measurement results of corrosion inhibitor film at different test cycles.

The cross-section and surface micromorphology of the inhibitor film at different test cycles shows in Figure 11. The two film layers are relatively complete in the fresh acid medium at 180 °C, which has a good protective effect on the substrate, but the local film layer is damaged, which leads to pitting corrosion on the surface of the sample. Although the thickness of the film in the 12 h test was significantly larger than that in the 4 h test, the compactness of the film was not significantly improved, and the degree of damage was more serious than that in the 4 h test.

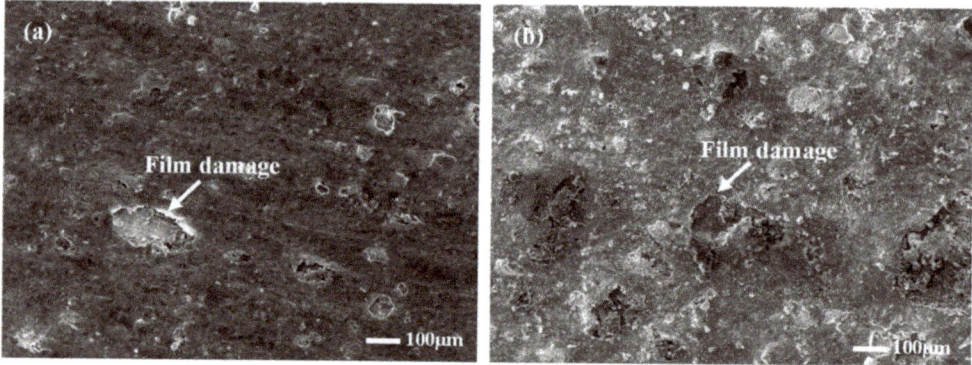

Figure 11. Micromorphology of the inner film of the corrosion inhibitor on the surface of the samples at different test cycles magnified by 100× ((**a**) test cycle 4 h; (**b**) test cycle 12 h).

Figures 12 and 13 display the statistical results of pitting depth and the surface microscopic corrosion morphology of samples after different test cycles. It can be seen from Figure 12a that the maximum pitting depth is 37 μm, the average pitting depth is 16 μm, and the depth of the corrosion pit is mainly concentrated at 11~15 μm at 180 °C for 4 h. The maximum pitting depth is 64 μm, the average pitting depth is 23 μm, and the depth of the corrosion pit is mainly concentrated in 21~25 μm at 180 °C for 12 h (see Figure 12b).

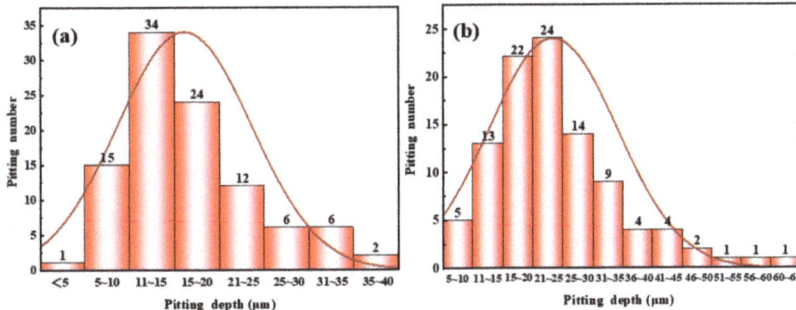

Figure 12. Normal distribution of pitting depth on the surface of samples at different test cycles (**a**) 4 h; (**b**) 12 h.

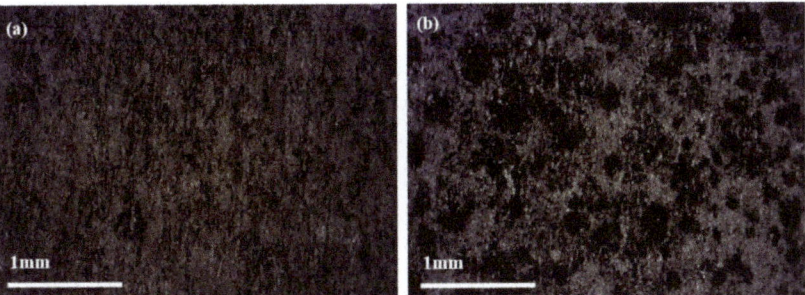

Figure 13. Micromorphology of sample surface magnified by 100× under optical microscope at different test cycles: (**a**) 4 h, (**b**) 12 h.

3.3.3. Corrosion Inhibitor Layer Analysis

In the acid solution system with high temperature and high concentration, the common organic corrosion inhibitor is seriously desorbed at high temperature, and its adsorption capacity differs greatly from that of austenite and ferrite, which is prone to selective corrosion and difficult to play a good protective role. The corrosion inhibitor in this study constructs a double-layer membrane adsorption structure, which uses inorganic substances with good stability at high temperature to form a complex with the metal matrix to enhance the binding ability with the metal matrix and maintain the high-temperature stability of the film. However, the inorganic film is not dense enough; organic film is used to supplement, as shown in Figure 14a. The microscopic morphology of the film cross-section after adding corrosion inhibitor in the 180 °C acid solution is displayed in Figure 14b. It can be seen the inhibitor has a double-layer membrane structure, which is consistent with the expected results of the inhibitor mechanism. Inorganic salt has a good binding ability with metal matrix at high-temperature. However, the film is easy to crack at high-temperature and forms a channel for acid diffusion. The crack defect could be repaired by the organic film, making the corrosion inhibitor film more dense and better protective for the matrix.

The cross-section of the line scan results of the corrosion inhibitor film is shown in Figure 15. The outer film and the inner film is mainly composed of element C and Sb, respectively, which is consistent with Table 5. From the results of line scanning, it can be seen that the organic film and the inorganic salt film penetrate into each other and combine closely.

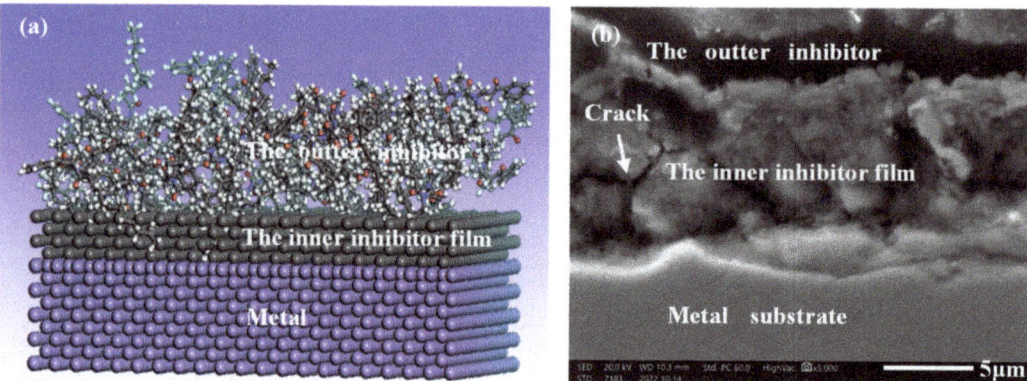

Figure 14. Structure of high-temperature acidizing corrosion inhibitor double-layer film ((**a**) schematic diagram, (**b**) section SEM).

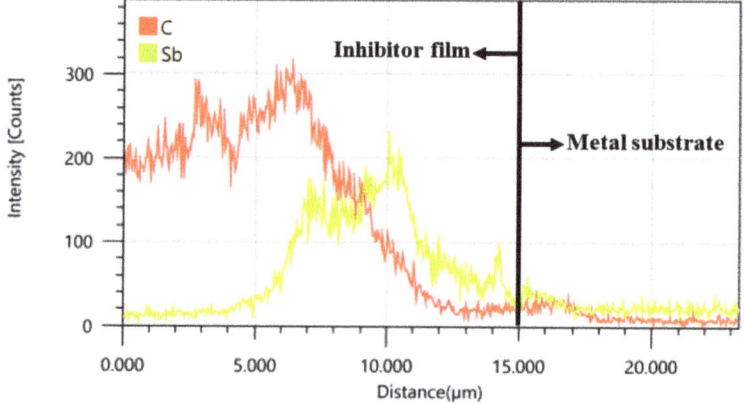

Figure 15. Line Scanning Results of Element Distribution of Corrosion Inhibitor Film.

Figure 16 shows the cross-section morphology of 2205 duplex stainless steel after testing in 180 °C acid solution with a corrosion inhibitor. No selective corrosion was observed. Figure 17 shows the XRD analysis results of the sample surface after the 180 °C test without corrosion inhibitor and with corrosion inhibitor. The peaks of the austenitic phase are much stronger than that of the ferritic phase in tests without the corrosion inhibitors. However, the peaks of the ferritic phase are as strong as those of the austenitic phase when the corrosion inhibitor is added. The results show that without adding a corrosion inhibitor, selective ferrite corrosion could be observed in high-temperature and high-concentration acid solutions, and selective corrosion is effectively inhibited after adding a corrosion inhibitor.

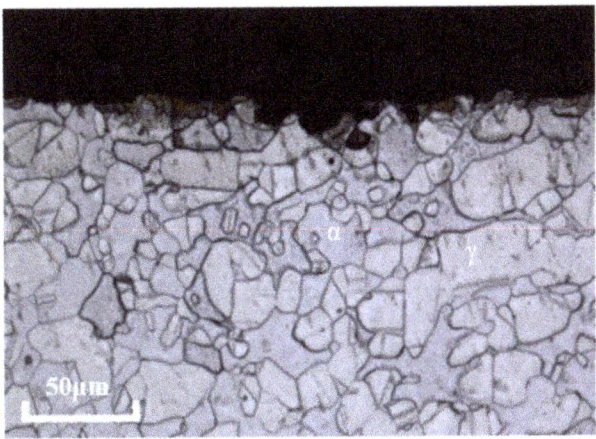

Figure 16. Cross-sectional morphology of 2205 duplex stainless steel sample without corrosion product film after testing in 180 °C acid solution with corrosion inhibitor.

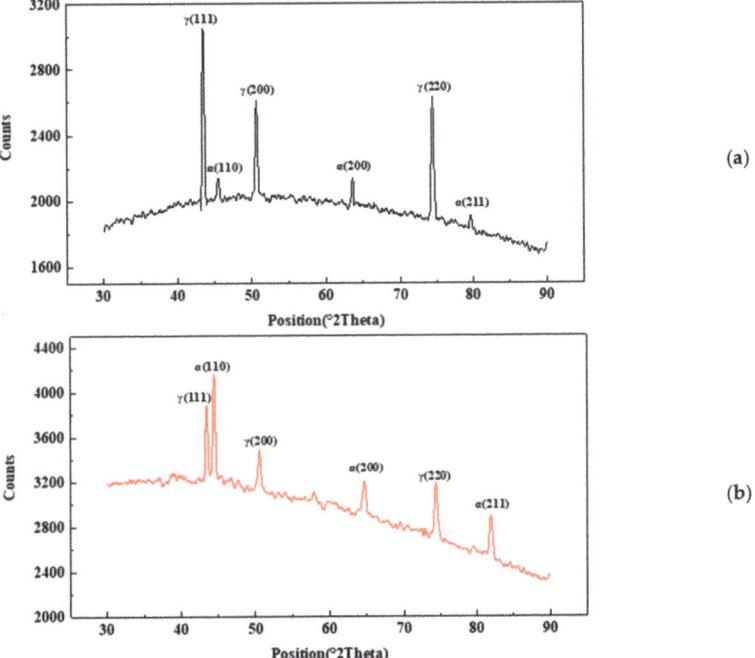

Figure 17. XRD analysis results of samples without and with corrosion inhibitor ((**a**) 180 °C without inhibitor, (**b**) 180 °C with inhibitor).

4. Discussion

Figure 18 shows the corrosion mechanism of duplex steel in a high-temperature and high-concentration acid solution system. Minor differences could be seen in the element composition of 2205 duplex stainless steel. Cr, Mo, Ni, and N tend to be concentrated in the ferrite and austenite phases, respectively [26,42,43]. The distribution of elements in the phase is a key factor affecting selective corrosion. The 2205 duplex stainless steel is in an active state when working with high-temperature and high-concentration hydrochloric acid and hydrochloric acid plus hydrofluoric acid [12,26]. Because the potential of Cr is more negative than that of Fe, Cr is more likely to lose electrons and form ions than Fe, and the corrosion of ferrite is more serious than that of austenite. The possible chemical reactions are:

anodic [44]:

$$Cr \rightarrow Cr^{3+} + 3e \tag{7}$$

$$Fe \rightarrow Fe^{2+} + 2e \tag{8}$$

cathodic [12]:

$$2H^+ + e \rightarrow H_2 \tag{9}$$

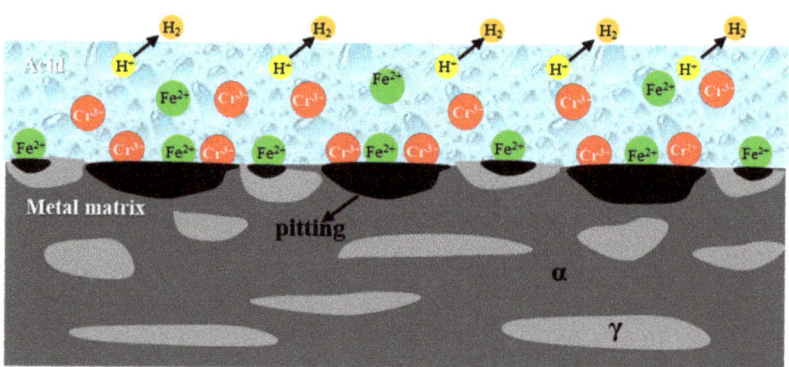

Figure 18. Corrosion mechanism of duplex stainless steel in high-temperature and high-concentration acid solution system.

Figure 19 shows the corrosion inhibition mechanism of acid corrosion inhibitors in high-temperature and high-concentration acid solution systems. The passive film of 2205 duplex stainless steel was dissolved in those harsh environments. Using high-temperature acid corrosion inhibitors can promote the steel to form a protective film on its surface, thus isolating the metal from the acid solution system and reducing the corrosion process. The SEM (see Figure 14) and EDS (see Figure 15) results all confirmed that the inhibitor film shows a double layer, and the inorganic film combined with the substrate forms a coordination bond, which enhances the stability of the inhibitor film at high-temperature. The outer layer is an organic layer. The quantum chemical calculation results show that the organic corrosion inhibitor molecule mainly takes the benzene ring and the O and N atoms in the molecule as the main adsorption sites. The O and N atoms have solitary electron pairs, which will coordinate with the empty orbit on the metal surface in the anode region to form an adsorption film [45] or adsorb at the defects of the inorganic film layer so as to improve the overall compactness of the film layer. The results of molecular dynamics simulation of the diffusion behavior of corrosion medium particles in the organic corrosion inhibitor film also show a good blocking effect on the corrosion media and improves the corrosion resistance of the double-layer film. The simulation result is confirmed by the EDS test results in Table 5, the content of Cl^- in the inner membrane is far lower than that of in the outer membrane, and the outer film provides a good barrier and supplement for the inner membrane. When the temperature rises from 140 °C~180 °C, the binding energy of the organic corrosion inhibitor decreases from 188.39 kcal·mol^{-1} to 180.75 kcal·mol^{-1}. The high-temperature corrosion simulation test results demonstrate that there is no obvious local corrosion on the surface of the sample at 140 °C, and there is obvious local corrosion on the surface of the test sample at 160–180 °C, and the pitting corrosion is deeper at higher temperatures.

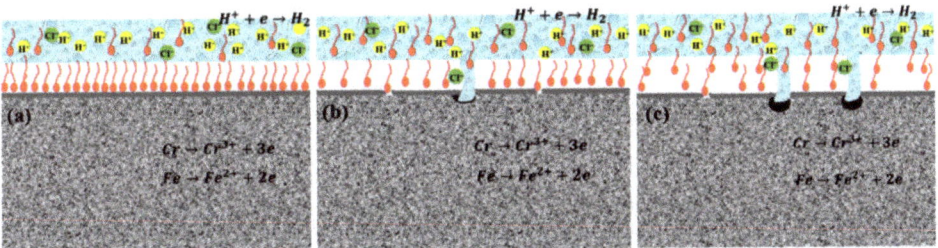

Figure 19. Schematic diagram of corrosion inhibition mechanism of acidizing corrosion inhibitor for duplex stainless steel in high-temperature and high-concentration acid solution system ((**a**) 140 °C, 4h (**b**) 180 °C, 4 h, (**c**) 180 °C, 12 h).

The corrosion inhibitor bilayer film is relatively dense when the temperature is low (see Figure 19a). The inorganic film is cracked or damaged with the increase in the experimental temperature, and the organic matter is adsorbed, which plays a supplementary role in the film layer (see Figure 19b). However, the binding energy of the organic matter and the adsorption property is reduced with the increase in temperature. Parts of the damaged film could not be repaired, leading to the base metal contact with the acid solution, then localized corrosion occurred (see Figure 19c). Although a thicker film could be formed with the further extension of test cycles, a loose inorganic film would be formed in the inner layer, and the organic film on the outer layer was also not intact. Hence, the local corrosion depth and scope were greatly increased (see Figure 19c). To sum up, the compactness of corrosion inhibitor film will be affected by the higher testing temperature and longer cycles.

5. Conclusions

In this paper, the corrosion inhibition performance of the self-designed ultra-high temperature acidizing corrosion inhibitor was studied by using quantum chemical calculation, MD simulation, SEM, EDS, and XRD. The main conclusions are summarized as follows:

(1) The 2205 duplex stainless steel ultra-high-temperature acidizing corrosion inhibitor showed excellent corrosion effect in the 180 °C 9 wt. % HCl + 1.5 wt. % of HF + 3 wt. % CH_3COOH acid system, the corrosion rate can be reduced to 12.1881 $g \cdot m^{-2} \cdot h^{-1}$ and is far below the commonly accepted 81 $g \cdot m^{-2} \cdot h^{-1}$, and no selective corrosion was observed.

(2) The ultra-high temperature acidizing corrosion inhibitor has a double-layer membrane structure, and the inorganic membrane bonded with the substrate is mainly composed of Sb element, and the outer layer is an organic membrane mainly composed of C element.

(3) The ultra-high temperature acidizing corrosion inhibitor had a good blocking effect on the diffusion of corrosive medium particles. With the increase in temperature, the binding energy of the corrosion inhibitor and substrate decreases, the blocking effect of corrosion inhibitor film on Cl^- also decreases, and pitting corrosion could easily be detected.

(4) With the increase in temperature and the extension of test time, the density of corrosion inhibitor film could be damaged, and the corrosion inhibition effect decreased, which was manifested as the increase in local corrosion depth and diameter.

(5) The developed acidizing corrosion inhibitor has great engineering application value to broaden the selection of materials for ultra-high-temperature and high-pressure wells and reduce the cost of oil and gas field string.

Author Contributions: Investigation, D.L., W.S., M.Z., H.C. and Z.L.; Writing—original draft, D.L.; Methodology, J.Z. (Junping Zhang); Writing—review & editing, J.Z. (Junping Zhang), C.Y. and J.Z. (Juantao Zhang); Formal analysis, L.F., W.L. and X.L.; Project administration, L.F.; Data curation, W.L. and X.L. All authors have read and agreed to the published version of the manuscript.

Funding: This research received no external funding.

Institutional Review Board Statement: Not applicable.

Informed Consent Statement: Not applicable.

Data Availability Statement: Not applicable.

Conflicts of Interest: The authors declare no conflict of interest.

References

1. Ming, L.; Yongcun, F.; Yun, G.; Jingen, D.; Cheng, H. Development status and prospect of key technologies for high temperature and high pressure drilling. *Pet. Sci. Bull.* **2021**, *6*, 228–244.
2. Jin-Cheng, R.; Jin, L.; Yu-Chun, S.; Liang-ying, C.; Ping, Z. Oil Testing Technology of L17 Ultra High Pressure Gas Well in Sichuan Basin. *Nat. Gas Oil* **2008**, *28*, 58–60.
3. Shadravan, A.; Amani, M. HPHT 101—What every engineer or geoscientist should know about high pressure high temperature wells. In Proceedings of the SPE Kuwait International Petroleum Conference and Exhibition, Kuwait City, Kuwait, 10–12 December 2012; Volume 2, pp. 917–943. [CrossRef]
4. Ueda, M.; Omura, T.; Nakamura, S.; Abe, T.; Nakamura, K.; Tomohiko, T.O.; Nice, P.I. *Development of 125ksi Grade HSLA Steel OCTG for Mildly Sour Environments. Corrosion*; NACE International: Houston, TX, USA, 2005.
5. Liu, M. Finite element analysis of pitting corrosion on mechanical behavior of E690 steel panel. *Anti-Corrosion Methods Mater.* **2022**, *69*, 351–361. [CrossRef]
6. Juanita, M.C. Design and Investigation of a North Sea Acid Corrosion Inhibition System. In *Corrosion 2006*; OnePetro: Richardson, TX, USA, 2006.
7. Iqbala, M.; Lyonb, B.; Benavidesa, E.; Ehsan, M.; Yunping, F.; Charles, M.; Kevin, J.; Christopher, E.; Pennell, K.D.; Keith, J. High temperature stability and low adsorption of sub-100 nm magnetite nanoparticles grafted with sulfonated copolymers on Berea sandstone in high salinity brine. *Physicochem. Eng. Asp.* **2017**, *520*, 257–267. [CrossRef]
8. Pang, H.; Chen, J.; Pang, X.; Liu, L.; Liu, K.; Xiang, C. Key factors controlling hydrocarbon accumulations in Ordovician carbonate reservoirs in the Tazhong area, Tarim basin, western China. *Mar. Pet. Geol.* **2013**, *43*, 88–101. [CrossRef]
9. Liu, M. Corrosion and Mechanical Behavior of Metal Materials. *Materials* **2023**, *16*, 973. [CrossRef] [PubMed]
10. Ali, G.; Bülent, K.; Nuri, O.; Kanca, E. The Investigation of Corrosion Behavior of Borided AISI 304 Austenitic Stainless Steel with Nanoboron Powder. *Prot. Met. Phys. Chem. Surf.* **2014**, *50*, 104–110.
11. Lo, K.; Shek, C.; Lai, J. Recent developments in stainless steel. *Mater. Sci. Eng.* **2009**, *65*, 39–104. [CrossRef]
12. Wang, Y.Q. Study on Corrosion Resistance of 2205 Duplex Stainless Steel in Underground Environment. Master's Thesis, Xi'an Shiyou University, Xi'an, China, 2018.
13. Liu, M. Effect of uniform corrosion on mechanical behavior of E690 high-strength steel lattice corrugated panel in marine environment: A finite element analysis. *Mater. Res. Express* **2021**, *8*, 066510. [CrossRef]
14. Guo, L.Q.; Li, M.; Shi, X.L.; Yan, Y.; Li, X.; Qiao, L. Effect of annealing temperature on the corrosion behavior of duplex stainless steel studied by in situ techniques. *Corros. Sci.* **2011**, *53*, 3733–3741. [CrossRef]
15. Ishiguro, Y.; Suzuki, T.; Miyata, Y.; Suzuki, T.; Kimura, M.; Sato, H.; Shimamoto, K. Enhanced corrosion-resistant stainless steel OCTG of 17Cr for sweet and sour environments. In Proceedings of the 68th NACE Annual Conference, Orlando, FL, USA, 17–21 March 2013.
16. Tsai, W.-T.; Tsai, K.-M.; Lin, C.-j. Selective corrosion in duplex stainless steel. In Proceedings of the CORROSION 2003, San Diego, CA, USA, 16–20 March 2003.
17. Lo, I.-H.; Fu, Y.; Lin, C.-J. Effect of electrolyte composition on the active-to-passive transition behavior of 2205 duplex stainless steel in H_2SO_4/HCl solutions. *Corros. Sci.* **2006**, *48*, 696–708. [CrossRef]
18. Tsai, W.-T.; Chen, J.-R. Galvanic corrosion between the constituent phases in duplex stainless steel. *Corros. Sci.* **2007**, *49*, 3659–3668. [CrossRef]
19. Lo, I.-H.; Tsai, W.-T. Effect of selective dissolution on fatigue crack initiation in 2205 duplex stainless steel. *Corros. Sci.* **2007**, *49*, 1847–1861. [CrossRef]
20. Fu, Y.; Lin, C. A study of the selective dissolution behaviour of duplex stainless steel by micro-electrochemical technique. *Acta Metall. Sin.* **2005**, *41*, 302–306.
21. Tsai, W.; Lo, I. Effects of potential and loading frequency on corrosion fatigue behaviour of 2205 duplex stainless steel. *Corrosion* **2008**, *64*, 155–163. [CrossRef]
22. Symniotis, E. Galvanic effects on active dissolution of duplex stainless steels. *Corrosion* **1990**, *46*, 2–12. [CrossRef]

23. Yau, Y.-H.; Streicher, M.A. Galvanic Corrosion of Duplex FeCr-10%Ni Alloys in Reducing Acids. *Corrosion* **1987**, *43*, 366–373. [CrossRef]
24. Symniotis, E. Dissolution mechanism of duplex stainless steels in the active-to-passive transition range and role of microstructure. *Corrosion* **1995**, *51*, 571–580. [CrossRef]
25. Sridhar, N.; Kolts, J. Effects of nitrogen on the selective dissolution of a duplex stainless steel. *Corrosion* **1987**, *143*, 646. [CrossRef]
26. Lee, J.; Fushimi, K.; Nakanishi, T.; Hasegawa, Y.; Park, Y. Corrosion behaviour of ferrite and austenite phases on super duplex stainless steel in a modified green-death solution. *Corros. Sci.* **2014**, *89*, 111–117. [CrossRef]
27. Ho, M.Y.; Geddes, J.; Barmatov, E.; Crawford, L.; Hughes, T. Effect of composition and microstructure of duplex stainless steel on adsorption behaviour and efficiency of corrosion inhibitors in 4 molar hydrochloric acid. Part I: Standard DSS 2205. *Corros. Sci.* **2018**, *137*, 43–52. [CrossRef]
28. Mizukami, H.; Yasui, A.; Hirano, S.; Ooba, T.; Yasui, A.; Sunaba, T.; Uno, M. Effect of Additives to Corrosion inhibitor for Duplex Stainless Steel in Acidizing. In Proceedings of the CORROSION Virtual Conference & Expo, Virtual, 19–30 April 2021.
29. De Mello Joia, C.J.B.; Brito, R.F.; Barbosa, B.C.; Pereira, A.Z.I. Performance of Corrosion Inhibitors for Acidizing Jobs in Horizontal Wells Completed with CRA Laboratory Tests. In Proceedings of the 56st NACE Annual Conference, Houston, TX, USA, 11–17 March 2021.
30. Ornek, C.; Engelberg, D.L. SKPFM measured Volta potential correlated with strain localisation in microstructure to understand corrosion susceptibility of cold-rolled grade 2205 duplex stainless steel. *Corros. Sci.* **2015**, *99*, 164–171. [CrossRef]
31. Du, J.; Yu, M.; Liu, P.; Fu, Y.; Xiong, G.; Liu, J.; Chen, X. Corrosion behavior of 2205 duplex stainless steel in acidizing stimulation solution for oil and gas wells at 200 °C. *Anti-Corros. Methods Mater.* **2022**, *69*, 149–159. [CrossRef]
32. Lin, Y.; Chen, X. Investigation of moisture diffusion in epoxy system: Experiments and molecular dynamics simulations. *Chem. Phys. Lett.* **2005**, *412*, 322–326. [CrossRef]
33. Yazdani, S.; Prince, L.; Vitry, V. Optimization of electroless Ni–B-nanodiamond coating corrosion resistance and understanding the nanodiamonds role on pitting corrosion behavior using shot noise theory and molecular dynamic simulation. *Diam. Relat. Mater.* **2023**, *134*, 109793. [CrossRef]
34. Hao, C.; Zhibin, F.; Zhijian, C.; Xuejie, Z.; Penghua, Z.; Jun, W. Effect of Cl^- and HSO_3^- on Corrosion Behavior of 439 Stainless Steel Used in Construction. *J. Chin. Soc. Corros. Prot.* **2022**, *42*, 493–500.
35. Liu, L.F.; Liu, J.X.; Zhang, J.; Lin-fa, L.; Li-Jun, X.; Gui-Min, J. Molecular dynamics simulation of the corrosive medium diffusion behavior inhibited by the corrosion inhibitor membranes. *Chem. J. Chin. Univ.* **2010**, *31*, 537. [CrossRef]
36. Wang, Z.W.; Li, B.; Lin, Q.B.; Hu, C.Y. Molecular dynamics simulation on diffusion of five kinds of chemical additives in polypropylene. *Packag. Technol. Sci.* **2018**, *31*, 277. [CrossRef]
37. Razaghi, Z.; Rezaei, M. Corrosion mechanism of sulfate, chloride, and tetrafluoroborate ions interacted with Ni-19wt% Cr coating: A combined experimental study and molecular dynamics simulation. *J. Mol. Liq.* **2020**, *319*, 114243. [CrossRef]
38. Damej, M.; Kaya, S.; EL Ibrahimi, B.; Lee, H.-S.; Molhi, A.; Serdaroğlu, G.; Benmessaoud, M.; Ali, I.; EL Hajjaji, S.; Lgaz, H. The corrosion inhibition and adsorption behavior of mercaptobenzimidazole and bis-mercaptobenzimidazole on carbon steel in 1.0 M HCl: Experimental and computational insights. *Surf. Interfaces* **2021**, *24*, 101095. [CrossRef]
39. Qiao, G.M.; Ren, Z.J.; Zhang, J.; Hu, S.; Yan, Y.; Ti, Y. Molecular Dynamics Simulation of Corrosive Medium Diffusion in Corrosion Inhibitor Membrane. *Acta Phys. Chim. Sin.* **2010**, *26*, 3041–3046.
40. Liu, L.-f. *Theoretical Study on the Mechanism of the Inhibition Efficiency of 1-(2-hydroxyethyl)-2-alkyl Imidazoline Inhibitor*; China University of Petroleum (East China): Shandong, China, 2010.
41. Finšgar, M.; Jackson, J. Application of corrosion inhibitors for steels in acidic media for the oil and gas industry: A review. *Corros. Sci.* **2014**, *86*, 17–41. [CrossRef]
42. Merello, R.; Botana, F.; Botella, J.; Matres, M.; Marcos, M. Influence of chemical composition on the pitting corrosion resistance of non-standard low-Ni high-Mn–N duplex stainless steels. *Corros. Sci.* **2003**, *45*, 909–921. [CrossRef]
43. Bautista, A.; Alvarez, S.M.; Velasco, F. Corrosion of duplex stainless steel bars in acid Part 1: Effect of the composition, microstructure and anodic polarizations. *Mater. Corros.* **2015**, *66*, 348. [CrossRef]
44. Zhao, Y.; Qi, W.; Xie, J.; Chen, Y.; Zhang, T.; Xu, D.; Wang, F. Investigation of the failure mechanism of the TG-201 inhibitor:Promoting the synergistic effect of HP-13Cr stainless steel during the well completion. *Corros. Sci.* **2020**, *166*, 108448. [CrossRef]
45. Zhang, D.Q.; Cai, Q.R.; He, X.M.; Gao, L.; Kim, G. Corrosion inhibition and adsorption behavior of methionine on copper in HCl and synergistic effect of zine ions. *Mater. Chem. Phys.* **2009**, *114*, 612–617. [CrossRef]

Disclaimer/Publisher's Note: The statements, opinions and data contained in all publications are solely those of the individual author(s) and contributor(s) and not of MDPI and/or the editor(s). MDPI and/or the editor(s) disclaim responsibility for any injury to people or property resulting from any ideas, methods, instructions or products referred to in the content.

Article

Standard Deviation Effect of Average Structure Descriptor on Grain Boundary Energy Prediction

Ruoqi Dang and Wenshan Yu *

State Key Laboratory for Strength and Vibration of Mechanical Structures, Shaanxi Engineering Laboratory for Vibration Control of Aerospace Structures, School of Aerospace Engineering, Xi'an Jiaotong University, Xi'an 710049, China
* Correspondence: wenshan@mail.xjtu.edu.cn

Abstract: The structural complexities of grain boundaries (GBs) result in their complicated property contributions to polycrystalline metals and alloys. In this study, we propose a GB structure descriptor by linearly combining the average two-point correlation function (PCF) and standard deviation of PCF via a weight parameter, to reveal the standard deviation effect of PCF on energy predictions of Cu, Al and Ni asymmetric tilt GBs (i.e., Σ3, Σ5, Σ9, Σ11, Σ13 and Σ17), using two machine learning (ML) methods; i.e., principal component analysis (PCA)-based linear regression and recurrent neural networks (RNN). It is found that the proposed structure descriptor is capable of improving GB energy prediction for both ML methods. This suggests the discriminatory power of average PCF for different GBs is lifted since the proposed descriptor contains the data dispersion information. Meanwhile, we also show that GB atom selection methods by which PCF is evaluated also affect predictions.

Keywords: grain boundary; descriptor; pair distribution function; grain boundary energy; machine learning method

---

Citation: Dang, R.; Yu, W. Standard Deviation Effect of Average Structure Descriptor on Grain Boundary Energy Prediction. *Materials* **2023**, *16*, 1197. https://doi.org/10.3390/ma16031197

Academic Editors: Adam Grajcar and Frank Czerwinski

Received: 6 December 2022
Revised: 4 January 2023
Accepted: 13 January 2023
Published: 31 January 2023

Copyright: © 2023 by the authors. Licensee MDPI, Basel, Switzerland. This article is an open access article distributed under the terms and conditions of the Creative Commons Attribution (CC BY) license (https://creativecommons.org/licenses/by/4.0/).

1. Introduction

Grain boundaries (GBs) are one of the most commonly seen planar defects in polycrystalline metals and alloys. Due to local atomic distortions and inconsistent atomic arrangement, GBs play important roles in determining the mechanical, thermal and electric, etc., properties of materials [1,2]. For example, GBs may act as the dislocation and point defect sources or sinkers, and they may block the dislocation motion and absorb them; thus the strength and ductility of materials can be greatly changed [3]. For an idealized GB, from the geometrical point of view, it can be completely governed by five parameters, usually represented by misorientation and a normal GB plane [4]. Unfortunately, the structures, as well as the properties of the GB, are hard to completely determine. This is simply because a GB may have numerous states due to the point defect absorptions and emissions. It means the structures of a given GB may no longer be unique for a given energy [5–11]. Thus, the connection between structure and property of a GB, such as energy, volume and mechanical behavior, etc., are usually built via atomistic simulations by using molecular dynamics (MD) and density functional theory (DFT) methods [12,13], which is also of great significance for the macroscopic modeling of material behavior [14,15]. Technically speaking, it is possible to do so using MD and DFT, but also needs a heavy workload if such connections for a large number of GBs are expected.

The ML method has been applied in many research fields [16–20], and provides an efficient technique by which to link the structure-property of a GB, particularly to extract correlations from high-dimensional datasets [21–25], and has been successfully applied in predicting GB energies [21,23–27], point defect segregation energies [28,29], GB structures [30] and damages and deformations in GB [31,32]. Usually, an appropriate ML method is employed according to the datasets and the expected correlations. Regardless of these, a problem is how to mathematically describe the GB structure, which should contain the essential characteristics of GB structures. One of the representative examples is structure units (SUs) [33–36], which are usually used to describe GB structures, but are incapable of doing so for general GBs or GBs driven out of equilibrate states [11]. Furthermore, some studies have also focused on developing structural matrices [12,13,37]. We could indeed gain a unique insight from these studies. However, these descriptions cannot be readily applied in ML.

So far, some descriptors related to GB structures, which can be well used in ML so that the atomic structure-property relationships can be constructed, have been developed. Generally, these descriptors can be divided into two categories; i.e., local atom environment descriptor and descriptors for atom connectivity [21,22,38,39]. The former mostly considers the local atom environment, such as the atomic neighboring arrangement. One of the representative examples, developed by Banadaki and Patala [38,39], describes the local GB structure by using some polyhedral units in a way similar to SUs. In principle, this method can be used to represent any general GB. However, GB structures are usually not at a equilibrate state or subjected to deformation perturbations due to vacancies or self-interstitial diffusions, absorptions and emissions. It means this technique itself suffers from limitations. Banadaki et al. proposed the point-pattern matching algorithm to enhance the power of PU for describing GBs, particularly with local distortions [38]. Another example is the smooth overlap of atomic positions (SOAP) descriptor, which is essentially a combination of radial and spherical spectral bases, including spherical harmonics [22]. For the atom connectivity descriptors, they essentially correlate the positions of atoms in GB and in the vicinity of GB. Such descriptors usually vary smoothly upon the perturbations of atom positions [22]. An example is the pair correlation function (PCF) method due to Gomberg et al.'s study [21]. It was shown that the two-point correlation functions can reliably predict GB energies. Based on these descriptors, a variety of GB properties, such as energy, mobility and mechanical behavior, etc., can be successfully predicted by using ML methods.

Using ML methods to predict GB properties, a better quantitative prediction usually requires that a descriptor should carry the information of the GB structure as much as possible [38]. An extreme case is to consider all atoms composed of the GB structure by evaluating the neighboring atom distribution for each GB atom [22]. This not only drastically increases the dimension of data, but may also lead to data dimension inconsistency between GBs. To avoid such a dilemma, a straightforward method is to further compute the average quantity [21,38]. This way, the descriptor can be seen as an average structure representation (ASR) [22]. From the statistics, ASR does not contain the information of data scatter, such as the standard deviation. Moreover, it is still unclear how standard deviation affects the prediction.

In this study, we establish 464 asymmetric GB models for Cu, Al and Ni metals and relax all GB models using MD. We discuss the prediction of GB energies by comparing two ML methods, i.e., principal component analysis (PCA) -based linear regression [40] and recurrent neural networks (RNN) [41,42], in an effort to answer the following questions:

Based on two-point correlation functions, i.e., the PCF method proposed by Gomberg et al. [21], we introduce a new GB structure descriptor by linearly combining PCF and its standard distribution PCF_{std} via a parameter. How does such a descriptor affect the prediction?

GB atoms can be selected from GB using common neighbor analysis (CNA) [43] and centro-symmetric parameter (CSP) methods [44]. It can be imaged that a different number of GB atoms could be selected for two methods when setting different critical values. Will this affect the prediction?

Comparing two ML methods, what happens to the prediction when considering the GB atom selection method and the GB structure descriptor? Can we predict GB energy without clearly distinguishing the tilt axis of the GB?

The content of this paper is organized as follows. Firstly, we introduce the GB models and the establishment of the GB structure descriptor. Secondly, the PCA-based linear regression of energies of Cu, Al and Ni GBs according to the tilt axis of GB considering full data and partition data is discussed. Thirdly, the effect of standard deviation of PCF on the RNN-based prediction of GB energies is discussed. Finally, the prediction comparisons between the two ML methods and conclusions of this study are made.

2. Methodology and GB Structure Descriptor

We consider a total number of 464 asymmetric tilt GBs (ATGBs) with misorietations $\Sigma 3, \Sigma 5, \Sigma 9, \Sigma 11, \Sigma 13$ and $\Sigma 17$ as the dataset for the subsequent GB energy prediction study. Each GB model is a bi-crystal composed of two grains with specified orientations. To construct it with periodic boundary conditions (PBCs) applicable, crystalline orientations of two grains are needed. Usually, a given GB misorientation Σ can be defined by the overlapped lattices of two crystals with one of them rotated around a specified axis ρ with a certain angle θ. Namely, Σ is equivalent to (ρ, θ), which can also be represented by a rotation matrix $\mathbf{R}_\Sigma$. Take $\Sigma = 3$ as an example, (ρ, θ) related to $\Sigma 3$ equals ([110], 70.53°) [33], and the corresponding unit cell of $\Sigma 3$ coincidence site lattice (CSL) is spanned by [1$\bar{1}$0], [111] and [11$\bar{2}$]. Then, the normal GB $\mathbf{m}$ of a series of $\Sigma 3$ ATGBs expressed in one grain are linear combinations of [111] and [11$\bar{2}$], i.e., i [111] + j [11$\bar{2}$] for different integers i and j. GBs normally expressed in another grain can be obtained by $\mathbf{R}_{\Sigma 3}.\mathbf{m}$. The angle between $\mathbf{m}$ and [111] defines the inclination angle, denoted as ϕ. Using $\mathbf{m}$ and $\mathbf{R}_{\Sigma 3}.\mathbf{m}$, along with the tilt axis [1$\bar{1}$0], the crystal orientations of two grains of all $\Sigma 3$ ATGBs can be defined. By following this approach, asymmetric GBs of all other misorientations can be readily created.

For $\Sigma 3, \Sigma 5, \Sigma 9, \Sigma 11, \Sigma 13$ and $\Sigma 17$, lattice symmetry requires the inclination angle ϕ varying from 0° to 90° for $\Sigma 3, \Sigma 9$ and $\Sigma 11$, with ϕ varying from 0° to 45° for $\Sigma 5, \Sigma 13$ and $\Sigma 17$, respectively. PBCs are imposed within the GB plane for all bicrystal models. Two grains (i.e., Grains A and B) terminate with free surfaces in the direction perpendicular to the GB plane by setting two 10Å thick vacuum spaces on the top and bottom ends of the bicrystal model, as schematically shown in Figure S1 in Supplementary Materials. This allows us to release the stress possibly produced in the z direction during the GB structure optimization. Embedded atom method (EAM) potentials [45,46] are used to model the atomic interactions in Al [47], Cu [48] and Ni [47]. We relax all 464 ATGBs via the conjugate gradient (CG) method using LAMMPS [49] and compute the average energies of all GBs. Tables S1 and S2 in Supplementary Materials list the variations in atom numbers and energies for $\Sigma 3, \Sigma 5, \Sigma 9, \Sigma 11, \Sigma 13$ and $\Sigma 17$ GB models of each metal. Atomic structures are visualized using Ovito [50].

With all GBs relaxed, we are in a position to introduce the descriptor by which the GB structure and structure differences between GBs can be described and distinguished. Herein, we employ the pair correlation function (PCF) method proposed Gomberg et al. [21] as a GB structure descriptor. In doing so, a primary concern is how many atoms in GB should be considered when evaluating the average PCF (PCF$_{mean}$(r)). In other words, an appropriate method for selecting GB atoms should include a certain number of atoms in the vicinity of the GB carrying structure information, but exclude other atoms. Herein, we consider common neighbor analysis (CNA) [43] and centro-symmetric parameter (CSP) methods [44]. For the CSP method, two CSP critical values are considered (i.e., CSP > 0.1 and 0.5). As exemplified in Figure S2 in Supplementary Material, the three methods identify a different number of GB atoms. Such effects on GB energy predictions will be discussed in the following section.

According to the approach of Gomberg et al. [21], the PCF of a given GB is computed by averaging the radial distribution function $g_a(r)$ of all N_{GB} GB atoms selected out of the GB using CNA, CSP$_{0.1}$ and CSP$_{0.5}$, which can be expressed as

$$\text{PCF}_{\text{mean}}(r) = \frac{1}{N_{GB}} \sum_{\alpha \in GB} g_\alpha(r) \tag{1}$$

where the radial distribution function $g_\alpha(r)$ of a GB atom can be calculated by considering all of its N_{in} neighboring atoms within a specified cut-off radius for three metals.

$$g_\alpha(r) = \sum_{k=1}^{N_{in}} \frac{K^e\left(r - \|R_\alpha^k\|, h^e\right)}{4\pi r^2 n_0} \tag{2}$$

where a kernel function K^e with bandwidth h^e is used to smoothen the radial distribution function. n_0 is the atom density in the FCC lattice and $\|R_\alpha^k\|$ is the distance between atom a and its kth neighboring atom. The parameters needed in Equation (2) are listed in Table S3 in Supplementary Material.

Figure 1a compares PCF_{mean} of the Al single crystal with results taken from [21]. Good agreement validates our algorithm for computing PCF_{mean}. In fact, PCF_{mean} is an averaged radial distribution function (RDF) curve of each GB atom, by which the PCF data fluctuations of different GB atoms cannot be well considered, as evidenced by the variation in standard deviation of PCF ($\text{PCF}_{\text{std}}(r)$) for three GBs in Cu in Figure 1b. In order to incorporate the data fluctuation into the averaged PCF, we further propose a PCF_{comb} by combining $\text{PCF}_{\text{mean}}(r)$ and $\text{PCF}_{\text{std}}(r)$ as

$$\text{PCF}_{comb}(r) = (1-\zeta)\frac{\text{PCF}_{meam}(r)}{\max(\text{PCF}_{meam}(r))} + \zeta \frac{\text{PCF}_{std}(r)}{\max(\text{PCF}_{std}(r))} \tag{3}$$

where parameter ζ is introduced to weigh the portions of $\text{PCF}_{\text{mean}}(r)$ and $\text{PCF}_{\text{std}}(r)$ in PCF_{comb}. PCF_{comb} is reduced to PCF by letting $\zeta = 0$. As an example, Figure 1c,d shows the $\text{PCF}_{\text{comb}}(r)$ of $\Sigma 5(310)$ and $\Sigma 9(114)$ Cu GBs for three values of ζ. Clearly, the variation trends of $\text{PCF}_{\text{comb}}(r)$ changes as ζ varies. In the following, how the variation of ζ influences the prediction will be discussed. The PCF_{comb} curve of each GB is further represented as 512 discrete points, serving as the input data for the ML methods.

In the following, we adopt two ML methods to predict GB energies, i.e., principal component analysis (PCA)-based linear regression [40] and recurrent neural networks (RNN) [41,42]. PCA is usually implemented in two steps, a dimensionality-reduction of data and regression based on the principle component, which are essentially the eigenvalues of the covariance matrix of raw data. Therefore, the regression of PCA is achieved only using a few principle components. The principle components for regression are selected by considering the explained variance percentage of each principle component. However, RNNs do not require dimensionality reduction. The training and prediction are performed by using raw data. To quantitatively compare the predictions, mean absolute error (MAE) and mean relative error (MRE) are assessed via

$$MAE = \frac{1}{n}\sum_{i=1}^{n}\left|\gamma_i^{Pred} - \gamma_i^{MD}\right| \tag{4}$$

$$MAE = \frac{1}{n}\sum_{i=1}^{n}\left|\gamma_i^{Pred} - \gamma_i^{MD}\right|/\gamma_i^{MD} \tag{5}$$

where γ_i^{Pred} and γ_i^{MD} are GB energies predicted by ML methods and computed via MD.

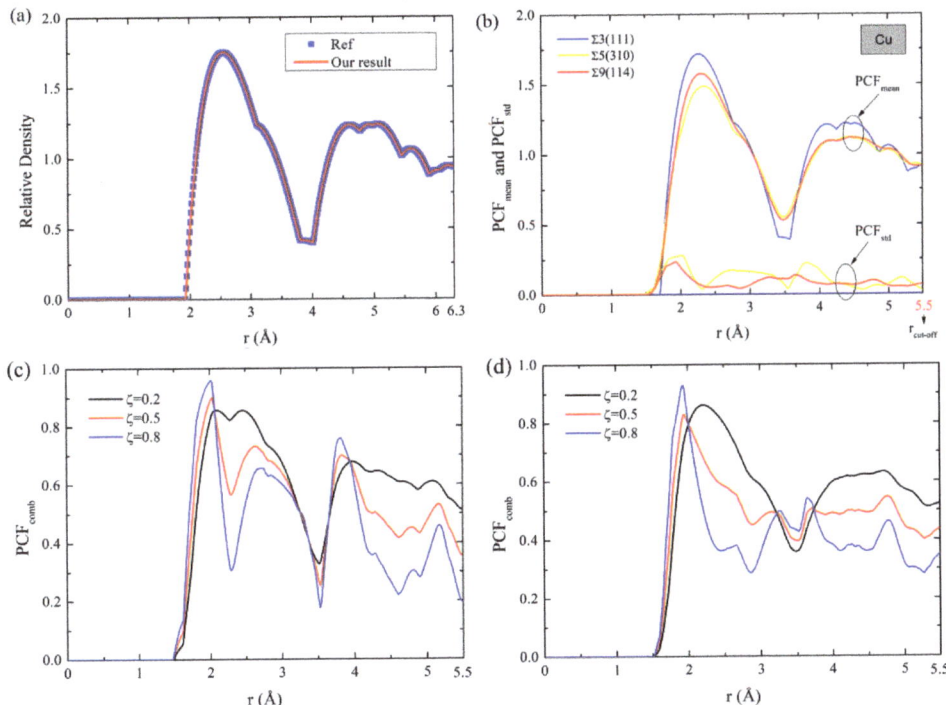

Figure 1. (a) PCF of Al single crystal compared with the results taken from [16], (b) PCF$_{mean}$ and PCF$_{std}$(r) of Cu GBs $\Sigma5(310)$, $\Sigma9(114)$ and $\Sigma3(111)$. PCF$_{comb}$(r) of Cu GBs (c) $\Sigma5(310)$ and (d) $\Sigma9(114)$ for $\zeta = 0.2$, 0.5 and 0.8. Results of figure (b) for Al and Ni are shown in Figure S3 in Supplementary Material.

3. PCA-Based Prediction

To implement PCA, we need to determine which principle components will be used in the regression. To do so, we analyze the explained variance percentage of the first ten principle components for Cu, as shown in Figure S4 in Supplementary Materials. It turns out that the explained variation of the first PC is up to 93%, while those of the other nine PCs are lower than 3%. This suggests that only the first few PCs accounting for higher explained variance percentages retain most of the original data, while the rest only keep a small amount of the data. It is therefore unnecessary to consider many PCs in the subsequent GB energy regression. Because of this, only the first three PCs, e.g., PC$_1$, PC$_2$ and PC$_3$, are used in the regression. With the multiple linear regression method, GB energy can be written as

$$\gamma_{GB} = a.PC_1 + b.PC_2 + c.PC_3 + d \qquad (6)$$

where a, b, c and d are fitting parameters. PC$_1$, PC$_2$ and PC$_3$ are obtained from data training, which is dependent on the dataset. The dataset in this study consists of GBs with <100> and <110> tilts axes. Thus, there are two possible ways to obtain PCs by reducing the dimensionality of data when considering all data together and two data subsets corresponding to <100> and <110> tilt axes, denoted as *full data* and *partitioned data* methods, respectively. Thus, two sorts of PCs PC$_i$ through training different datasets can be obtained.

The PCs obtained from data training are actually a representation of data in a lower dimensional space. Although these PCs retain most of the original data, it is still challenging to impart each PC with possible physical interpretability. By following the approach by

Gomberg et al. [14], it is possible to correlate each PC to a geometrical parameter of GB by interpolating each PC as a function of the geometrical parameter. Herein, such a geometrical parameter is considered to be inclination angle ϕ related to asymmetric GBs for a specified Σ. In this study, PC_1, PC_2 and PC_3 are assumed to be cubic polynomial interpolation functions of ϕ.

$$PC_i(\varphi) = A_i \varphi^3 + B_i \varphi^2 + C_i \varphi + D_i \tag{7}$$

where A_i, B_i, C_i and D_i are fitting parameters. Such cubic polynomial interpolation can well characterize the variation of calculated PC_i vs ϕ for most GBs and PCs. From the above analysis, there are four ways of predicting GB energies in terms of different approaches of obtaining PCs, denoted as *full data, full data-fitting, partitioned data* and *partitioned data-fitting* methods, respectively. For the partitioned data, Equation (4) corresponds to three metals, shown in Equations (S2)–(S8) in Supplementary Material. The parameters in Equation (5) are listed in Tables S4–S6 in Supplementary Material.

For comparison, Figure 2 exemplifies energy predictions of Σ3 and Σ5 asymmetric copper GBs related to <110> and <100> tilt axes. In order to assess the prediction improvement due to the data partition, MAEs of all predictions are calculated, as shown in Figure 2. For Σ3 GBs, MAEs for PCi- and PCi(ϕ)-based linear fittings using full data are 64.57 mJ/m^2 and 113.77 mJ/m^2, but those for partitioned data are 45.49 mJ/m^2 and 87.42 mJ/m^2. For Σ5 GBs, MAEs for PCi- and PCi(ϕ)-based linear fittings using full data are 25.58 mJ/m^2 and 29.62 mJ/m^2, but those for partitioned data are 4.29 mJ/m^2 and 10.66 mJ/m^2. From further inspection of the variation of MAEs due to the data partition, MAEs for PCi- and PCi(ϕ)-based linear fittings are reduced by ~30% and ~23%, respectively, but, for Σ3 GBs, they are up to ~83% and ~64%. This suggests that a better prediction can be achieved by separately considering <110> and <100> GB datasets, which is particularly more prominent for <100> GBs. From the MAEs results of PCi and PCi(ϕ) linear fittings for Σ3 and Σ5 GBs, PCi(ϕ) linear fittings indeed lead to a larger MAE than PCi linear fittings, which is understandable since PCi(ϕ) is approximately obtained from cubic interpolation. Nevertheless, these results show that PCi is capable of being correlated with inclination angle ϕ.

As previously mentioned, γ_i^{Pred} should be dependent on ζ, therefore, both MAE and MRE are functions of ζ. As an example, Table 1 shows MAEs and MREs of all Σs for $\zeta = 0.5$. Comparing the MAE and MRE predictions for <100> and <110> GBs, both MAE and MRE are lower for <100> GBs, which further demonstrates that better predictions can be obtained for <100> GBs. In fact, this can be explained by considering the structure differences of <110> and <100> GBs. It is known that SUs for <100> GBs are composed of some [100] dislocations [2,51,52]. This brings simpler and mutually similar structures to <100> GBs. However, <110> GBs are composed of SUs much more complicated than those of <100> GBs [2,33,53–55]. Therefore, the structures of two <100> GBs may be quite different from each other. Thus, predictions for <100> GBs are better than those for their <110> counterparts, as also evidenced by the results of Al and Ni (see Figures S11, S12, S16 and S17 in Supplementary Material).

Table 1. MAE and MRE of Cu GB energy predictions. Note that this table lists the prediction for PCF_{comb} computed for CNA-based GB atom selection and ζ is taken as 0.5.

Error	Σ3	Σ9	Σ11	Σ5	Σ13	Σ17
MAE (mJ/m^2)	99.72	50.86	70.48	31.70	34.22	64.19
MRE (%)	42.54	6.40	11.89	3.14	3.48	6.30

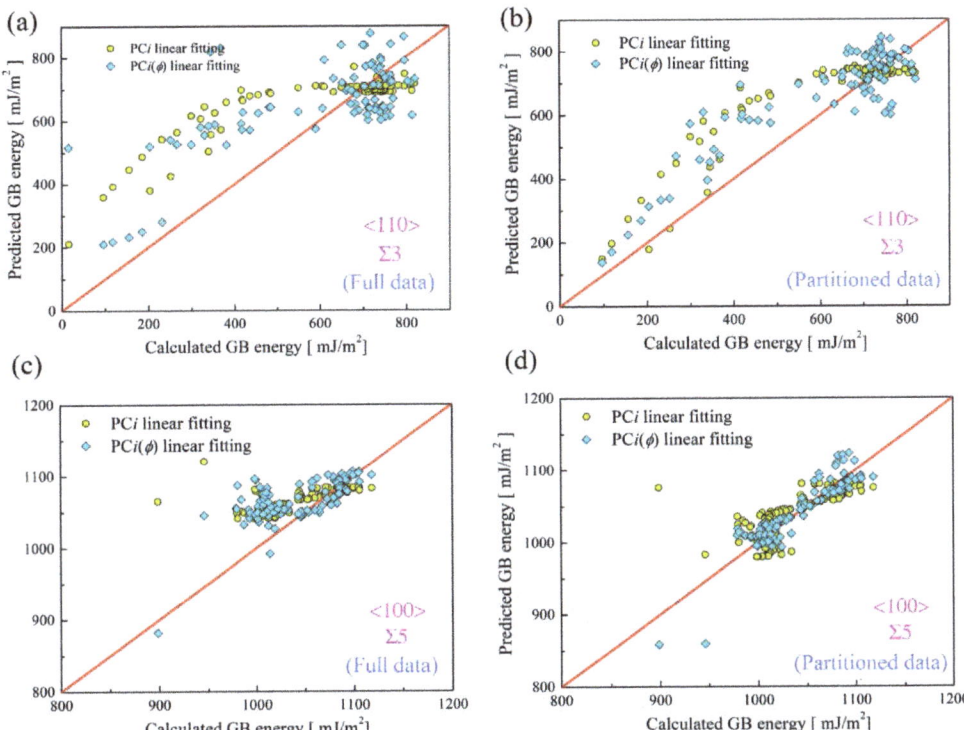

Figure 2. Energy predictions for (**a**,**b**) Σ3 and (**c**,**d**) Σ5 asymmetric copper GBs. Dimensionality reductions are performed based on full data (Figures (**a**,**c**)) and partitioned data (Figures (**b**,**d**)). Note that this figure exemplifies the predictions using PCF_{comb} computed for CNA-based GB atom selection and $\zeta = 0.5$. Results of Σ9, Σ11, Σ13 and Σ17 GBs based on full data and partitioned data are shown in Figures S5–S8 in Supplementary Material.

In order to compare the effects of the GB atom selection method (i.e., CNA, CSP = 0.1 and CSP = 0.5) on the prediction, Figure 3a,b exemplify the MRE of <110> and <100> Cu GBs vs. ζ. From Figure 3, with increasing ζ, the MRE of <110> GBs keep increasing; however, that of <110> GBs keep decreasing. Finally, MREs of both <110> and <100> GBs reach plateaus. Further inspection of Figure 3 reveals that the minimum values of MRE for <110> and <100> GBs for CAN and CSP0.1 methods correspond to $\zeta = 0.0$ and 1.0, respectively. For the CSP0.5 method, the minimum values of MRE for <110> and <100> GBs are $\zeta = 0.2$ and 0.1. Moreover, considering CAN, CSP0.1 and CSP0.5 alone, MREs also differ at $\zeta = 0.0$, but their general variation trends are similar. Therefore, it can be seen that a better prediction not only requires an appropriate GB atom selection method, but an appropriate value of ζ. In fact, the MREs of Al and Ni are also dependent on ζ, as seen from Figures S15 and S20 in Supplementary Material.

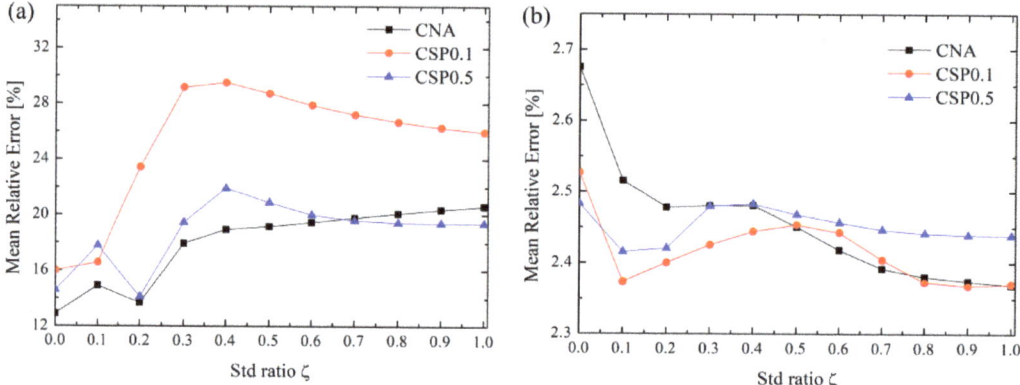

Figure 3. MRE of Cu GBs for three methods of selecting GB atoms (**a**) <110> and (**b**) <100> tilt axis. Results of Al and Ni are shown in Figures S15 and S20 in Supplementary Material.

4. RNN-Based Predictions

In this section, we discuss the predictions using the RNN method. Due to the dimension reduction in the PCA method, some data is lost. Moreover, a better prediction can be achieved provided that GB types are distinguished based on the tilt axis; i.e., <100> and <110>. In comparison to PCA, on the other hand, the RNN method is highly nonlinear. Considering all of these factors, we do not distinguish GB types for each metal; i.e., the prediction is performed using the full data for each metal. For each metal, 10-fold cross validation is performed, with each fit being performed on a training set consisting of 70% of the total training set selected at random, with the remaining 30% used as a holdout set for testing. This yields good convergence of MAE, as shown in Figure S21 in Supplementary Materials. Figure 4 shows the prediction results of the RNN method considering three GB atom selection methods. A preliminary comparison between PCA and RNN, as shown in Figures 2 and 4, reveals that the RNN method gives a better prediction. This is not surprising due to the higher nonlinearity of the RNN method.

We further evaluated the MAE of RNN predictions for three GB atom selection methods and three metals, as shown in Figure 5. Clearly, with increasing ζ, there is a sudden drop in the MAE. Meanwhile, such a drop in CNA, CSP0.1 and CSP0.5 for the same metal almost occurs at the same value of ζ. Moreover, MAEs for Cu, Al and Ni suddenly drop by ~75%, ~75% and ~70% at $\zeta_{crit} \approx 0.3$, 0.6 and 0.7. After the sudden drops, all curves approach plateaus with nearly the same MAE, regardless of the three GB atom selection methods. This evidences the significant dependence of ζ in RNN prediction, and also implies that considerable errors will be caused in RNN prediction when letting $\zeta = 0$. To avoid such errors, the standard deviation of PCFs (PCF_{std}) must be incorporated into the descriptor. Moreover, the ζ_{crit} of the three metals differ, which implies that the prediction accuracy can be enhanced only when more data scatter information of PCF is considered. ζ_{crit} of Cu is the smallest, while that of Ni is the largest. It suggests that data scatter of the average PCF for Cu is the lowest, but that of Ni is the highest.

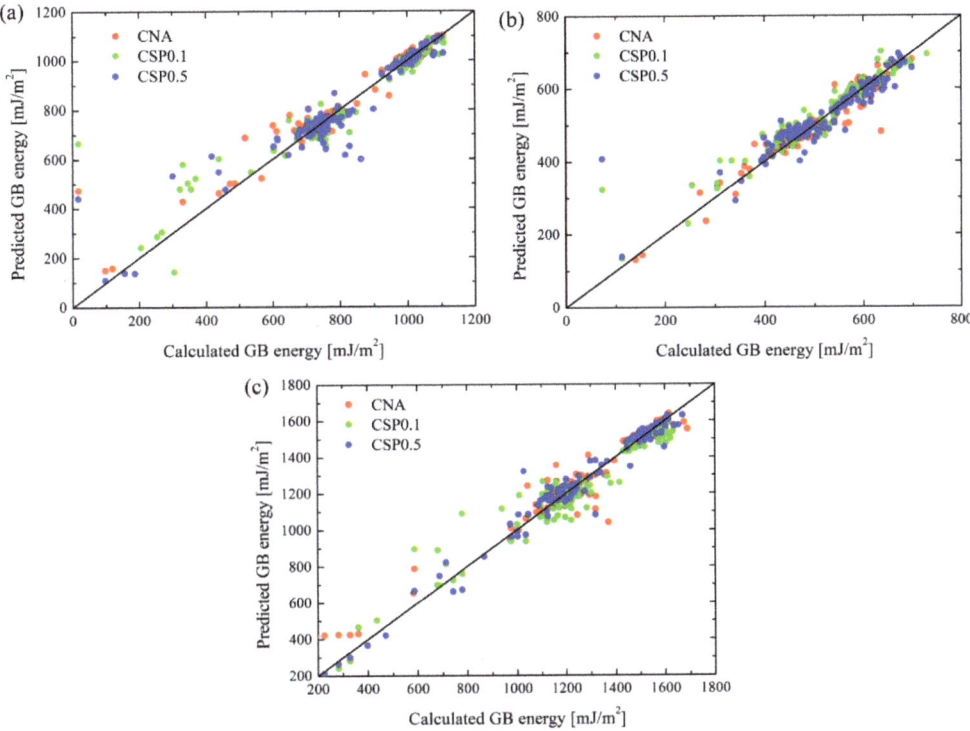

Figure 4. A comparison of RNN-based predictions and MD results for (**a**) Cu, (**b**) Al and (**c**) Ni. Note that predictions are made by considering the parameter ζ corresponding to the minimum MAE of three GB atom selection methods, as shown in Figure 5.

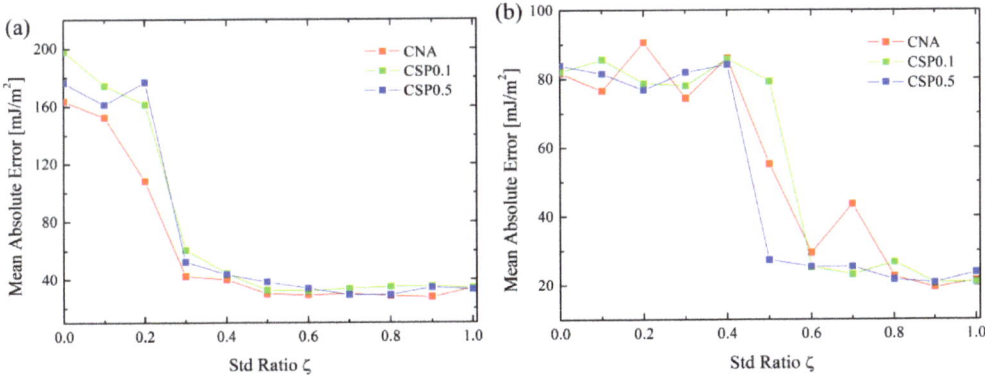

Figure 5. *Cont.*

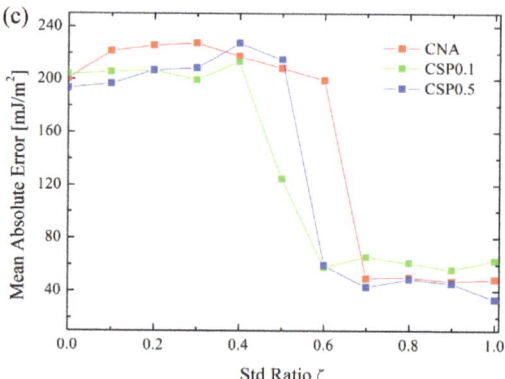

Figure 5. MAE of RNN prediction vs. ζ of three GB atom selection methods for (**a**) Cu, (**b**) Al and (**c**) Ni.

5. Discussions

In this study, we predicted GB energies by using PCA-based linear regression and RNN. Figure 6 compares the GB energy prediction properties of two ML methods. From Figure 6a, RNN gives a better prediction than the PCA method. For PCA-based linear regression, linear fitting parameters of GB energy are obtained by dividing a full dataset into two separate datasets according to GBs of <100> and <110> tilt axis for three metals. In doing so, we attempted to weaken the effects of the mutual interference due to <100> and <110> GBs when obtaining linear fitting parameters. Indeed, such a way of treating a dataset for PCA-based predictions decreases the prediction errors (Figure 2). We also tried to further obtain the linear fitting parameters by expressing them as cubic polynomial interpolation functions of an inclination angle of GB ϕ, instead of obtaining them from data training. The purpose of doing so was to obtain an empirical GB energy prediction function and intensify the interpretability of ML prediction, which may be impossible for the RNN method. Such a method may work for PCA-based predictions, as evidenced in Figure 2.

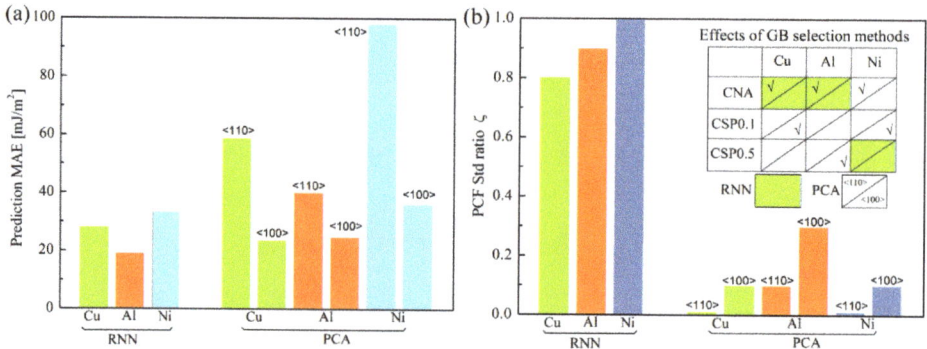

Figure 6. (**a**) Lowest MAE and (**b**) the variation of PCF standard deviation ratio ζ at the lowest MAE for PCA and RNN predictions. Inset in figure (**b**) shows which GB selection methods yield the lowest MAEs for PCA and RNN predictions.

An appropriate GB structure descriptor is vital in ML-based GB energy prediction, as whether the descriptor contains the essential information of GB structures or not determines the prediction accuracy. In fact, PCF as a GB structure descriptor [16], compared with those of polyhedral units [34,35], is much easier use. However, the definition of PCF shows that the GB structure is described by an average function. It is believed that the GB structure may not be well described without considering the higher order moment of RDF data from the statistical point of view. Motivated by this, we further incorporated the standard deviation of PCF into the PCF function to further extend the descriptor of the GB structure (Equation (2)). Figure 6b shows that predictions using PCA and RNN are significantly dependent on the parameter and GB atom selection methods for three metals, also suggesting the necessity of considering PCF_{std} in GB structure descriptors.

6. Conclusions

In this paper, we studied the GB energies prediction properties of Cu, Al and Ni using two ML methods; i.e., PCA-based linear regression and RNN. We considered the asymmetric GBs $\Sigma 3$, $\Sigma 5$, $\Sigma 9$, $\Sigma 11$, $\Sigma 13$ and $\Sigma 17$ of <110> and <100> types. Atomistic models were constructed and relaxed using the MD method. By extending the PCF-based GB structure descriptor and using three methods of selecting GB atoms, we compared the prediction of two ML methods. The main conclusions of this study were drawn as:

For the three metals, the lowest MAE can be obtained when ζ is greater than 0.8 for RNN, while that should be smaller than 0.3 for PCA-based linear regression. This indicates the dependence of GB descriptors on the ML method. Meanwhile, PCF as an average function and GB structure descriptor needs to consider the PCF_{std}, by which ML prediction accuracy can be improved. The GB structure descriptor in the form of average structure representation (ASR) may need to further take into account the standard deviation of ASR.

In comparison to RNN, it is indeed possible to intensify or realize the interpretability of ML prediction by using the PCF-based linear regression method, though how to generalize the fitting method of the linear regression function when considering a dataset of different GB misorientations still needs to be addressed.

For a specific ML method, the MAE of the prediction is determined by multiple factors, such as the GB atom selection method and a portion of PCF standard deviation. A better quantitative descriptor of GB structure is a trade-off between computation cost and complexity. It is expected that prediction accuracy can be enhanced by combining those comprehensive descriptors together. Moreover, it will be interesting to examine the performance of GB structure descriptors if we consider GBs of mixed types.

Supplementary Materials: The following supporting information can be downloaded at: https://www.mdpi.com/article/10.3390/ma16031197/s1, Figure S1. Schematic of the 3D bicrystal GB model. Figure S2. Three methods of selecting GB at-oms in a $\Sigma 3$ Cu GB. Red atoms in the center of model are those selected out of whole model based on different methods. Figure S3. PCF(r) and PCFcomb(r) of GBs $\Sigma 5(310)$, $\Sigma 9(114)$ and $\Sigma 3(111)$ in Al and Ni. Figure S4. the explained variance percentage for first-ten principle components by considering full data set of Cu. Figure S5. PCA based prediction results of Cu <110> GBs using full data. Regression is performed using PCi based on dimensionality reduction and further interpolation as a function of ϕ, respectively. Figure S6. PCA based prediction results of Cu <100> GBs using full data. Regression is performed using PCi based on dimensionality reduction and further interpolation as a function of ϕ, respectively. Figure S7. PCA based prediction results of Cu <110> GBs using partitioned <110> data. Regression is performed using PCi based on dimensionality reduction and further interpolation as a function of ϕ, respectively. Figure S8. PCA based prediction results of Cu <100> GBs using portioned <100> data. Regression is performed using PCi based on dimensionality reduction and further interpolation as a function of ϕ, respectively. Figure S9. Comparison of Computed PCi with cubic interpolated PCi as a function of inclination angle for <110> type Cu GBs. Figure S10. Comparison of Computed PCi with cubic interpolated PCi as a function of inclination angle for <100> type Cu GBs. Figure S11. PCA based prediction results of Al <110> GBs using partitioned <110> data. Regression is performed using PCi based on dimensionality reduction and further interpolation as a function of ϕ, respectively. Figure S12. PCA based prediction results of Al <100> GBs using portioned <100> data. Regression is performed using PCi based on dimensionality reduction and further interpolation as a function of ϕ, respectively. Figure S13. Comparison of Computed PCi with cubic interpolated PCi as a function of inclination angle for <110> type Al GBs. Figure S14. Comparison of Computed PCi with cubic interpolated PCi as a function of inclination angle for <100> type Al GBs. Figure S15. MRE of Al

GBs for three methods of selecting GB atoms. Figure S16. PCA based prediction results of Ni <110> GBs using partitioned <110> data. Regression is performed using PCi based on dimensionality reduction and further interpolation as a function of ϕ, respectively. Figure S17. PCA based prediction results of Ni <100> GBs using portioned <100> data. Regression is performed using PCi based on dimensionality reduction and further interpolation as a function of ϕ, respectively. Figure S18. Comparison of Computed PCi with cubic interpolated PCi as a function of inclination angle for <100> type Ni GBs. Figure S19. Comparison of computed PCi with cubic interpolated PCi as a function of inclination angle for <100> type Ni GBs. Figure S20. MRE of Cu GBs for three methods of selecting GB atoms. Figure S21. MAE convergence curve for RNN prediction of Cu GBs. Table S1. Some additional information of all GB models. Table S2. Variation range of GB energies for all GBs of each metal (mJ/m^2). Table S3. Some material constants and parameter used for computing PCF. Table S4. Cubic interpolation of PCi as a function of inclination angle for all Cu GBs. Table S5. Cubic interpolation of PCi as a function of inclination angle for all Al GBs. Table S6. Cubic interpolation of PCi as a function of inclination angle for all Ni GBs.

Author Contributions: R.D.: Methodology, Investigation, Writing—original draft; W.Y.: Investigation, Supervision, Funding acquisition. All authors have read and agreed to the published version of the manuscript.

Funding: This research was funded by the support of NSFC (Grant No: 11872049 and 12090032).

Institutional Review Board Statement: Not applicable.

Data Availability Statement: The data that supports the findings of this study are available in the supplementary material of this article.

Conflicts of Interest: The authors declare no conflict of interest.

References

1. Rohrer, G.S. Grain boundary energy anisotropy: A review. *J. Mater. Sci.* **2011**, *46*, 5881. [CrossRef]
2. Cahn, J.W.; Mishin, Y.; Suzuki, A. Coupling grain boundary motion to shear deformation. *Acta Mater.* **2006**, *54*, 4953. [CrossRef]
3. Mishin, Y.; Asta, M.; Li, J. Atomistic modeling of interfaces and their impact on microstructure and properties. *Acta Mater.* **2010**, *58*, 1117. [CrossRef]
4. Cantwell, P.R.; Tang, M.; Dillon, S.J.; Luo, J.; Rohrer, G.S.; Harmer, M.P. Grain boundary complexions. *Acta Mater.* **2014**, *62*, 1. [CrossRef]
5. Suzuki, A.; Mishin, Y. Interaction of point defects with grain boundaries in fcc metals. *Interface Sci.* **2003**, *11*, 425. [CrossRef]
6. Bai, X.M.; Vernon, L.J.; Hoagland, R.G.; Voter, A.F.; Nastasi, M.; Uberuaga, B.P. Role of atomic structure on grain boundary-defect interactions in Cu. *Phys. Rev. B* **2012**, *85*, 204103. [CrossRef]
7. Tschopp, M.A.; Solanki, K.N.; Gao, F.; Sun, X.; Khaleel, M.A.; Horstemeyer, M.F. Probing grain boundary sink strength at the nanoscale: Energetics and length scales of vacancy and interstitial absorption by grain boundaries in a-Fe. *Phys. Rev. B* **2012**, *85*, 064108. [CrossRef]
8. Siegel, R.W.; Chang, S.M.; Balluffi, R.W. Vacancy Loss at Grain-Boundaries in Quenched Polycrystalline Gold. *Acta Met. Mater.* **1980**, *28*, 249. [CrossRef]
9. Dollar, M.; Gleiter, H. Point-Defect Annihilation at Grain-Boundaries in Gold. *Scr. Met.* **1985**, *19*, 481. [CrossRef]
10. Frolov, T.; Olmsted, D.L.; Asta, M.; Mishin, Y. Structural phase transformations in metallic grain boundaries. *Nat. Commun.* **2013**, *4*, 1899. [CrossRef]
11. Tucker, G.J.; McDowell, D.L. Non-equilibrium grain boundary structure and inelastic deformation using atomistic simulations. *Int. J. Plasticity* **2011**, *27*, 841. [CrossRef]
12. Olmsted, D.L.; Foiles, S.M.; Holm, E.A. Survey of computed grain boundary properties in face-centered cubic metals: I. Grain boundary energy. *Acta Mater.* **2009**, *57*, 3694. [CrossRef]
13. Olmsted, D.L.; Holm, E.A.; Foiles, S.M. Survey of computed grain boundary properties in face-centered cubic metals—II: Grain boundary mobility. *Acta Mater.* **2009**, *57*, 3704. [CrossRef]
14. Liu, M. Effect of uniform corrosion on mechanical behavior of E690 high-strength steel lattice corrugated panel in marine environment: A finite element analysis. *Mater. Res. Express* **2021**, *8*, 066510. [CrossRef]
15. Liu, M. Finite element analysis of pitting corrosion on mechanical behavior of E690 steel panel. *Anti-Corros. Methods Mater.* **2022**, *69*, 351. [CrossRef]
16. Elsheikh, A.H.; Sharshir, S.W.; Abd Elaziz, M.; Kabeel, A.E.; Guilan, W.; Haiou, Z. Modeling of solar energy systems using artificial neural network: A comprehensive review. *Sol. Energy* **2019**, *180*, 622. [CrossRef]
17. Elsheikh, A.H.; Elaziz, M.A.; Das, S.R.; Muthuramalingam, T.; Lu, S. A new optimized predictive model based on political optimizer for eco-friendly MQL-turning of AISI 4340 alloy with nano-lubricants. *J. Manuf. Process.* **2021**, *67*, 562. [CrossRef]

18. Elsheikh, A.H.; Katekar, V.P.; Muskens, O.L.; Deshmukh, S.S.; Elaziz, M.A.; Dabour, S.M. Utilization of LSTM neural network for water production forecasting of a stepped solar still with a corrugated absorber plate. *Process Saf. Environ. Prot.* **2021**, *148*, 273. [CrossRef]
19. Elsheikh, A.H.; Abd Elaziz, M.; Vendan, A. Modeling ultrasonic welding of polymers using an optimized artificial intelligence model using a gradient-based optimizer. *Weld. World* **2022**, *66*, 27. [CrossRef]
20. Elsheikh, A.H.; Shanmugan, S.; Sathyamurthy, R.; Kumar Thakur, A.; Issa, M.; Panchal, H.; Muthuramalingam, T.; Kumar, R.; Sharifpur, M. Low-cost bilayered structure for improving the performance of solar stills: Performance/cost analysis and water yield prediction using machine learning. *Sustain. Energy Technol. Assess.* **2022**, *49*, 101783. [CrossRef]
21. Gomberg, J.A.; Medford, A.J.; Kalidindi, S.R. Extracting knowledge from molecular mechanics simulations of grain boundaries using machine learning. *Acta Mater.* **2017**, *133*, 100. [CrossRef]
22. Rosenbrock, C.W.; Homer, E.R.; Csányi, G.; Hart, G.L.W. Discovering the building blocks of atomic systems using machine learning: Application to grain boundaries. *NPJ Comput. Mater.* **2017**, *3*, 29. [CrossRef]
23. Ye, W.; Zheng, H.; Chen, C.; Ong, S.P. A Universal Machine Learning Model for Elemental Grain Boundary Energies. *Scr. Mater.* **2022**, *218*, 114803. [CrossRef]
24. Homer, E.R.; Hart, G.L.W.; Braxton Owens, C.; Hensley, D.M.; Spendlove, J.; Serafin, L.H. Examination of computed aluminum grain boundary structures and energies that span the 5D space of crystallographic character. *Acta Mater.* **2022**, *234*, 118006. [CrossRef]
25. Song, X.; Deng, C. Atomic energy in grain boundaries studied by machine learning. *Phys. Rev. Mater.* **2022**, *6*, 043601. [CrossRef]
26. Snow, B.D.; Doty, D.D.; Johnson, O.K. A Simple Approach to Atomic Structure Characterization for Machine Learning of Grain Boundary Structure-Property Models. *Front. Mater.* **2019**, *6*, 120. [CrossRef]
27. Guziewski, M.; Montes de Oca Zapiain, D.; Dingreville, R.; Coleman, S.P. Microscopic and Macroscopic Characterization of Grain Boundary Energy and Strength in Silicon Carbide via Machine-Learning Techniques. *ACS Appl. Mater. Interfaces* **2021**, *13*, 3311. [CrossRef]
28. Huber, L.; Hadian, R.; Grabowski, B.; Neugebauer, J. A machine learning approach to model solute grain boundary segregation. *NPJ Comput. Mater.* **2018**, *4*, 64. [CrossRef]
29. Wagih, M.; Larsen, P.M.; Schuh, C.A. Learning grain boundary segregation energy spectra in polycrystals. *Nat. Commun.* **2020**, *11*, 6376. [CrossRef]
30. Homer, E.R.; Hensley, D.M.; Rosenbrock, C.W.; Nguyen, A.H.; Hart, G.L.W. Machine-Learning Informed Representations for Grain Boundary Structures. *Front. Mater.* **2019**, *6*, 168. [CrossRef]
31. Vieira, R.B.; Lambros, J. Machine Learning Neural-Network Predictions for Grain-Boundary Strain Accumulation in a Polycrystalline Metal. *Exp. Mech.* **2021**, *61*, 627–639. [CrossRef]
32. Zhang, S.; Wang, L.; Zhu, G.; Diehl, M.; Maldar, A.; Shang, X.; Zeng, X. Predicting grain boundary damage by machine learning. *Int. J. Plasticity* **2022**, *150*, 103186. [CrossRef]
33. Tschopp, M.A.; McDowell, D.L. Structural unit and faceting description of Sigma 3 asymmetric tilt grain boundaries. *J. Mater. Sci.* **2007**, *42*, 7806. [CrossRef]
34. Sutton, A.P.; Vitek, V. On the Structure of Tilt Grain Boundaries in Cubic Metals I. Symmetrical Tilt Boundaries. *Philos. Trans. R. Soc. London. Ser. A Math. Phys. Sci.* **1983**, *309*, 1.
35. Sutton, A.P.; Vitek, V. On the Structure of Tilt Grain Boundaries in Cubic Metals II. Asymmetrical Tilt Boundaries. *Philos. Trans. R. Soc. London. Ser. A Math. Phys. Sci.* **1983**, *309*, 37.
36. Sutton, A.P.; Vitek, V. On the Structure of Tilt Grain Boundaries in Cubic Metals. III. Generalizations of the Structural Study and Implications for the Properties of Grain Boundaries. *Philos. Trans. R. Soc. London. Ser. A Math. Phys. Sci.* **1983**, *309*, 55.
37. Francis, T.; Chesser, I.; Singh, S.; Holm, E.A.; De Graef, M. A geodesic octonion metric for grain boundaries. *Acta Mater.* **2019**, *166*, 135. [CrossRef]
38. Patala, S. Understanding grain boundaries—The role of crystallography, structural descriptors and machine learning. *Comp. Mater. Sci.* **2019**, *162*, 281. [CrossRef]
39. Banadaki, A.D.; Patala, S. A three-dimensional polyhedral unit model for grain boundary structure in fcc metals. *NPJ Comput. Mater.* **2017**, *3*, 13. [CrossRef]
40. Jolliffe, I.T. *Principal Component Analysis*, 2nd ed.; Springer: New York, NY, USA, 2002.
41. Aggarwal, C.C. *Neural Networks and Deep Learning: A Textbook*; Springer: New York, NY, USA, 2018.
42. Nielsen, M. *Neural Networks and Deep Learning*; Determination Press: Los Angeles, CA, USA, 2015.
43. Faken, D.; Jónsson, H. Systematic analysis of local atomic structure combined with 3D computer graphics. *Comp. Mater. Sci.* **1994**, *2*, 279. [CrossRef]
44. Kelchner, C.L.; Plimpton, S.J.; Hamilton, J.C. Dislocation nucleation and defect structure during surface indentation. *Phys. Rev. B* **1998**, *58*, 11085. [CrossRef]
45. Daw, M.S.; Baskes, M.I. Embedded-Atom Method—Derivation and Application to Impurities, Surfaces, and Other Defects in Metals. *Phys. Rev. B* **1984**, *29*, 6443. [CrossRef]
46. Daw, M.S.; Foiles, S.M.; Baskes, M.I. The embedded-atom method: A review of theory and applications. *Mater. Sci. Rep.* **1993**, *9*, 251. [CrossRef]

47. Voter, A.F.; Chen, S.P. Accurate Interatomic Potentials for Ni, Al and Ni3Al. In *MRS Proceedings*; Materials Research Society: Warrendale, PA, USA, 1986.
48. Mishin, Y.; Farkas, D.; Mehl, M.J.; Papaconstantopoulos, D.A. Interatomic potentials for monoatomic metals from experimental data and ab initio calculations. *Phys. Rev. B* **1999**, *59*, 3393. [CrossRef]
49. Plimpton, S. Fast Parallel Algorithms for Short-Range Molecular Dynamics. *J. Comp. Phys.* **1995**, *117*, 1. [CrossRef]
50. Alexander, S. Visualization and analysis of atomistic simulation data with OVITO–the Open Visualization Tool. *Model Simul. Mater. Sci. Eng.* **2010**, *18*, 015012.
51. Sansoz, F.; Molinari, J.F. Incidence of atom shuffling on the shear and decohesion behavior of a symmetric tilt grain boundary in copper. *Scr. Mater.* **2004**, *50*, 1283. [CrossRef]
52. Vitek, V.; Sutton, A.P.; Gui Jin, W.; Schwartz, D. On the multiplicity of structures and grain boundaries. *Scr. Met.* **1983**, *17*, 183. [CrossRef]
53. Spearot, D.E.; Tschopp, M.A.; Jacob, K.I.; McDowell, D.L. Tensile strength of < 100 > and < 110 > tilt bicrystal copper interfaces. *Acta Mater.* **2007**, *55*, 705.
54. Tschopp, M.A.; Tucker, G.J.; McDowell, D.L. Structure and free volume of < 110 > symmetric tilt grain boundaries with the E structural unit. *Acta Mater.* **2007**, *55*, 3959.
55. Spearot, D.E.; Jacob, K.I.; McDowell, D.L. Nucleation of dislocations from [0 0 1] bicrystal interfaces in aluminum. *Acta Mater.* **2005**, *53*, 3579. [CrossRef]

Disclaimer/Publisher's Note: The statements, opinions and data contained in all publications are solely those of the individual author(s) and contributor(s) and not of MDPI and/or the editor(s). MDPI and/or the editor(s) disclaim responsibility for any injury to people or property resulting from any ideas, methods, instructions or products referred to in the content.

Editorial

Corrosion and Mechanical Behavior of Metal Materials

Ming Liu

Center for Advancing Materials Performance from the Nanoscale (CAMP-Nano), State Key Laboratory for Mechanical Behavior of Materials, Xi'an Jiaotong University, Xi'an 710049, China; liuming0313@xjtu.edu.cn

Many high-strength metal-related materials and structures work under the coupling condition of harsh corrosion environments and complex loading [1–17], and related failure cases have been reported extensively all over the world. Hence, it is absolutely essential to investigate the corrosion and mechanical behavior of metal materials, aspects which mainly include corrosion fatigue [3–5], stress corrosion cracking [6–8], erosion corrosion [9], hydrogen-induced cracking [10], wear corrosion [18], etc. From the point view of materials and structures, failure can be caused by the unique mechanical and corrosive environment during their service life. The research methods of most forward environmental fractures [4,19,20] and the new mechanical analysis techniques for structures could all be useful in the study of those particular failure behaviors. Hence, this Special Issue, entitled "Corrosion and Mechanical Behavior of Metal Materials", will mainly concentrate on how high-strength metal materials and structures work under the conditions of corrosion and complex loading.

The aim of this Special Issue is to discover the current state of the new methods, novel ideas, and advanced techniques of the related issues that link to the corrosion and mechanical behavior of metal materials. A wide range of research findings on different topics has been helpful in contributing to this Special Issue. The emphasis of these topics covers fundamental science and scientific problems that exist in engineering, experimental studies, analysis tools, numerical approaches, and design receipts. This Special Issue has the ambition to inspire and to disseminate the latest knowledge on the corrosion and mechanical behavior of metal materials and structures, laying the foundation for new ideas covering a range of topics for young researchers as well as leading experts in materials science and engineering and civil engineering.

The published papers covered in the topic area of this Special Issue encompass the corrosion fatigue characteristics of high-strength bridge steel, i.e., the degradation characteristics of galvanized and Galfan high-strength steel wire under marine corrosion and fatigue loading [21] and evaluating the corrosion fatigue degradation of the elastic center buckle of the short suspender of a suspension bridge under traffic loading [22]. The probabilistic seismic performance analysis of a corroded, reinforced concrete column and a corroded elastic bridge bearing was carried out by using the analytical model of the material degradation phenomenon. The seismic vulnerability of an aging bridge system was obtained by considering the failure functions of several related components [23]. Crucial attention was paid to the effect of severe plastic deformation on the corrosion behavior of a tantalum–tungsten alloy [24]; the severely deformed crystallographic orientations in the tantalum–tungsten alloy could be greatly weakened by an electrochemical process and could reduce the corrosion rate. The pre-exposure SCC (PESCC) of a ZK60 alloy induced by preliminary immersion in a NaCl-containing solution was systematically studied in one paper [25], and it was argued that the hydrogen stored within the corrosion product layer and the corrosion solution was responsible for the formation of these two zones. Meanwhile, the corrosion resistance of dilute Fe–Al alloys could be improved by preheating a nanoscale Al_2O_3 protective layer in a H_2 atmosphere [26]. Similarly, a study on the corrosion behavior of a high-strength CuNi alloy in a harsh environment is also included within the scope of this Special Issue [27].

Citation: Liu, M. Corrosion and Mechanical Behavior of Metal Materials. *Materials* **2023**, *16*, 973. https://doi.org/10.3390/ma16030973

Received: 13 January 2023
Accepted: 19 January 2023
Published: 20 January 2023

Copyright: © 2023 by the author. Licensee MDPI, Basel, Switzerland. This article is an open access article distributed under the terms and conditions of the Creative Commons Attribution (CC BY) license (https://creativecommons.org/licenses/by/4.0/).

Conflicts of Interest: The author declares no conflict of interest.

References

1. Liu, M. Effect of uniform corrosion on mechanical behavior of E690 high-strength steel lattice corrugated panel in marine environment: A finite element analysis. *Mater. Res. Express* **2021**, *8*, 066510. [CrossRef]
2. Liu, M. Finite element analysis of pitting corrosion on mechanical behavior of E690 steel panel. *Anti-Corros. Methods Mater.* **2022**, *28*, 7527–7536. [CrossRef]
3. Calvo-García, E.; Valverde-Pérez, S.; Riveiro, A.; Álvarez, D.; Román, M.; Magdalena, C.; Badaoui, A.; Moreira, P.; Comesaña, R. An Experimental Analysis of the High-Cycle Fatigue Fracture of H13 Hot Forging Tool Steels. *Materials* **2022**, *15*, 7411. [CrossRef] [PubMed]
4. Chen, W.; Lu, W.; Gou, G.; Dian, L.; Zhu, Z.; Jin, J. The Effect of Fatigue Damage on the Corrosion Fatigue Crack Growth Mechanism in A7N01P-T4 Aluminum Alloy. *Metals* **2023**, *13*, 104. [CrossRef]
5. Liu, M.; Luo, S.J.; Shen, Y.; Lin, X.Z. Corrosion fatigue crack propagation behaviour of S135 high-strength drill pipe steel in H_2S environment. *Eng. Fail. Anal.* **2019**, *97*, 493–505. [CrossRef]
6. Yoo, Y.-R.; Choi, S.-H.; Kim, Y.-S. Effect of Laser Peening on the Corrosion Properties of 304L Stainless Steel. *Materials* **2023**, *16*, 804. [CrossRef]
7. Jiang, X.; Li, G.; Tang, H.; Liu, J.; Cai, S.; Zhang, J. Modification of Inclusions by Rare Earth Elements in a High-Strength Oil Casing Steel for Improved Sulfur Resistance. *Materials* **2023**, *16*, 675. [CrossRef]
8. Kang, C.-Y.; Chen, T.-C.; Tsay, L.-W. Effects of Micro-Shot Peening on the Stress Corrosion Cracking of Austenitic Stainless Steel Welds. *Metals* **2023**, *13*, 69. [CrossRef]
9. Lyu, L.; Qiu, X.; Yue, H.; Zhou, M.; Zhu, H. Corrosion Behavior of Ti3SiC2 in Flowing Liquid Lead–Bismuth Eutectic at 500 °C. *Materials* **2022**, *15*, 7406. [CrossRef]
10. Rudskoi, A.I.; Karkhin, V.A.; Starobinskii, E.B.; Parshin, S.G. Modeling of Hydrogen Diffusion in Inhomogeneous Steel Welded Joints. *Materials* **2022**, *15*, 7686. [CrossRef]
11. Al-Huri, M.A.; Al-Osta, M.A.; Ahmad, S. Finite Element Modelling of Corrosion-Damaged RC Beams Strengthened Using the UHPC Layers. *Materials* **2022**, *15*, 7606. [CrossRef] [PubMed]
12. Wu, W.; Qin, L.; Cheng, X.; Xu, F.; Li, X. Microstructural evolution and its effect on corrosion behavior and mechanism of an austenite-based low-density steel during aging. *Corros. Sci.* **2023**, *212*, 110936. [CrossRef]
13. Chen, Y.; Chen, Q.; Pan, Y.; Xiao, P.; Du, X.; Wang, S.; Zhang, N.; Wu, X. A Chemical Damage Creep Model of Rock Considering the Influence of Triaxial Stress. *Materials* **2022**, *15*, 7590. [CrossRef] [PubMed]
14. Yang, X.; Yang, Y.; Sun, M.; Jia, J.; Cheng, X.; Pei, Z.; Li, Q.; Xu, D.; Xiao, K.; Li, X. A new understanding of the effect of Cr on the corrosion resistance evolution of weathering steel based on big data technology. *J. Mater. Sci. Technol.* **2022**, *104*, 67–80. [CrossRef]
15. Zemková, M.; Minárik, P.; Jablonská, E.; Veselý, J.; Bohlen, J.; Kubásek, J.; Lipov, J.; Ruml, T.; Havlas, V.; Král, R. Concurrence of High Corrosion Resistance and Strength with Excellent Ductility in Ultrafine-Grained Mg-3Y Alloy. *Materials* **2022**, *15*, 7571. [CrossRef]
16. Tao, J.; Xiang, L.; Zhang, Y.; Zhao, Z.; Su, Y.; Chen, Q.; Sun, J.; Huang, B.; Peng, F. Corrosion Behavior and Mechanical Performance of 7085 Aluminum Alloy in a Humid and Hot Marine Atmosphere. *Materials* **2022**, *15*, 7503. [CrossRef]
17. Zhao, D.; Ye, F.; Liu, B.; Du, H.; Unigovski, Y.B.; Gutman, E.M.; Shneck, R. Effect of Surface Dissolution on Dislocation Activation in Stressed FeSi6.5 Steel. *Materials* **2022**, *15*, 7434. [CrossRef]
18. Lu, C.-J.; Yeh, J.-W. Improved Wear and Corrosion Resistance in TiC-Reinforced SUS304 Stainless Steel. *J. Compos. Sci.* **2023**, *7*, 34. [CrossRef]
19. Dorado, S.; Arias, A.; Jimenez-Octavio, J.R. Biomechanical Modelling for Tooth Survival Studies: Mechanical Properties, Loads and Boundary Conditions—A Narrative Review. *Materials* **2022**, *15*, 7852. [CrossRef]
20. Zhang, F.; Wu, Z.; Zhang, T.; Hu, R.; Wang, X. Microstructure Sensitivity on Environmental Embrittlement of a High Nb Containing TiAl Alloy under Different Atmospheres. *Materials* **2022**, *15*, 8508. [CrossRef]
21. Zhao, Y.; Su, B.; Fan, X.; Yuan, Y.; Zhu, Y. Corrosion Fatigue Degradation Characteristics of Galvanized and Galfan High-Strength Steel Wire. *Materials* **2023**, *16*, 708. [CrossRef]
22. Zhao, Y.; Guo, X.; Su, B.; Sun, Y.; Li, X. Evaluation of Flexible Central Buckles on Short Suspenders' Corrosion Fatigue Degradation on a Suspension Bridge under Traffic Load. *Materials* **2023**, *16*, 290. [CrossRef] [PubMed]
23. Liu, X.; Zhang, W.; Sun, P.; Liu, M. Time-Dependent Seismic Fragility of Typical Concrete Girder Bridges under Chloride-Induced Corrosion. *Materials* **2022**, *15*, 5020. [CrossRef]
24. Ma, G.; Zhao, M.; Xiang, S.; Zhu, W.; Wu, G.; Mao, X. Effect of the Severe Plastic Deformation on the Corrosion Resistance of a Tantalum–Tungsten Alloy. *Materials* **2022**, *15*, 7806. [CrossRef] [PubMed]
25. Merson, E.; Poluyanov, V.; Myagkikh, P.; Merson, D.; Vinogradov, A. Effect of Air Storage on Stress Corrosion Cracking of ZK60 Alloy Induced by Preliminary Immersion in NaCl-Based Corrosion Solution. *Materials* **2022**, *15*, 7862. [CrossRef]

26. Li, C.; Freiberg, K.; Tang, Y.; Lippmann, S.; Zhu, Y. Formation of Nanoscale Al2O3 Protective Layer by Preheating Treatment for Improving Corrosion Resistance of Dilute Fe-Al Alloys. *Materials* **2022**, *15*, 7978. [CrossRef]
27. Gao, X.; Liu, M. Corrosion Behavior of High-Strength C71500 Copper-Nickel Alloy in Simulated Seawater with High Concentration of Sulfide. *Materials* **2022**, *15*, 8513. [CrossRef]

Disclaimer/Publisher's Note: The statements, opinions and data contained in all publications are solely those of the individual author(s) and contributor(s) and not of MDPI and/or the editor(s). MDPI and/or the editor(s) disclaim responsibility for any injury to people or property resulting from any ideas, methods, instructions or products referred to in the content.

Article

Corrosion Fatigue Degradation Characteristics of Galvanized and Galfan High-Strength Steel Wire

Yue Zhao [1], Botong Su [1,*], Xiaobo Fan [2], Yangguang Yuan [3] and Yiyun Zhu [1]

1. School of Civil Engineering and Architecture, Xi'an University of Technology, Xi'an 710048, China; zhaoyue@chd.edu.cn (Y.Z.); zyyun@xaut.edu.cn (Y.Z.)
2. Xi'an Municipal Engineering Design & Research Institute Co., Ltd., Xi'an 710068, China; 18717393234@163.com
3. School of Architecture and Civil Engineering, Xi'an University of Science and Technology, Xi'an 710054, China; yuanyg31@163.com
* Correspondence: 2210721161@stu.xaut.edu.cn

Citation: Zhao, Y.; Su, B.; Fan, X.; Yuan, Y.; Zhu, Y. Corrosion Fatigue Degradation Characteristics of Galvanized and Galfan High-Strength Steel Wire. *Materials* 2023, 16, 708. https://doi.org/10.3390/ma16020708

Academic Editor: Ming Liu

Received: 1 December 2022
Revised: 29 December 2022
Accepted: 9 January 2023
Published: 11 January 2023

Copyright: © 2023 by the authors. Licensee MDPI, Basel, Switzerland. This article is an open access article distributed under the terms and conditions of the Creative Commons Attribution (CC BY) license (https://creativecommons.org/licenses/by/4.0/).

Abstract: Cables are the main load-bearing components of a cable bridge and typically composed of high strength steel wires with a galvanized coating or Galfan coating. Galfan steel wire has recently started to be widely used because of its better corrosion resistance than galvanized steel wire. The corrosion characteristics of the coating and the difference in the corrosion fatigue process of the two types of steel wire are unclear. To further improve the service performance and maintenance of cable bridges, this study investigated the corrosion characteristics of galvanized steel wire and Galfan steel wire through accelerated corrosion tests and established a time-varying model of uniform corrosion and pitting corrosion of high-strength steel wire. Then, a long-span suspension bridge was taken as the research object, and the corrosion fatigue degradation of the two kinds of steel wire under a traffic load was analyzed on the basis of traffic monitoring data. The results showed that the uniform corrosion of the two types of steel wire conformed to an exponential development trend, the corrosion coefficient of galvanized steel wire conformed to the normal distribution, and the corrosion coefficient of Galfan steel wire conformed to the Cauchy distribution. The maximum pitting coefficient distribution of the two kinds of steel wire conformed to the generalized extreme value distribution. The location parameters and scale parameters of the two distributions showed an exponential downward trend with the increase of corrosion duration. When the traffic intensity was low, the corrosion characteristics of the steel wire was the main factor affecting its service life, and the average service life of Galfan steel wire was significantly higher than that of galvanized steel wire. Under a dense traffic flow, the service life of the steel wire was mainly controlled by the traffic load, and the service life of Galfan steel wire was slightly improved. Effective anti-corrosion measures are a key factor for improving the service life of steel wire.

Keywords: high strength steel wire; corrosion fatigue; uniform corrosion; pitting corrosion; traffic load

1. Introduction

High-strength steel wire is the key load-bearing component of cable bearing bridges, such as cable-stayed bridges and suspension bridges, and will degrade during operation [1]. The reliability of a bridge in operation is deeply influenced by the corrosion degradation of its components [2]. The corrosion problem of cable systems has aroused extensive consideration from scholars. The main types of cable structure are parallel wire rope and steel strand; both are composed of single steel wire. The earliest cable steel wires were all-steel wires without coating. The cable was wrapped with a protective sleeve outside and filled with barrier materials inside, but the anti-corrosion effect was poor, as proven in practice. To resist the corrosion of environmental factors, the cable components began to have a coating on the surface of the high-strength steel wire, to isolate the corrosion medium. Early cable bearing bridges mostly used galvanized high-strength steel wire, and relevant research has mainly focused on galvanized steel wire. The corrosion development law of galvanized coating is relatively clear. In recent years, Galfan coating steel wire with better corrosion resistance has gradually started to be widely used. Its corrosion resistance is better than that of galvanized steel wire, but the corrosion characteristics of this coating lack systematic research. The deterioration of the steel wire is the result of the combined effect of corrosion and fatigue. The corrosion characteristics of a steel wire coating directly affect the corrosion state of the steel substrate and the subsequent fatigue crack growth. The difference of corrosion fatigue properties between the two types of steel wires is unclear. To further improve the service performance and operation and maintenance level of a cable load-bearing structure, this work intended to conduct an experimental study on the corrosion characteristics of galvanized steel wire and Galfan steel wire, analyze the degradation characteristics of the two types of high-strength steel wire, and discuss the influence of corrosion and fatigue on the degradation of high-strength steel wire.

The degradation of high-strength steel wire is the result of the combined effect of corrosion and fatigue. Simple corrosion and fatigue degradation of steel wire is relatively slow. The pitting corrosion pits produced along with the uniform corrosion of steel wire provide conditions for the initiation of fatigue cracks, thus greatly reducing the fatigue life of steel wire; the fatigue strength of the steel wire decreases with the increase of corrosion [3,4]. Betti et al. conducted in-depth research on the deterioration mechanism of the high-strength steel wire of a suspension bridge, studied the corrosion evolution of galvanized and non-galvanized steel wire under different environmental conditions through an accelerated cyclic corrosion test, and pointed out that the uneven change of the steel wire section along the length reduced the elongation of the steel wire [5]. Nakamura, Suzumura, and others researched the influence of reagent concentration, ambient temperature, and humidity on the corrosion rate of galvanized steel wire through experiments, given the loss rate of the galvanized layer of galvanized steel wire. They pointed out that the main reason for the deterioration of the properties of corroded steel wire is the reduction of elongation, torsional strength, and fatigue strength [6–9]. Lan et al. conducted an acid salt spray test and fatigue test on high-strength steel wire and fitted the corrosion fatigue life of steel wire based on the Weibull distribution. The change trend of the fatigue life of steel wire and cable components with the development of the corrosion process is basically consistent, and the fatigue life of a stay cable decreases significantly as the corrosion degree of the steel wire increases [10]. The above research confirmed that the fatigue life of steel wire decreases due to corrosion, from practical engineering and laboratory research. Jiang et al. and Wang et al. used solutions to create corrosive environments and studied the effects of different solutions, solution concentrations, stress amplitudes, and load frequencies on fatigue life. The corrosion fatigue performance of steel wire in acidic environments was the worst, and electrochemical reaction greatly reduced the life of the steel wire [11,12]. Sun established a corrosion fatigue degradation model of steel wire based on fracture mechanics and compared it with test results, which proved that the proposed model could better simulate the evolution of corrosion fatigue of steel wire [13]. Li et al. established improved uniform corrosion and pitting models for high-strength steel wire, verified that the maximum pitting factor obeys a Gumbel distribution based on an accelerated corrosion test, fitted relevant parameters, and studied corrosion fatigue through finite element simulation [14]. Jiang et al. measured the corroded steel wire 3D profile and proposed that the pitting depth of steel wire follows a normal distribution and that the location and scale parameters increase with the degree of corrosion. A method for predicting the residual life of corroded steel wire based on 3D measurement and AFGROW software was established [15]. The basic process and principles of corrosion fatigue degradation have been confirmed by scholars.

The above research mainly focused on the corrosion characteristics of galvanized steel wire. Compared with galvanized steel wire, galvanized aluminum high-strength steel wire has been gradually applied to engineering construction in recent years. However, the research on its corrosion behavior characteristics is relatively scarce. Xue et al. [16] studied the corrosion fatigue behavior of Galfan coating, and Cao et al. [17] studied the effect of Nd on the corrosion behavior of Zn-5Al (wt.%) alloy in neutral 3.5wt.%NaCl solution using electrochemical impedance spectroscopy. The addition of Nd can improve the corrosion resistance of Zn-5Al alloy. The above studies are important references for the study of Galfan coating corrosion resistance performance. Nonetheless, the development of Galfan coating corrosion resistance and pitting has not been systematically studied, and the difference in the service life between the two kinds of steel wires is unclear.

Systematic studies on the corrosion resistance of Galfan wire coating and the difference between its corrosion fatigue and the fatigue of galvanized wire are few. The effect on improving the service performance of cable structures in engineering applications is unclear. In this study, the corrosion resistance of galvanized and Galfan high-strength steel wires was studied using an accelerated corrosion test. On the basis of the test results, a uniform corrosion development model of the two kinds of steel wires was established, and a dynamic distribution model of the maximum pitting coefficient was established using Gumbel distribution. The time-varying characteristics of the scale and location parameters of the maximum pitting coefficient distribution for the two types of steel wire are given. On the basis of the traffic load monitoring data of a bridge during operation, the corrosion fatigue degradation characteristics of the steel wires of the cable components in the service period were analyzed, which can provide a reference for the design and maintenance of the components of bridge structure cables.

2. Accelerated Corrosion Experiment

Corrosion tests can generally be divided into two categories: one is the traditional corrosion tests under natural conditions. The real corrosion conditions of test objects can be obtained by directly exposing the test samples to the real environment, but the time cost is high and the test cycle is too long. The other is accelerated corrosion tests under a laboratory environment. By putting the test samples into a corrosion chamber and using a salt spray environment, atmospheric pressure, temperature, and other factors to accelerate corrosion, the test time is greatly shortened. With the rapid development of bridge component materials and the increasing demand for corrosion-resistance research, the salt spray test has become the most commonly used method for cable corrosion research. Based on the specification "Corrosion Test in Artificial Atmospheres—Salt Spray Test"(GB/T 10125-2012) [18], a neutral salt spray test was selected to study the corrosion characteristics of galvanized and Galfan high-strength steel wire. The corrosion atmosphere was formed using a salt spray test chamber, and an accelerated corrosion effect was achieved by combining temperature and air pressure. Given the few parameters of the accelerated corrosion test, the concentration and continuity of the salt spray, temperature, and pressure in the test chamber were mainly guaranteed during the test.

The accelerated corrosion time and number of test pieces are presented in Table 1. With reference to existing research results, the planned test duration for galvanized steel wire was 510 h, and the test duration for Galfan steel wire was 1445 h. Each group of galvanized steel wires had 5 test pieces, and the first 10 groups of Galfan steel wire had 5 steel wires. To ensure the accuracy of data, each group of Galfan steel wires had 10 steel wires. Before placing the steel wire test pieces, all steel wires were weighed and numbered, and then the test pieces were placed in a salt spray test chamber for artificial atmospheric corrosion. The scheduled steel wire test pieces were taken out in batches, according to the planned corrosion time period, and then the corrosion products were removed using a combined chemical method and physical method, according to the corrosion product removal specification "Corrosion of Metals and Alloys–Removal of Corrosion Products from Corrosion Test Specimens" (GB/T 16545) [19]. The pickling solution was a saturated solution of NH_3CH_2COOH, where 1000 mL solution was prepared by mixing 250 g NH_3CH_2COOH and distilled water. The specific steps were as follows: (1) Put the sample into the saturated NH_3CH_2COOH solution (pickling solution) at 20~25 °C and soak it for 10 min. (2) Rub off the residual corrosion products with abrasive paper. (3) Successively put it into water and alcohol 5 times for cleaning. (4) Wipe it with a towel and dry it after cleaning. After each group of steel wire samples was taken out at specific times, the corrosion products were removed and weighed according to the above steps.

Table 1. Accelerated corrosion period of high-strength steel wires.

Galvanized Steel Wire		Galfan Steel Wire			
Corrosion Duration (h)	Amount of Test Pieces	Corrosion Duration (h)	Amount of Test Pieces	Corrosion Duration (h)	Amount of Test Pieces
24	5	24	5	606	10
48	5	48	5	686	10
96	5	96	5	784	10
168	5	168	5	848	10
216	5	216	5	944	10
264	5	264	5	1016	10
336	5	336	5	1088	10
384	5	384	5	1192	10
438	5	438	5	1389	10
510	5	510	5	1445	10

2.1. Specimens of High-Strength Steel Wire

The high-strength steel wire used in the test was provided by a cable manufacturer. The material parameters of the steel wire samples were tested, and the results are given in Table 2. According to the mass and density of the coating, the thickness of the zinc coating was 28.05 μm, and the thickness of the Galfan coating was 29.53 μm. Galvanized steel wire coatings consisted of pure zinc, and Galfan steel wire consisted of 5% aluminum zinc alloy and a small amount of mixed rare earth elements. The chemical composition of the steel wire coating is shown in Table 3.

Table 2. Steel wire sample parameters.

Type of Steel Wire	Diameter (mm)	Out of Roundness	Tensile Strength (MPa)	Yield Strength (MPa)	Elastic Modulus (MPa)	Coating Quality (g/m^2)
Galvanized steel wire	5.34	0	1895	1760	2.08×10^5	336
Galfan steel wire	5.25	0	1926	1775	2.08×10^5	337

Table 3. Mass percentages of microelements of the high-strength steel wires coating (%).

Material Composition	Zn	Al	Si	Mn	P	S	Cr
Galvanized Coating	≥99	/	/	/	/	/	/
Galfan Coating	≤95	≤5	0.24	0.86	0.009	0.002	0.17

The steel wire was cut to make an experimental sample, and the sample length was about 20 cm. Before the test, all steel wire samples were cleaned, dried, numbered, measured, and weighed one by one. After the preparation, they were put into the salt spray box.

2.2. Test Device and Accelerated Corrosion Medium

The test materials required for the neutral salt spray test included: steel wire sample, salt spray box, electronic balance, C_2H_5OH, NaCl, and pickling agent. The quality of the steel wire was measured using a high-precision electronic balance. The reagent grade used in the test was chemically pure. The parameters of the relevant instruments and chemical reagents are provided in Table 4.

Table 4. Instruments and chemical reagents.

Instrument Name	Electronic Balance	C_2H_5OH	NaCl	Pickling Agent
Parameter grade	0.0001 g	Analytically pure	Analytically pure	Analytically pure

A Zhongte LX120 multi-functional climate and environment salt fog test chamber was adopted for the test salt fog chamber, which can realize the simulation of a salt fog environment, high temperature, and high pressure conditions, as well as the coupled effect of different conditions. The technical parameters are listed in Table 5. The experimental equipment and materials are shown in Figure 1.

Table 5. Technical parameters of the salt spray box.

Internal Dimensions (mm)	Test Room Temperature (°C)	Pressure Barrel Temperature (°C)	Compressed Air Force (kgf/cm^2)
1200 × 1000 × 500	NSS ACSS 35 ± 1 CASS 50 ± 1	NSS ACSS 47 ± 1 CASS 63 ± 1	1.00 ± 0.01

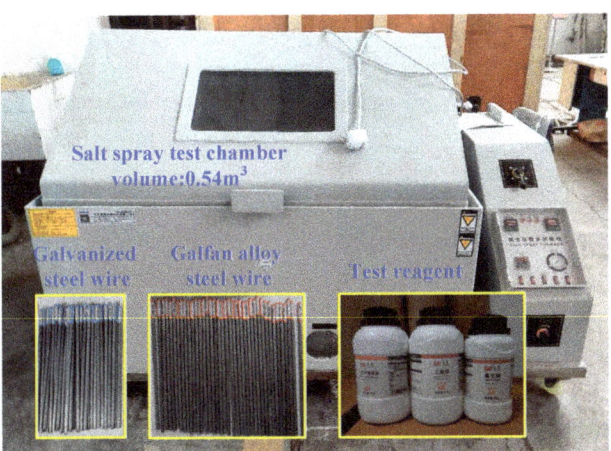

Figure 1. Equipment and materials for the salt spray test device.

According to the requirements of the specification for the neutral salt spray test and to test the reproducibility of the test equipment results, a steel reference test verification was carried out. Four defect-free CR4 grade cold-rolled carbon steel plates with a thickness of 1 mm were selected. After cleaning, the back of the samples was protected with a film, and a sample was placed at the four corners of the salt spray box for 48 h, as shown in Figure 2. The specific test parameters were set according to the specifications. The volume of the salt spray box was 0.6 m^3. No solution accumulated on the top of the box on the sample, and the spray was always uniform. The pH value of the spray solution collected by the collector was in the range 6.5–7.2. The test temperature was 35 ± 1 °C, and the concentration of NaCl was 50 g/L. The settling rate of the salt spray met the specifications.

The reference sample was taken out immediately after the test, and ammonium acetate solution and a mechanical cleaning method were used to remove the corrosion products. The mass loss per unit area obtained after weighing is presented in Table 6. The loss of each test piece was within the range of 70 ± 20 g/m^2 required by the specification, indicating that the equipment operated normally and met the operational requirements of the neutral salt spray test. The steel wire corrosion tests were then carried out according to the specification requirements. The test solution was NaCl solution with a concentration of

50 g/L ± 5 g/L, the pH value of spray solution was 6.5 to 7.2, and the temperature was 35 ± 2 °C.

Figure 2. Steel reference test.

Table 6. Mass loss of the cold-rolled carbon steel sheet.

Specimen Number	Specimen 1	Specimen 2	Specimen 3	Specimen 4
Mass loss (g/m^2)	75	78	77	80

3. Corrosion Phenomenon

The box was opened, and the corresponding test pieces were taken out, according to the time specified in the test plan. They were cleaned and dried in strict accordance with the corrosion product removal specification. Figure 3 shows the corrosion morphology of a single Galfan steel wire at different corrosion stages. The degree of corrosion of the steel wire gradually increased with time, and the corrosion resistance of the Galfan steel wire was much better than that of the galvanized steel wire. The early corrosion of the steel wire was reflected in the loss of luster of the coating, accumulation of salt on the surface of the steel wire, and the gradual production of uniformly distributed white accumulation products on the surface of the steel wire.

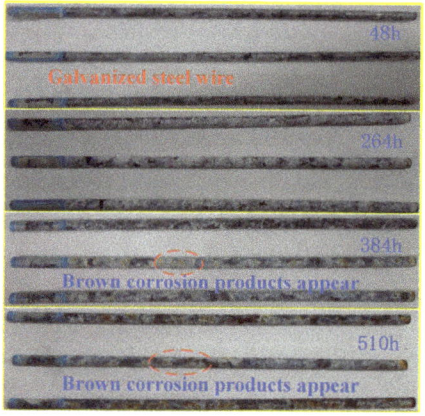 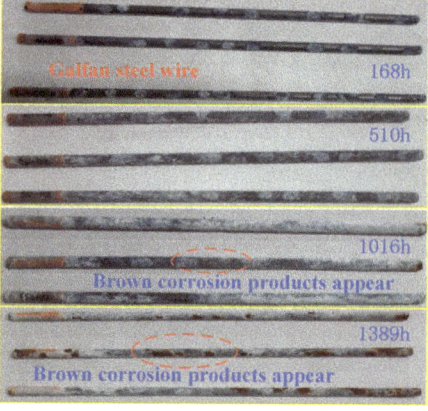

Figure 3. Corrosion morphology in the different stages of corrosion.

The galvanized steel wire and the Galfan steel wire were completely covered by white corrosion products from the 264 h and 510 h, respectively. After the corrosion products were removed, part of the coating had lost luster, indicating that the coating had been corroded at this time and part of the iron matrix had begun to be exposed. At 384 h and 1016 h, some corrosion spots had appeared on the surface of the steel wire, indicating that the iron matrix under the partial coating of the steel wire had started to corrode at this time, producing reddish brown corrosion products. At 510 h, many reddish brown corrosion products had appeared at the middle and end of the galvanized steel wire, and the reddish brown corrosion products had connected into sheets. At 1389 h, the same situation had occurred to the Galfan steel wire. At this time, uneven pits had appeared on the steel wire surface after the corrosion products had been removed. The corrosion of steel wire can be generally divided into two parts: The first part is the corrosion of the surface coating. When the corrosion depth exceeds the coating thickness, the corrosion of the second part of the steel wire matrix begins. Owing to the protection of oxidation products formed after the coating corrosion, the corrosion rate of the steel wire matrix decreases.

4. Corrosion Process

4.1. Uniform Corrosion

The development of the uniform corrosion of steel wire is mainly affected by two factors: the corrosion time, and the uniform corrosion rate. As the uniform corrosion depth is not easy to obtain directly, it is generally described using the volumetric method, weight-loss method, or other methods. In this study, the weight-loss method was used to describe the uniform corrosion of steel wire, which is given as Equation (1). With reference to the specification for removal of corrosion products, the chemical substances generated after steel wire corrosion can be removed without damaging the metal matrix, and the quality loss of metal in the corrosive environment can be accurately measured, to evaluate the degree of corrosion of steel wire.

$$\psi = \frac{m_0/l_0 - m_1/l_1}{m_0/l_0} \times 100\% \tag{1}$$

where ψ represents the loss rate of steel wire mass, m_0 represents the quality of steel wire before corrosion, l_0 represents the length of steel wire before corrosion, m_1 represents the quality of steel wire after corrosion, and l_1 represents the length of steel wire after corrosion.

Owing to the large slenderness ratio of the steel wire specimen, the sectional area of both ends of the steel wire is small and the calculation constant is large, so the length change caused by corrosion can be ignored. That is $l_0 = l_1$. Therefore, Equation (1) can be rewritten as Equation (2):

$$\psi = \frac{m_0 - m_1}{m_0} \tag{2}$$

According to Equation (3), the mass loss of steel wire is converted into the coating and corrosion depth of steel wire du.

$$du = \frac{\psi}{A\rho} = \frac{m_0 - m_1}{\pi D \rho l_0} \tag{3}$$

where A is the surface area of steel wire, $A = \pi D l_0$, D is the initial diameter of steel wire, ρ is the material density, and l_0 is the length of steel wire.

Figures 4 and 5 show the average mass loss per unit length of all steel wire samples, indicating that the overall distribution of test results was relatively regular. The change of steel wire mass loss at each stage was stable, without obvious mutations, indicating that the test effect was good. The mass loss of Galfan steel wire increased obviously with the increase of corrosion time. During the entire corrosion period, the increasing trend of corrosion quality was close to an exponential change, and the corrosion rate gradually slowed down.

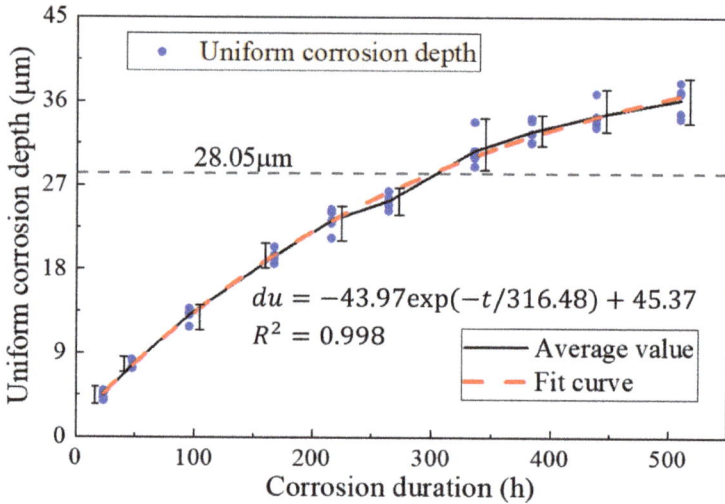

Figure 4. Average mass loss per unit length of galvanized steel wire.

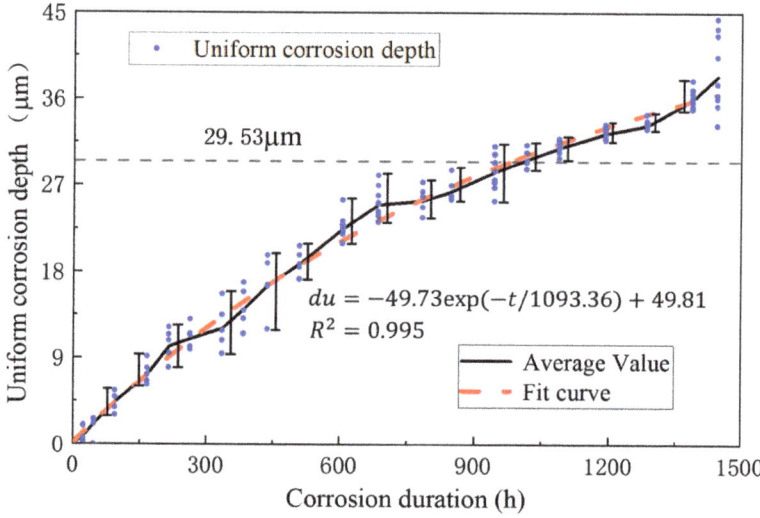

Figure 5. Average mass loss per unit length of Galfan steel wire.

Owing to the influence of the coating thickness and corrosion randomness, the corrosion rate of different steel wires varies, so the steel wire corrosion coefficient was introduced to describe the randomness. Figures 6 and 7 show the fit distribution law of the two kinds of steel wires. The corrosion coefficient of the galvanized steel wire conformed to the normal distribution $N(\mu, \sigma^2)$ with $\mu = 1.00$ and $\sigma = 0.0483$, whereas the Galfan steel wire conformed to the Cauchy distribution $C(\gamma, x_0)$ with $x_0 = 1.00$ and $\gamma = 0.0391$.

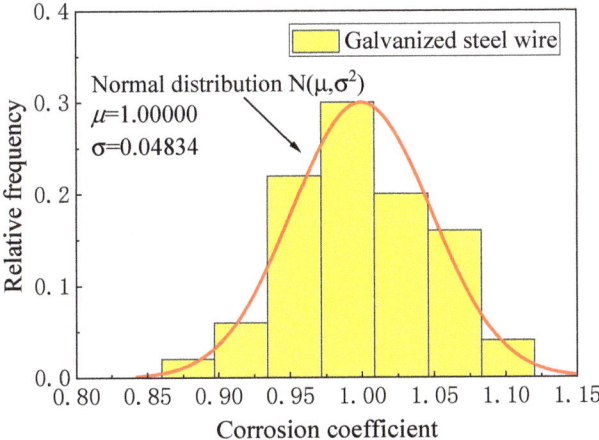

Figure 6. Corrosion coefficient distribution of galvanized steel wire.

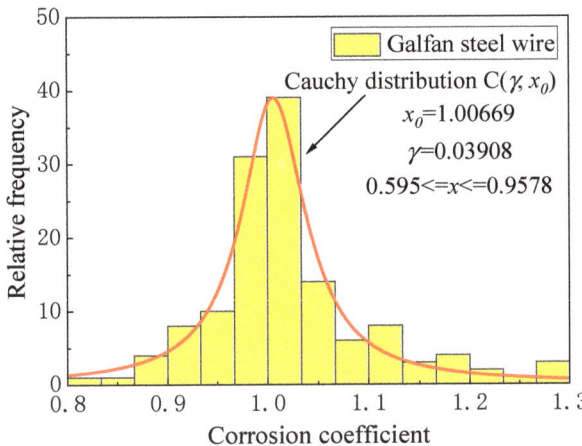

Figure 7. Corrosion coefficient distribution of Galfan steel wire.

Therefore, the uniform corrosion rate of the two kinds of steel wire can be expressed as in Equation (4).

$$\begin{cases} du_{Zn}(t) = \psi_{Zn} * (-43.97 * exp(-t/316.48) + 45.37) & \text{Galvanized steel wire} \\ du_{Gal}(t) = \psi_{Gal} * (-49.73 * exp(-t/1093.36) + 49.81) & \text{Galfan steel wire} \end{cases} \quad (4)$$

where t is the corrosion duration, ψ_{Zn} conforms to the normal distribution $N(\mu, \sigma^2)$ with $\mu = 1.00$, $\sigma = 0.04834$, and ψ_{Gal} conforms to the Cauchy distribution $C(\gamma, x_0)$ with $x_0 = 1.00669$, $\gamma = 0.03908$.

4.2. Pitting Corrosion

The key factor causing cracks in corrosion fatigue is pitting corrosion, which is accompanied by uniform corrosion. When a passivation or film forms on the metal material surface, a small and deep corrosion pit is generated on the substrate surface after the protective layer is consumed. Owing to the stress concentration effect, the stress at the edge of the corrosion pit is far greater than the overall stress level. When the depth of the corrosion pit increases to a certain extent, it becomes a crack. A crack usually occurs at the position with the greatest pitting corrosion, so the deepest pitting corrosion determines the working state of the steel wire and is a key analysis point in corrosion fatigue analysis.

Figure 8 shows the surface morphology of the steel wire after cleaning. The pits on the surface of the steel wire are obvious. To further determine the distribution characteristics of the pitting corrosion of the different types of steel wire, the steel wire indication was detected using a 3D shape scanner, and a 3D model of the steel wire surface was established based on the regression of 2D scanning results. As the steel wire specimen was not completely straight, the surface profile presented an irregular curve shape as a whole. To eliminate the influence of the steel wire specimen's own curved shape on the test results, the small window moving average automatic baseline correction method was used to estimate the baseline corresponding to the measured profile. According to the difference between the measured contour coordinates and the baseline coordinates, the pitting depth on the axial length of the steel wire could be determined.

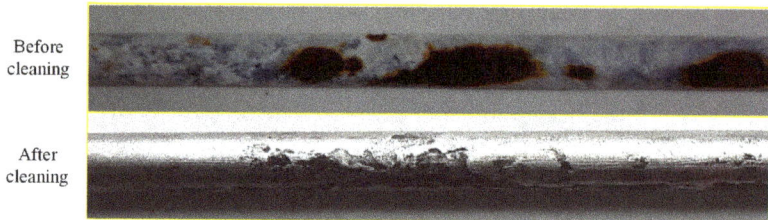

Figure 8. Surface morphology of steel wire.

The pitting depth could be calculated using the uniform corrosion depth and maximum pitting factor [20]. Figures 9 and 10 show the 3D surface regression profile of the steel wire surface. According to the 40 measured surface profiles and regression analysis results, the block maximum value method was used to obtain a sample of the maximum pitting factor in each exposure period. The analysis accuracy of the block maximum method is directly affected by the selected block size, and the sample size of the block maximum should be sufficiently large. After comprehensive consideration, this study determined that the block size for calculating the pitting factor was 10 mm. The maximum pitting depth was taken every 10 mm along the length of the steel wire, and the maximum pitting factor was calculated according to Equation (5).

$$G(t) = da(t)/du(t) \tag{5}$$

where $G(t)$ is the maximum pitting factor, $da(t)$ is the maximum pitting depth, and $du(t)$ is the uniform corrosion depth in the same period.

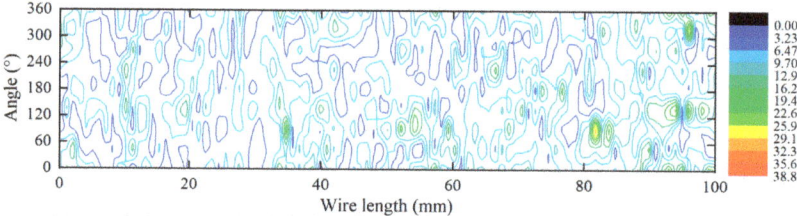

Figure 9. Regression of the 3D corrosion morphology of galvanized steel wire.

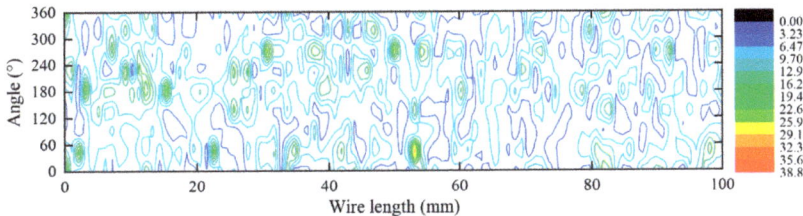

Figure 10. Regression of the 3D corrosion morphology of Galfan steel wire.

The existing research ignored the time-varying characteristics of the maximum pitting factor and posited that the maximum pitting factors in different periods have the same distribution characteristics. Figures 11 and 12 show the fitting results of the maximum pitting factor of steel wire in certain periods. The maximum pitting factor has obvious time-varying characteristics. Gumbel distribution was used to fit the maximum pitting factor. The relationship between the pitting system and location parameters, scale parameters, and corrosion time can be expressed as in Equation (6).

$$Z_p = exp\left\{-exp\left[-\left(\frac{x-\mu(t)}{\sigma(t)}\right)\right]\right\} \qquad (6)$$

where $\mu(t)$ and $\sigma(t)$ are the location parameters and scale parameters corresponding to the accelerated corrosion duration t, respectively.

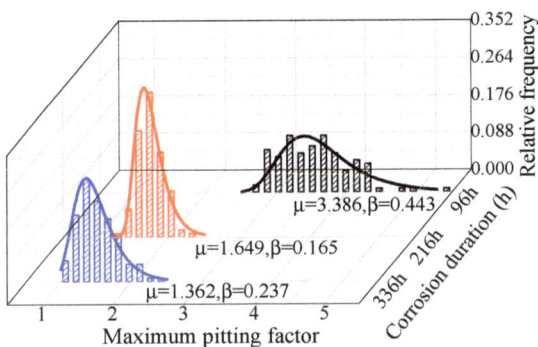

Figure 11. Fitting results of maximum pitting factor of galvanized steel wire.

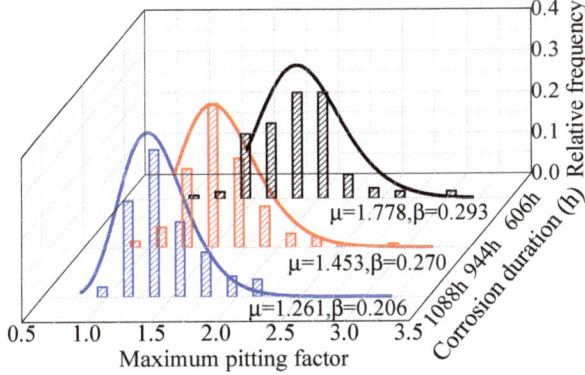

Figure 12. Fitting results of maximum pitting factor of Galfan steel wire.

For galvanized steel wire, the pitting factor was larger at the early stage of corrosion; μ was 3.386 at 96 h. The pitting factor decreased significantly with the increase of time at the later stages; μ was 1.649 at 216 h. The change amplitude of Galfan steel wire was smaller than that of the galvanized steel wire but also decreased with the increase of time; μ was 1.778 at 606 h and 1.261 at 1088 h. The increase of the pitting depth was limited, and estimating the corrosion process of the degraded steel wire according to a single distribution law may underestimate the service state of steel wire. Furthermore, the fitting analysis of the location parameters and scale parameters of the pitting distribution function in different corrosion stages showed that the distribution parameters of the two coatings conformed to the exponential change law, which could be calculated according to Equations (7) and (8).

$$\begin{cases} \mu_{Zn}(t) = 9.8698 \exp(-t/59.2058) + 1.3595 \\ \sigma_{Zn}(t) = 1.3967 \exp(-t/58.0061) + 0.2190 \end{cases} \quad \textit{Galvanized steel wire} \quad (7)$$

$$\begin{cases} \mu_{Gal}(t) = 18.7549 \, exp(-t/71.0266) + 1.7746 \\ \sigma_{Gal}(t) = 4.5949 \, exp(-t/55.5349) + 0.3375 \end{cases} \quad \textit{Galfan steel wire} \quad (8)$$

where $\mu_{Zn}(t)$, $\sigma_{Zn}(t)$, $\mu_{Gal}(t)$, and $\sigma_{Gal}(t)$, respectively, represent the location parameters and scale parameters of the maximum pitting coefficient distribution between the zinc coating and Galfan coating corresponding to the accelerated corrosion duration t.

5. Corrosion Fatigue of High-Strength Steel Wire

5.1. Corrosion Fatigue Degradation Model

When the protection system of cable components is damaged because the air contains water, salt, and other substances, the corrosion factors of the external environment enter the cable body and cause steel wire corrosion. The corrosion of steel wire generally goes through the following three stages [21]: (1) uniform corrosion and pitting corrosion, (2) pitting crack development, and (3) corrosion fatigue crack growth. Uniform corrosion refers to uniform corrosion and detachment of the steel wire surface, which belongs to pure chemical reaction and directly causes the reduction of the steel wire diameter. The reduction of diameter is approximately equal along the length of steel wire. Owing to the non-uniformity of the material, the diameter of the steel wire decreases, accompanied by pitting corrosion randomly distributed on the surface of the steel wire. Uniform corrosion includes the corrosion consumption of the steel wire coating and the time needed for uniform corrosion and pitting of the steel wire matrix. The specific development rules of uniform corrosion and pitting were obtained through accelerated corrosion tests in this study.

On the basis of the unit area A_0 of pitting calculation, Equation (9) can be used to obtain the maximum pitting coefficient distribution of A_k in any area [22,23]. When estimating the maximum pitting coefficient per unit area, Saint Venant's principle must be considered. The minimum length of the analysis unit should be greater than twice the diameter of the steel wire.

$$\mu_k = \mu_0 + \frac{1}{\sigma_0} ln(\frac{A_k}{A_0}), \sigma_k = \sigma_0 \quad (9)$$

where A_k is the surface area of the analysis target, and A_0 is the surface area of unit area. μ_0 and σ_0 are the location parameters and scale parameters of the corresponding fitting analysis results by exponential function.

Y. Kondo pointed out that the initiation of fatigue cracks is caused by pitting, and the transition from pitting depth to fatigue cracks is related to the stress amplitude of the steel wire. When the stress amplitude is large, the cracks easily occur in smaller pits, and vice versa. The development of pitting pits into fatigue cracks can be described based on the theory of fracture mechanics. The stress intensity factor K is introduced to describe the strength of the stress field near the crack tip. The generation of cracks is mainly affected by the alternating stress field at this location. The amplitude of the stress intensity factor ΔK can be calculated according to Equation (10).

$$\Delta K = F_a\left(\frac{a}{b}\right)\Delta\sigma_a\sqrt{\pi a} + F_b\left(\frac{a}{b}\right)\Delta\sigma_b\sqrt{\pi a} \tag{10}$$

where a is the crack depth, b is the diameter of steel wire, $\Delta\sigma_a$ is the equivalent axial stress amplitude, and $\Delta\sigma_b$ is the equivalent axial stress amplitude. $F\left(\frac{a}{b}\right)$ is calculated using Equation (11) [24].

$$\begin{cases} F_a\left(\frac{a}{b}\right) = 0.92 \cdot \frac{2}{\pi} \cdot \sqrt{\frac{2b}{\pi a} \cdot \tan\frac{\pi a}{2b}} \cdot \frac{0.752 + 1.286\left(\frac{a}{b}\right) + 0.37\left(1 - \sin\frac{\pi a}{2b}\right)^3}{\cos\frac{\pi a}{2b}} \\ F_b\left(\frac{a}{b}\right) = 0.92 \cdot \frac{2}{\pi} \cdot \sqrt{\frac{2b}{\pi a} \cdot \tan\frac{\pi a}{2b}} \cdot \frac{0.923 + 0.199\left(1 - \sin\frac{\pi a}{2b}\right)^4}{\cos\frac{\pi a}{2b}} \end{cases} \tag{11}$$

where $F_a\left(\frac{a}{b}\right)$ denotes axial stress, $F_b\left(\frac{a}{b}\right)$ denotes bending stress, a is the crack depth, and b is the diameter of steel wire.

The corrosion cracking of steel wire is generated from pitting to cracks. When the stress intensity factor at the pitting reaches the threshold value for crack growth of 2.8 MPa·m$^{1/2}$ [25], the pitting development is transformed into the crack development stage. The Paris formula is the most widely used method for the growth rate analysis of metal corrosion fatigue cracking, and the crack growth rate is expressed as the fatigue crack growth rate [26].

$$\frac{da}{dt} = C(\Delta K)^m N \tag{12}$$

where ΔK is the effective stress intensity factor amplitude, and C and m are the parameters of the Paris criterion.

To comprehensively consider the impact of the daily traffic flow intensity level on the crack development rate, a crack depth development model was established based on the proportion of daily traffic flow operations, as given in Equation (13).

$$\begin{cases} a_i = \Delta a + a_{i-1} \\ \Delta a = C \sum n_q \left[\sum e_j (\Delta K_{qj})^m N_{qj}\right] \end{cases} \tag{13}$$

where a_i is the depth of the crack at time i, Δa is the increment of the crack, e_j is the operating time of traffic flow with different intensities, $\sum e_j = 24$ h, and ΔK_j and N_j are the stress intensity factor range and the number of cycles.

Mayrbaurl pointed out that the critical relative crack depth conforms to the logarithmic normal distribution, with an average value of 0.390 and a coefficient of variation of 0.414. Based on the tests, the maximum critical relative depth was 0.5, which can be used as the judgment standard for steel wire failure.

5.2. Protype Bridge and Traffic Load

To further analyze the corrosion fatigue degradation of the two types of high-strength steel wire under a load, this study took a long-span suspension bridge as the research background and took high-strength steel wire suspenders as the research object, to investigate the corrosion fatigue degradation of galvanized and Galfan high-strength steel wire under the same conditions. The main bridge structure of the bridge is a single span 838 m steel concrete composite girder suspension bridge. The main cable span is (250 + 838 + 215) m. The vertical layout is shown in Figure 13. Steel concrete composite girders are used as stiffening girders, and the steel beams are combined with concrete bridge decks through shear studs. The section layout is shown in Figures 14 and 15. The full width of the stiffening girder is 33.2 m. The steel longitudinal beams on both sides are connected by steel cross beams. The center height is 2.8 m. The center distance between the webs of the longitudinal beams on both sides is 26.0 m. The bridge deck is a reinforced concrete bridge deck with a full width of 25.0 m and a thickness of 0.22 m. The standard spacing of suspender lifting points is 16 m. There are two suspenders for each lifting point, with 204 suspenders in total for the whole bridge, and 151 ϕ5 mm high-strength steel wires included in each suspender.

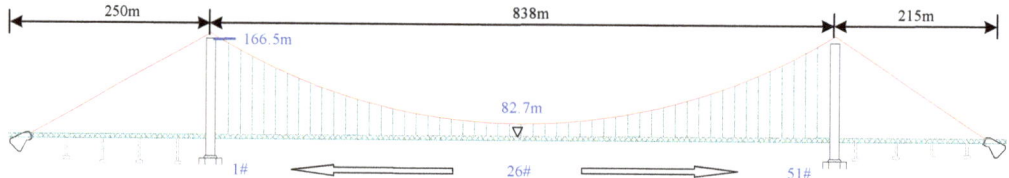

Figure 13. Elevation Layout of the Bridge.

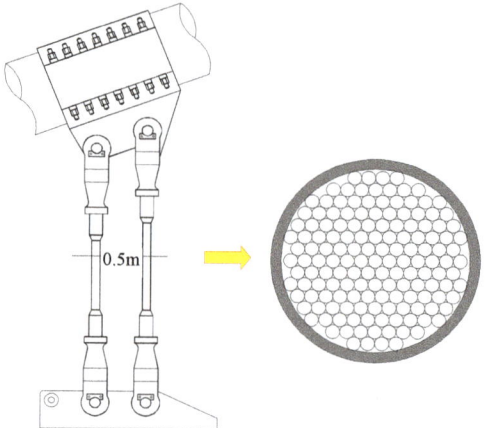

Figure 14. Schematic Diagram of a Suspender.

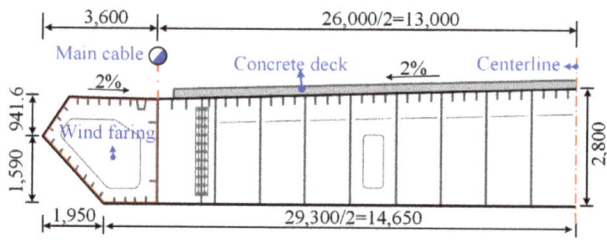

Figure 15. Section of the Main Girder.

Stochastic traffic flow simulation is the mainstream method for long-span bridge operation evaluation. After the degradation of the suspender steel wire changes from pitting to cracking, it enters the main stage of crack development. The crack growth speed is mainly affected by the stress response of the suspender under the traffic load. Based on the measured traffic flow in a certain area, obtained by the traffic load monitoring system, this study analyzed and selected the data of representative periods, to simulate the formation of traffic flow for loading, so as to obtain the corrosion fatigue degradation of suspender steel wire under different levels of traffic loading.

The traffic data collected by a weigh-in-motion system (WIM) included the vehicle data of the region from 1 March to 31 March 2015, including the vehicle wheelbase, axle load, vehicle speed, and other details [27]. Figure 16 shows the hourly traffic volume results obtained from the collected data by time and lane statistics. The daily traffic volume changed greatly. On the basis of the action area of the traffic flow with obvious changes in traffic volume, the traffic flow level was divided into three intensity levels: dense, moderate, and sparse. The traffic flow intensive periods were mainly concentrated at 9:00–11:00 and 13:00–17:00, and the sparse flow period was from 21:00 to 7:00. The vehicle traffic characteristics at different intensity levels were fitted to obtain the distribution characteristics of the traffic volume and vehicle flow speed under the three intensity levels, as shown in Figures 17 and 18.

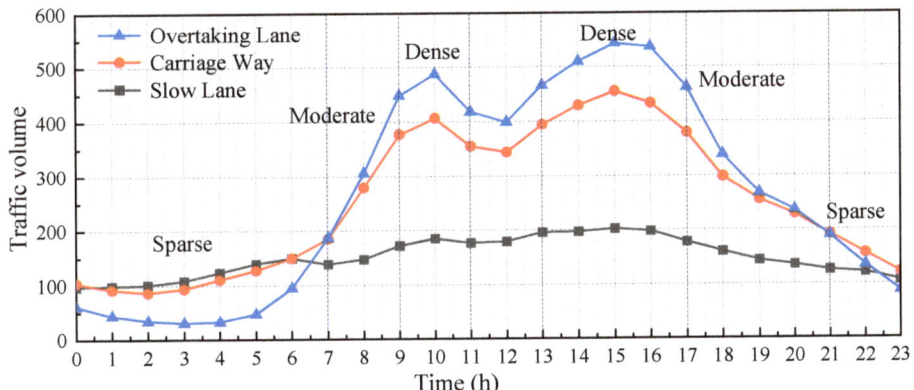

Figure 16. Change trend of daily average hourly traffic volume.

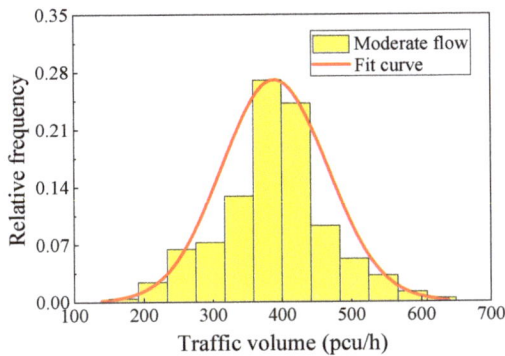

Figure 17. Traffic volume (moderate flow).

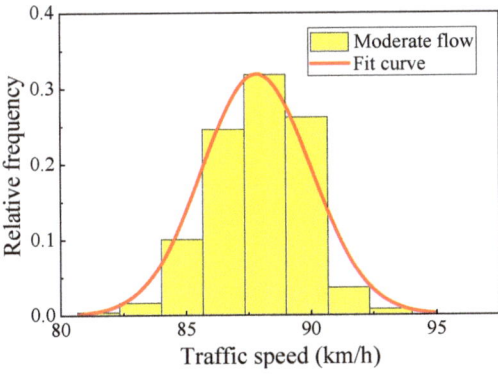

Figure 18. Vehicle flow speed (moderate flow).

Based on the statistical results of traffic volume and traffic flow speed, the average spacing of vehicles could be determined according to Equation (14). The key parameters of traffic flow are given in the Table 7, and the characteristics of traffic flow parameters could then be determined. In combination with the random traffic flow sampling method based on the Monte Carlo method, vehicle samples were obtained [28], so as to select the traffic flow in typical representative periods for a load effect analysis. The specific traffic flow simulation process is shown in Figure 19.

$$Q = K \cdot V \qquad (14)$$

where Q denotes traffic volume, K denotes traffic density, and V denotes traffic speed.

Table 7. Key parameters of traffic flow.

	Key Parameter	Passing Lane	Carriage-Way	Slow Lane
Dense flow	Traffic volume (pcu/h)	572	474	209
	Traffic speed (km/h)	93.36	88.4	70.76
Moderate flow	Traffic volume (pcu/h)	451	378	184
	Traffic speed (km/h)	94.12	87.48	67.98
Sparse flow	Traffic volume (pcu/h)	175	186	132
	Traffic speed (km/h)	92.12	81.80	67.98

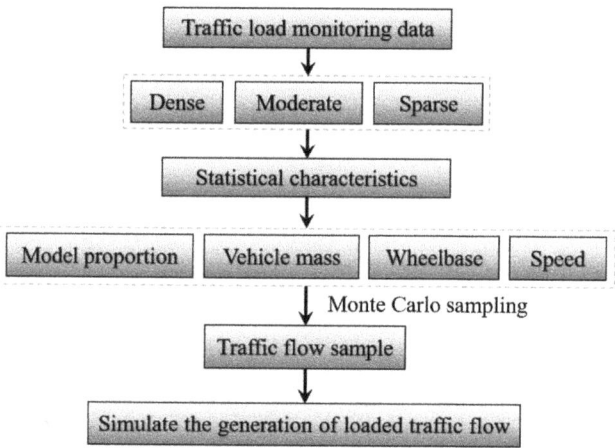

Figure 19. Vehicle flow simulation.

5.3. Numerical Analysis

Based on the above vehicle flow load simulation method, the dense, moderate, and sparse evacuation flows obtained from the sampling were loaded into the bridge finite element model. The vehicle bridge coupling analysis system established on the basis of finite element analysis software ANSYS in a previous study was used to obtain the bridge suspender response. For suspension bridges, short suspenders in the middle of the span are the most easily damaged suspenders, because of their short length and the large bending stress caused by the relative displacement between the main cable and the main girder. As the link element cannot directly give the bending stress, Wyatt's theoretical formula was introduced to calculate the bending stress, according to the axial stress of the suspender and the angle generated by the relative movement between the main cable and the stiffening girder [29].

$$\sigma_b = \tan\theta \cdot \sqrt{\sigma_a E} \qquad (15)$$

where σ_a is axial stress, E is elasticity modulus of steel wire, θ is the angle caused by the relative movement between the main cable and the stiffening girder.

Figures 20 and 21 show the stress response results of short suspenders in the mid span. The response of the suspenders under the overall traffic flow is affected by the number of vehicles and the distance between vehicles, and the response under a dense flow is most obvious. However, during driving, a driver spontaneously maintains a safe distance, to ensure safety requirements, which is usually greater than the distance between adjacent suspenders. Therefore, the extreme value of the axial force response of suspenders under sparse or moderate flow may exceed the extreme value of axial force under a dense flow, owing to the impact of single vehicle weight at the time of traffic flow. The bending stress is mainly affected by the continuous superposition of traffic flow effects, and the overall response level and extreme value of bending stress under a dense flow are the most obvious.

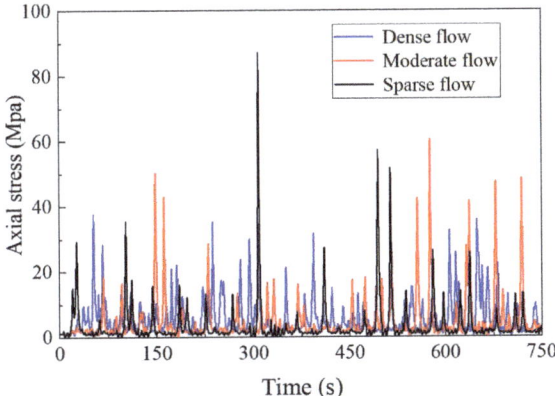

Figure 20. Axial stress response.

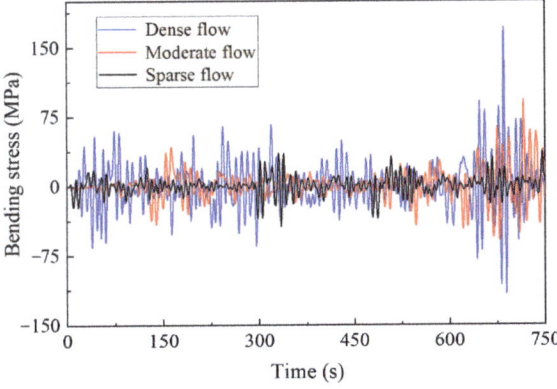

Figure 21. Bending stress response.

Basing on the obtained stress responses of suspender steel wires under different traffic flows, and combined with the uniform corrosion and pitting corrosion laws of steel wires obtained from the tests, the corrosion fatigue degradation of steel wires was analyzed. Based on the proportion of traffic flow at different levels determined from the traffic flow monitoring data, the corrosion fatigue of the two types of steel wire under the combined action of traffic flow was simulated by sampling. Figures 22 and 23 show the change rules of the uniform corrosion rate, pitting rate, and crack development rate of steel wire. The initial rate of uniform corrosion and pitting corrosion was fast and then rapidly decreased. On the contrary, the crack growth rate gradually increased, and the rate increased with time. The crack growth rate of galvanized steel wire exceeded the corrosion rate after seven years of corrosion, whereas that of galvanized aluminum steel wire exceeded the corrosion rate 10 years later. Although the time difference was small, the corrosion rate of galvanized aluminum steel wire was lower, the relative crack growth rate was also significantly lower than that of galvanized steel wire, which is conducive to improving the service life of steel wire.

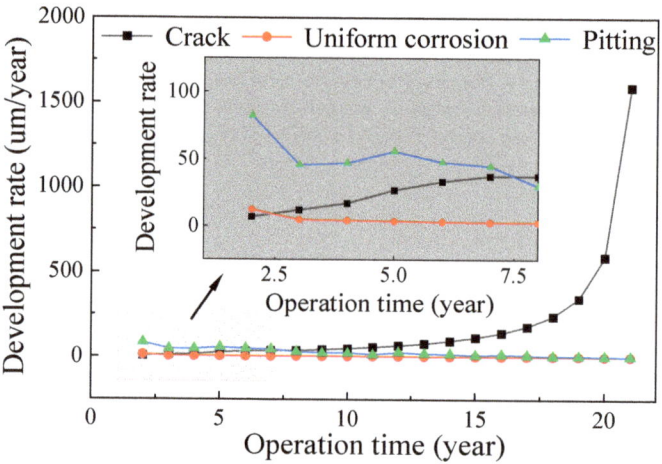

Figure 22. Corrosion rate of galvanized steel wire.

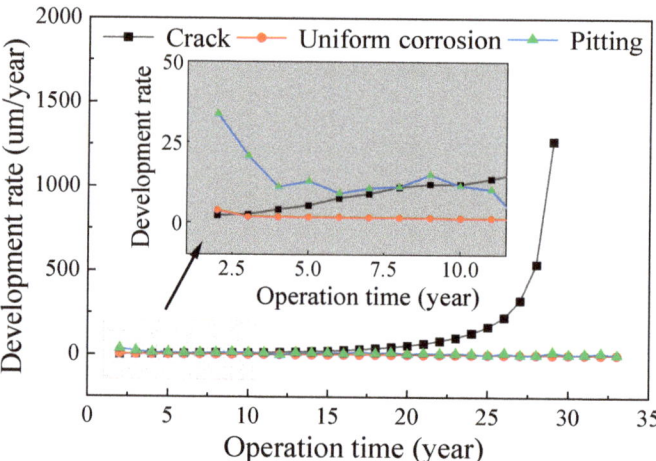

Figure 23. Corrosion rate of Galfan steel wire.

Figures 24 and 25 show the distribution law of steel wire crack depth at different stages. With the increase of service time, the average crack depth of the whole steel wire increased and the development rate increased, which is consistent with the analysis results in the above figure. However, owing to the randomness of the pitting corrosion and crack development, the STD value of crack distribution was large at the end of service, and the discreteness of wire life became obvious.

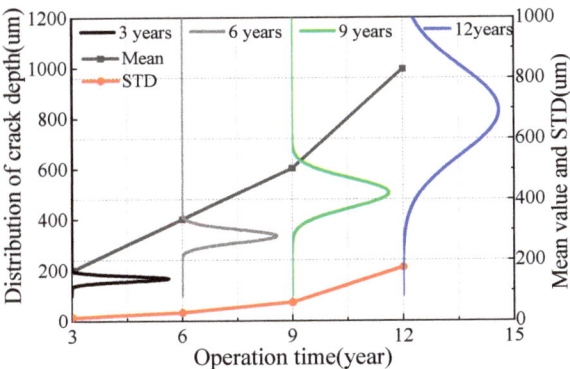

Figure 24. Crack depth distribution of galvanized suspender steel wire.

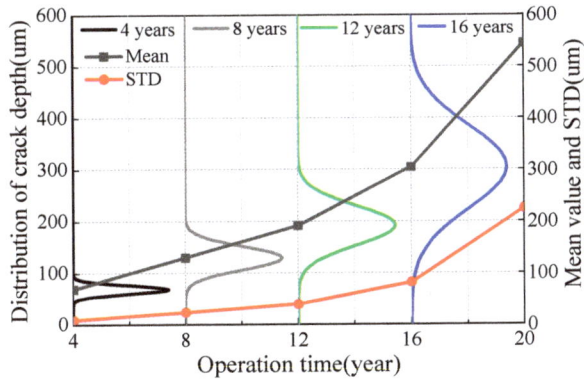

Figure 25. Crack depth distribution of Galfan suspender steel wire.

Figure 26 shows the life distribution of the two types of steel wire under a comprehensive traffic flow. The average life of Galfan steel wire was significantly higher than that of the galvanized steel wire, being 28.47 years and 17.24 years, respectively. Figures 27 and 28 show the corrosion fatigue life of the two kinds of coated steel wires under different strengths of traffic flow. The strength level of the traffic flow was the decisive factor affecting the steel wire degradation. The durability of Galfan steel wire was obviously better than that of galvanized steel wire, and the overall life of the steel wire was longer. Nevertheless, the average life of galvanized steel wire was 11.13 years, whereas that of the Galfan steel wire was 16.92 years under dense traffic flows. The corrosion fatigue life of steel wire was not significantly increased. Under a sparse traffic flow, the service life of steel wire was better than expected and was significantly higher than the design life of general cable structures (25 years). Under rarefaction flow, the average life of the two kinds of steel wires could reach 35 years and 55 years, respectively. When the proportion of sparse traffic flow and moderate traffic flow is relatively high during bridge operation, the slow corrosion fatigue degradation of Galfan steel wire at this stage could greatly improve the service life of steel wire. However, when the traffic flow intensity level is generally high, obtaining good results using Galfan steel wire would be difficult.

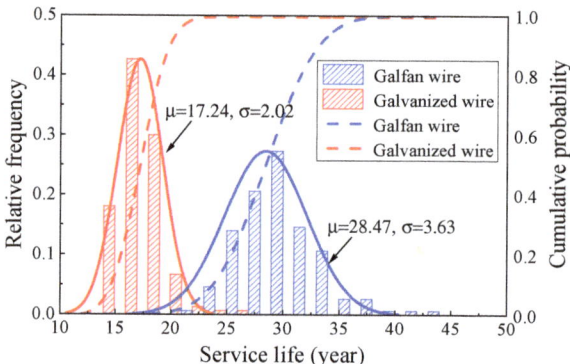

Figure 26. Life distribution of steel wire under comprehensive traffic flow.

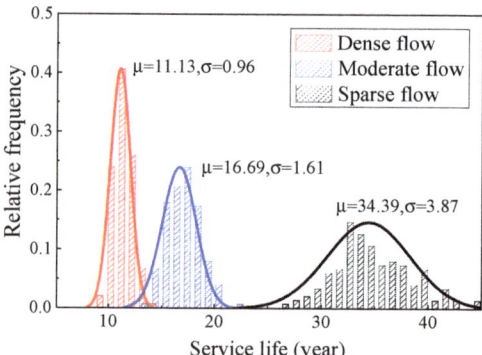

Figure 27. Service life of galvanized suspender wire.

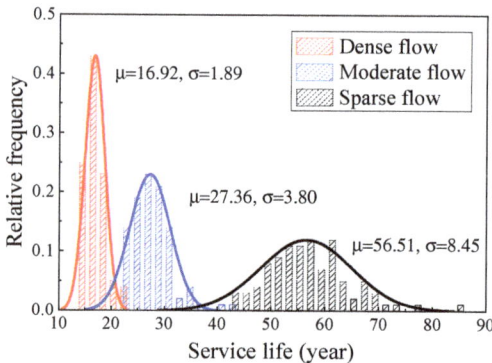

Figure 28. Service life of Galfan suspender steel wire.

6. Conclusions

This paper took the difference in the corrosion fatigue degradation characteristics of galvanized and Galfan high-strength steel wire coatings as the research goal. It analyzed the difference of uniform corrosion and pitting corrosion of two coatings under the same conditions through an accelerated corrosion test. It then constructed a dynamic distribution model of uniform corrosion and pitting corrosion during coating corrosion. On the basis of an accelerated corrosion test and the traffic load monitoring data of a large bridge, the corrosion fatigue degradation of two kinds of steel wires under traffic load during operation was investigated. The following conclusions were obtained:

(1) The macro morphology of the corrosion of high-strength steel wire had obvious stage change characteristics. At the early stage of corrosion, the steel wire coating could effectively protect the iron matrix, and the corrosion products were free of Fe oxides. When the corrosion developed to a certain stage, brown corrosion products began to appear; on the galvanized steel wire at 386 h and on the Galfan steel wire at 1016 h. The surface coating began to be consumed, and the steel wire Fe matrix started to corrode. At this time, the steel wire surface a reddish brown rust appeared. The corrosion of steel wire can be generally divided into two parts. The first part is the corrosion of the surface coating. When the corrosion depth exceeds the coating thickness, corrosion of the second part of the steel wire matrix starts. After the corrosion products are removed, uneven pits appeared on the steel wire surface. The corrosion resistance of Galfan steel wire is obviously better than that of galvanized steel wire.

(2) The development law of steel wire corrosion depth was fitted and analyzed. In addition, a uniform corrosion development model and pitting corrosion probability model of galvanized and Galfan steel wire were established. With the extension of corrosion time, the uniform corrosion depth of zinc coating and Galfan coating conformed to the exponential increase trend. The development trend of the two coatings was similar, and the corrosion rate gradually slowed down with the increase of time. The corrosion coefficient of galvanized steel wire conformed to normal random distribution, whereas that of Galfan steel wire conformed to Cauchy distribution. The section distribution of the maximum pitting coefficient did not reject the Gumbel distribution. The location and scale parameters of the maximum pitting coefficient distribution in the two coating intervals showed an exponential downward trend with the increase of corrosion duration.

(3) The early rate of uniform corrosion and pitting corrosion was fast, and then it decreased rapidly, whereas the crack growth rate gradually increased. With the increase of time, the rate increased. The average value of the overall crack depth of the steel wire also increased, and the development rate continued to increase. The STD value of the crack distribution in the late service period was large, and the discreteness of the steel wire life became obvious. Based on the analysis of the corrosion fatigue history of steel wire under different strength traffic loads, the corrosion characteristics of steel wire was the main factor affecting its service life when the traffic strength level was low. The average service life of Galfan steel wire was significantly higher than that of galvanized steel wire. However, under a dense traffic flow with high strength, the service life of steel wire was mainly controlled by the traffic load, and the service life of Galfan steel wire increased slightly. Effective anti-corrosion measures are the key to improving the service life of steel wire.

It should be mentioned that the corrosion fatigue properties of coated steel wires were investigated through accelerated corrosion tests. However, the period of the accelerated corrosion test was relatively short, which is quite different from the real environment in service. In a subsequent study, the corrosion rate and characteristics of the steel wire need to be investigated using field exposure tests and cable components in service, which could better determine the corrosion characteristics and effects of traffic load on the corrosion performance of steel wire coatings.

Author Contributions: Conceptualization, Y.Z. (Yue Zhao); Methodology, Y.Z. (Yue Zhao); Software, X.F.; Validation, X.F.; Formal analysis, B.S.; Writing—original draft, B.S.; Writing—review & editing, Y.Z. (Yiyun Zhu); Supervision, Y.Y. All authors have read and agreed to the published version of the manuscript.

Funding: This research was funded by the Natural Science Basic Research Program of Shaanxi (Program No. 2022JQ-336), and the open fund of Shaanxi Provincial Key Laboratory (Chang'an University) of Highway Bridges and Tunnels (Program No. 300102212509).

Data Availability Statement: Some or all data, models, or codes that support the findings of this study are available from the corresponding author upon reasonable request.

Conflicts of Interest: The authors declare no conflict of interest.

References

1. Meng, E.C.; Yao, G.W.; Yu, Y.L.; Gu, L.F.; Zhong, L. Influence factor analysis on the mechanical behavior of Galvanized steel wire under service environment. *J. Build. Mater.* **2020**, *23*, 934–940.
2. Liu, X.; Zhang, W.; Sun, P.; Liu, M. Time-Dependent Seismic Fragility of Typical Concrete Girder Bridges under Chloride-Induced Corrosion. *Materials* **2022**, *15*, 5020. [CrossRef] [PubMed]
3. Roffey, P. The Fracture Mechanisms of Main Cable Wires from the Forth Road Suspension. *Eng. Fail. Anal.* **2013**, *31*, 430–441. [CrossRef]
4. Liu, M. Finite element analysis of pitting corrosion on mechanical behavior of E690 steel panel. *Anti-Corros. Methods Mater.* **2022**, *28*, 7527–7536. [CrossRef]
5. Betti, R.; West, A.C.; Vermaas, G.; Cao, Y. Corrosion and Embrittlement in High-Strength Wires of Suspension Bridge Cables. *J. Bridg. Eng.* **2005**, *10*, 151–162. [CrossRef]
6. Nakamura, S.I.; Suzumura, K.; Tarui, T. Mechanical Properties and Remaining Strength of Corroded Bridge Wires. *Struct. Eng. Int.* **2004**, *14*, 50–54. [CrossRef]
7. Nakamura, S.-I.; Suzumura, K. Hydrogen embrittlement and corrosion fatigue of corroded bridge wires. *J. Constr. Steel Res.* **2008**, *65*, 269–277. [CrossRef]
8. Suzumura, K.; Nakamura, S.I. Environmental Factors Affecting Corrosion of Galvanized Steel Wires. *J. Mater. Civ. Eng.* **2004**, *16*, 1–7. [CrossRef]
9. Liu, M. Effect of uniform corrosion on mechanical behavior of E690 high-strength steel lattice corrugated panel in marine environment: A finite element analysis. *Mater. Res. Express* **2021**, *8*, 066510. [CrossRef]
10. Lan, C.; Xu, Y.; Liu, C.; Li, H.; Spencer, B. Fatigue life prediction for parallel-wire stay cables considering corrosion effects. *Int. J. Fatigue* **2018**, *114*, 81–91. [CrossRef]
11. Jiang, J.H.; Ma, A.B.; Weng, W.F.; Fu, G.H.; Zhang, Y.F.; Liu, G.G.; Lu, F.M. Corrosion fatigue performance of pre-split steel wires for high strength bridge cables. *Fatigue Fract. Eng. Mater. Struct.* **2009**, *32*, 769–779. [CrossRef]
12. Wang, S.; Zhang, D.; Chen, K.; Xu, L.; Ge, S. Corrosion fatigue behaviors of steel wires used in coalmine. *Mater. Des.* **2014**, *53*, 58–64. [CrossRef]
13. Sun, B. A Continuum Model for Damage Evolution Simulation of the High Strength Bridge Wires Due to Corrosion Fatigue. *J. Constr. Steel Res.* **2018**, *146*, 76–83. [CrossRef]
14. Li, S.; Xu, Y.; Li, H.; Guan, X. Uniform and Pitting Corrosion Modeling for High-Strength Bridge Wires. *J. Bridg. Eng.* **2014**, *19*, 04014025. [CrossRef]
15. Jiang, C.; Wu, C.; Jiang, X. Experimental study on fatigue performance of corroded high-strength steel wires used in bridges. *Constr. Build. Mater.* **2018**, *187*, 681–690. [CrossRef]
16. Xue, S.; Shen, R.; Chen, W.; Shen, L. The corrosion-fatigue measurement test of the Zn-Al alloy coated steel wire. *Structures* **2020**, *27*, 1195–1201. [CrossRef]
17. Cao, Z.; Kong, G.; Che, C.; Wang, Y. Influence of Nd addition on the corrosion behavior of Zn-5%Al alloy in 3.5wt.% NaCl solution. *Appl. Surf. Sci.* **2017**, *426*, 67–76. [CrossRef]
18. GB/T 10125-2012; Corrosion Test in Artificial Atmosphere—Salt Spray Test. Standards Press of China: Beijing, China, 2013.
19. GB/T 16545-2015; Corrosion of Metals and Alloys—Removal of Corrosion Products from Corrosion Test Specimens. Standards Press of China: Beijing, China, 2015.
20. Valor, A.; Caleyo, F.; Alfonso, L.; Rivas, D.; Hallen, J.M. Stochastic modeling of pitting corrosion: A new model for initiation and growth of multiple corrosion pits. *Corros. Sci.* **2007**, *49*, 559–579. [CrossRef]
21. Qiao, Y.; Miao, C.Q.; Sun, C.Z. Evaluation of corrosion fatigue life for corroded wire for cable-supported bridge. *J. Civ. Archit. Environ. Eng.* **2017**, *39*, 115–121.
22. Stewart, M.G. Mechanical behaviour of pitting corrosion of flexural and shear reinforcement and its effect on structural reliability of corroding RC beams. *Struct. Saf.* **2009**, *31*, 19–30. [CrossRef]
23. Steward, M.G.; Al-Harthy, A. Pitting corrosion and structural reliability of corroding RC structures: Experimental data and probabilistic analysis. *Reliab. Eng. Syst. Saf.* **2008**, *93*, 373–382. [CrossRef]
24. Mayrbaurl, R.M.; Camo, S. Cracking and fracture of suspension bridge wire. *J. Bridge Eng.* **2001**, *6*, 645–650. [CrossRef]
25. Li, S.L.; Zhu, S.; Xu, Y.L.; Chen, Z.W.; Li, H. Long-term condition assessment of suspenders under traffic loads based on structural health monitoring system: Application to Tsing Ma Bridge. *Struct. Control Health Monit.* **2013**, *19*, 82–101. [CrossRef]
26. Shi, P.; Mahadevan, S. Damage tolerance approach for probabilistic pitting corrosion fatigue life prediction. *Eng. Fract. Mech.* **2001**, *68*, 1493–1507. [CrossRef]
27. Han, W.S.; Liu, X.D.; Gao, G.Z.; Xie, Q.; Yuan, Y. Site-specific extra-heavy truck load characteristics and bridge safety assessment. *J. Aerosp. Eng.* **2018**, *31*, 04018098.1–04018098.12. [CrossRef]

28. Zhao, Y.; Huang, P.M.; Long, G.X.; Yuan, Y.; Sun, Y. Influence of Fluid Viscous Damper on the Dynamic Response of Suspension Bridge under Random Traffic Load. *Adv. Civ. Eng.* **2020**, *2020*, 1857378. [CrossRef]
29. Wyatt, T.A. Secondary Stress in Parallel Wire Suspension Cables. *J. Struct. Div.* **1960**, *86*, 37–59. [CrossRef]

Disclaimer/Publisher's Note: The statements, opinions and data contained in all publications are solely those of the individual author(s) and contributor(s) and not of MDPI and/or the editor(s). MDPI and/or the editor(s) disclaim responsibility for any injury to people or property resulting from any ideas, methods, instructions or products referred to in the content.

Article

Evaluation of Flexible Central Buckles on Short Suspenders' Corrosion Fatigue Degradation on a Suspension Bridge under Traffic Load

Yue Zhao [1], Xuelian Guo [2,*], Botong Su [1], Yamin Sun [3] and Xiaolong Li [2]

[1] School of Civil Engineering and Architecture, Xi'an University of Technology, Xi'an 710048, China
[2] School of Highway, Chang'an University, Xi'an 710064, China
[3] School of Architecture and Civil Engineering, Xi'an University of Science and Technology, Xi'an 710054, China
* Correspondence: guoxuelian@chd.edu.cn

Abstract: Suspenders are the crucial load-bearing components of long-span suspension bridges, and are sensitive to the repetitive vibrations caused by traffic load. The degradation of suspender steel wire is a typical corrosion fatigue process. Although the high-strength steel wire is protected by a coating and protection system, the suspender is still a fragile component that needs to be replaced many times in the service life of the bridge. Flexible central buckles, which may improve the wind resistance of bridges, are used as a vibration control measure in suspension bridges and also have an influence on the corrosion fatigue life of suspenders under traffic load. This study established a corrosion fatigue degradation model of high-strength steel wire based on the Forman crack development model and explored the influence of flexible central buckles on the corrosion fatigue life of suspenders under traffic flow. The fatigue life of short suspenders without buckles and those with different numbers of buckles was analyzed. The results indicate that the bending stress of short suspenders is remarkably greater than that of long suspenders, whereas the corrosion fatigue life of steel wires is lower due to the large bending stress. Bending stress is the crucial factor affecting the corrosion fatigue life of steel wires. Without flexible central buckles, short suspenders may have fatigue lives lower than the design value. The utilization of flexible central buckles can reduce the peak value and equivalent stress of bending stress, and the improved stress state of the short suspender considerably extends the corrosion fatigue life of steel wires under traffic flow. However, when the number of central buckles exceeds two, the increase in number does not improve the service life of steel wire.

Keywords: suspender; high-strength steel wire; corrosion fatigue; flexible central buckle; bending stress

1. Introduction

With the rapid development of highway transportation, long-span suspension bridges are constructed across mountains, valleys, and rivers for their good mechanical characteristics and excellent spanning performance. The construction of early large-span suspension bridges was limited by experience and technology, and structural vibration control measures were relatively lacking, leading to obvious vibration responses under external load. The longitudinal vibration displacement of structures caused by external load may lead to the fatigue of expansion joints and other ancillary components. Traffic load has been proven to be one of the main reasons causing the longitudinal vibration displacement of structures. Such vibration may cause the fatigue of expansion joints and other ancillary components [1,2]. Suspenders are key load-bearing components of suspension bridges, of which the degradation is the result of the comprehensive action of corrosion and fatigue, and the corrosion accelerates the generation of fatigue cracks. The corrosion degradation of components seriously affects the reliability of bridge operation [3]. The propagation of fatigue cracks is easily affected by vibrations under traffic load; thus, it has been an important research goal to evaluate the stress and service life of short suspenders. Flexible central buckles were set up in the midspan for long-span suspension bridges recently to enhance wind resistance performance with low cost and construction convenience, but the studies on them were limited and mainly focused on wind resistance and the vibration characteristics of the structure itself. Actually, flexible central buckles also play a contributing role to the suspenders' response under traffic flow that may reduce the fatigue degradation of suspenders, but the influence mechanism on the structural vibration under traffic load remains unclear. Thus, the control effect of flexible central cables should be investigated to optimize the designation of the flexible central buckle.

Corrosion fatigue is the phenomenon of crack formation and propagation under the interaction of alternating load and a corrosive medium that leads to a reduction in fatigue resistance [4]. Scholars have studied the corrosion fatigue degradation process of high-strength steel wires in bridge engineering. The surface of the suspender steel wire is provided with a coating to enhance the corrosion resistance. The damage to the coating's passive film is accompanied by pitting corrosion. Roffey indicates that the pit corrosion of the steel wire develops into vertical cracks inside, resulting in the decline of the bearing capacity of the steel wire based on the inspection results of the Fourth Highway Bridge in Scotland [5]. Qiao Yan divided the corrosion fatigue process of steel wire into three stages—coating corrosion, corrosion pit development, and crack development—and gave a calculation method for the development time of each stage [6]. Valor proposed a random model for pitting distribution simulation which uses a non-uniform Poisson process to simulate the generation of pits and verified it with experiments [7]. Nakamura investigated the corrosion of steel wires in different environments for fatigue loading. The results show that the fatigue life of steel wire in a corrosive environment decreases significantly [8]. Suzumura studied the effects of reagent concentration, ambient temperature, and humidity on the corrosion rate of galvanized steel wire through experiments, and gave the loss rate of zinc coating on galvanized steel wire [9]. Although the durability of galvanized steel wire has been significantly improved, it still does not meet the engineering requirements. The corrosion resistance of the coating can be achieved by improving the properties of the coating, such as improving the adhesion and porosity and adding elements that can form a passive film; it is proven that the corrosion resistance can be improved by the oxidation of the Al element [10]. In recent years, Galfan steel wires have been gradually widely used. The evaluation method of steel wire has been well developed, but the existing models mainly use the Paris criterion to calculate the crack growth rate; the influence of the average load factor and the difference in crack growth rate caused by the change in traffic flow intensity are not considered. There is a deviation when using the parameters under the same stress ratio to calculate the crack growth life.

Furthermore, the axial stress and bending stress fluctuations caused by relative displacements between the girder and cables easily damage the short suspenders along with fatigue degradation [11,12]. To reduce the fatigue damage of short suspenders, appropriate vibration control facilities are utilized to control bridge vibration [13,14]. The central buckle is a vibration control measure for long-span suspension bridges, which includes a flexible central buckle and a rigid central buckle. Previous research focuses on the influence of rigid central buckles on the dynamic characteristics of bridges [15]. Wang analyzed the influence of rigid central buckles on the wind-induced buffeting response of long-span suspension bridges and pointed out that rigid central buckles can suppress buffeting vibration [16]. Wang investigated the working and mechanical characteristics of the rigid central buckle of the Runyang Yangtze River Bridge under vehicle load based on measured results and finite element modeling [17]. Liu investigated the effects of central clamps in the midspan (i.e., rigid central buckle) on the fatigue life of short suspenders, and the results revealed that short suspenders were more prone to fatigue than others because of large bending stress, and central clamps can effectively improve their lifespan [18]. In addition to a rigid central buckle, a flexible central buckle cable was set up in the midspan to enhance wind resistance performance. Wang studied the influence of flexible central buckles on the displacement of stiffening girders. The results showed that the flexible central buckle remarkably reduces the longitudinal amplitude of the stiffening girder and increases its vibration frequency [19]. The influence of flexible buckles on structural vibration under random traffic flow remains unclear. The control effect of flexible buckles under random traffic flow should be studied.

Traffic flow is an important vibration source in the suspender stress response. Suspension bridges are a flexible system and structural deformation is evident under the action of traffic flow, which varies with traffic density. Characteristics such as traffic flow parameters, vehicle type, and vehicle weight generally have random distribution [20,21]. The load effect of traffic flow can be well considered by a macro traffic flow simulation method [22,23]. Thus, in this study, the influence of flexible central buckles on the stress response and corrosion fatigue life of suspenders under traffic flow were analyzed by numerical modeling. First, the corrosion fatigue of high-strength steel wire based on the Forman criterion was established. Then, the response of suspenders with flexible central buckles was calculated with consideration of the load effect of traffic flow at different levels. Finally, the fatigue life of suspender steel wires and the influence of flexible central buckles were evaluated. This research can provide a reference for the design and maintenance of long-span suspension bridges.

2. Prototype Bridge

2.1. Bridge Information

This study takes the Zhixi Yangtze River Bridge as the research object. The bridge is a single-span steel–concrete composite girder suspension bridge. The section layout is shown in Figure 1. The span of the main cable is arranged as 250 + 838 + 215 m, and the sagittal span ratio of the midspan main cable is 1/10. The standard distance between the adjacent lifting points of stiffening girders is 16 m, and the suspender adopts a φ5.0 mm galvanized aluminum alloy (i.e., Galfan coating) high-strength steel wire. To improve the vibration resistance of the bridge, two flexible central buckles are set near both sides of the middle span of each main cable to form a cable–beam connection. The entire bridge has a total of eight central buckles. The stiffening girder adopts a steel–concrete composite structure in which the steel beam is combined with the concrete deck through shear nails. The half section of the stiffening girder is shown in Figure 2. The full width of the stiffening girder is 33.2 m, the center height is 2.8 m, and the central transverse spacing of the two main cables in the midspan is 26.0 m. The small longitudinal beams are arranged longitudinally at the center line of the girder and the top surface is flush with the top surface of the steel beam. The bridge deck is reinforced concrete with a full width of 25.0 m and a thickness of 0.22 m.

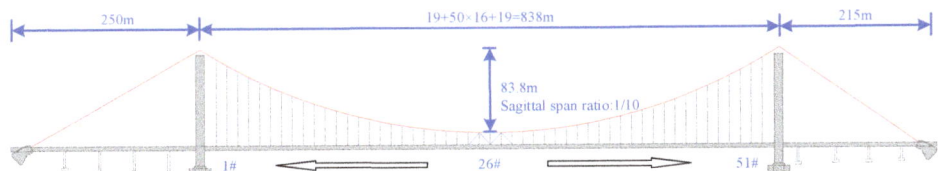

Figure 1. Layout of the prototype bridge.

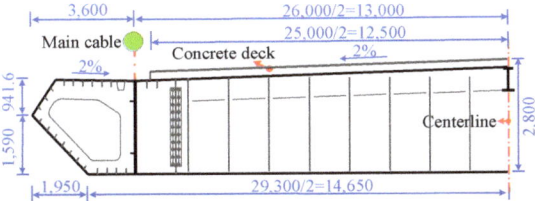

Figure 2. Half-section of the stiffening girder (mm).

As a common vibration control measure, central buckles are used to improve the vibration response of suspension bridges. These buckles are generally installed in the middle span; examples include the Runyang Yangtze River Bridge and the Sidu River Bridge, in which the rigid central buckle is installed in the middle span. Existing research indicates that the rigid central buckle can improve the structure frequency and reduce the longitudinal displacement response of the girder [24]. In the Zhixi Yangtze River Bridge, flexible central buckles that differ from traditional rigid central buckles are set in the middle of the main span of each main cable to coordinate with short suspenders, as shown in Figure 3. The flexible central buckle is composed of an inclined cable connected to a short suspender, forming a cable–girder connection to control the vibration response of the structure.

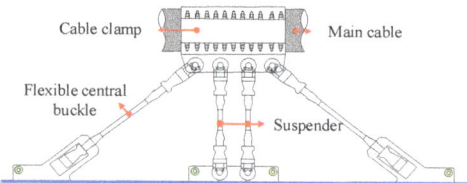

Figure 3. Flexible central buckle.

2.2. Finite Element (FE) Model

To simulate the structural characteristics, a three-girder model of a prototype bridge was established using ANSYS 18.0. The FE model is shown in Figure 4. The stiffening girder of the bridge is a steel-composite girder with an open section, and the longitudinal beams on both sides are the main bearing structures of the stiffening girder. Thus, the BEAM4 element was used to simulate the main stringer, small stringer, steel beam, and main tower. The LINK10 element was used to simulate the cable components. A total of 1836 BEAM4 elements for the girder, 82 BEAM4 elements for the pylon, and 279 LINK10 elements for the main cable and suspender were found. The bridge deck pavement contributes minimally to the stiffness of the stiffening girder; thus, only its mass was considered, and the stress stiffening of the LINK element was conducted in accordance with the measured cable force.

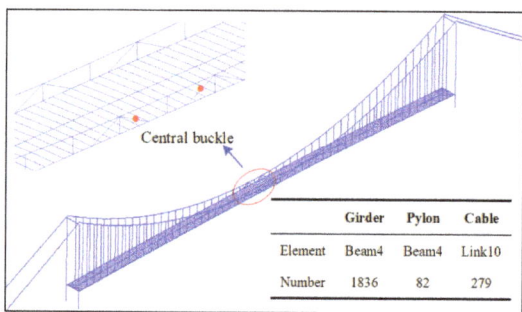

Figure 4. Bridge FE model.

The theoretical material properties and cable force vary from the actual state of the structure; thus, the FE model was modified according to the measured material properties and cable force in construction. Then, the structure frequency was calculated by the modal analysis module of ANSYS software, and the Block Lanczos feature solver based on the Lanczos algorithm was used in modal analysis. When calculating the natural frequencies of a certain range contained in the eigenvalue spectrum of a system, the Block Lanczos method is particularly effective for extracting modes. The frequencies of the FE model were compared with measured structure frequency to validate the FE model. The research team undertook the monitoring of the structural state during the bridge's construction. After construction, the actual vibration mode and frequency of the bridge were measured through the modal test analysis system. Figure 5 shows the modal analysis results and the test results; the modes of vibration are consistent with the test results. The first-order frequency L1 is 0.113 hz and smaller than other bridge types, which is determined by the flexibility characteristics of the suspension bridge. The error of L1 is 2.7%, and the maximum error is 5.2% in L2. The errors are within the acceptable range, which preliminarily proves the simulation effect of the FE model.

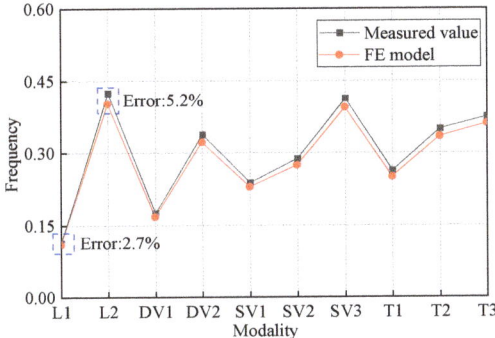

Figure 5. Comparison of structure frequency (L is lateral mode; DV is dissymmetry vertical mode; SV is symmetrical vertical mode; T is torsional mode).

Besides the modal test, the vehicle loading experiment was conducted to test its deformation performance under external load. Figure 6 shows the layout of the static and running tests of the bridge. The loading vehicle is a 35 t three-axle truck. The static test has four loading trucks in each row and eight rows in total. A comparison of the maximum girder vertical deflection of the static test is given in Table 1. The computing value and measured value are close, and the error is within 3%. The running test condition is that two 35 t loading vehicles drove through the bridge at a constant speed of 60 km/h, and the vertical dynamic deflection of the main girder in 1/2 L is measured. A comparison between the measured results and the FE model is shown in Figure 7; the results are in good agreement. The model can reflect the dynamic response of the bridge and satisfy the requirements of subsequent analysis under traffic flow.

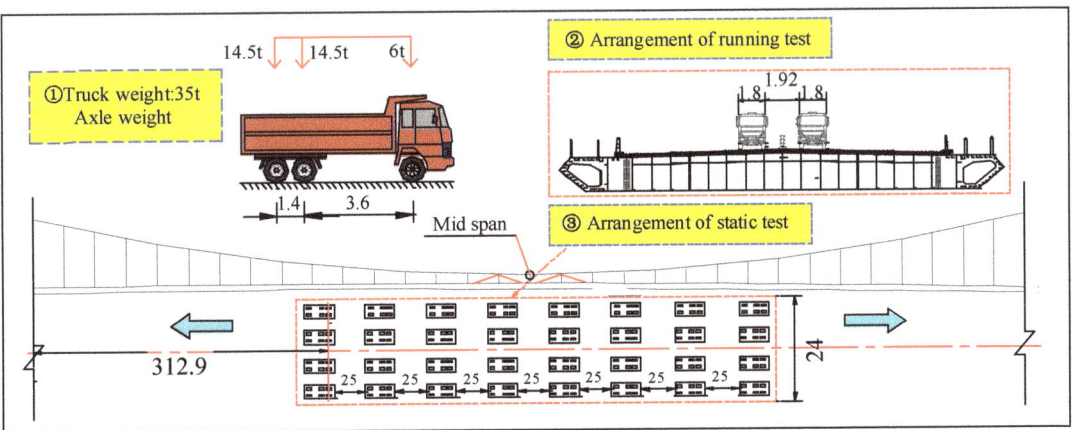

Figure 6. Load case of bridge static and running tests. (unit: m).

Table 1. Comparison of stiffening girder vertical deflection under static test.

	L/8	3L/8	L/2	5L/8	7L/8
Measured Value (mm)	175	606	1093	615	180
FE model (mm)	181	624	1120	630	185
Error (%)	0.03	0.03	−2.02	0.02	0.03

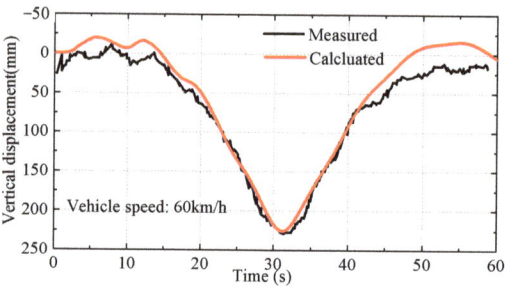

Figure 7. Comparison of midspan dynamic strain under running test.

3. Corrosion Fatigue of Suspender Steel Wire

3.1. Corrosion Fatigue Mechanism

The stress of long-span bridge suspenders is caused mostly by dead load; thus, the amplitude of stress change caused by vehicle load and other live loads is relatively small and is far lower than the fatigue limit of steel wire. Therefore, the degradation process is a typical corrosion fatigue process; that is, the corrosion defects on the steel wire surface develop into initial crack damage. The entire steel wire degradation process can be divided into stages of the development of corrosion and crack propagation, as shown in Figure 8. The tiny corrosion defects on the steel wire surface become the crack initiation site. When the corrosion defects transform into cracks, they continue to develop until destroyed under the action of load cycles. This degradation process can be simulated by the corrosion fatigue theory.

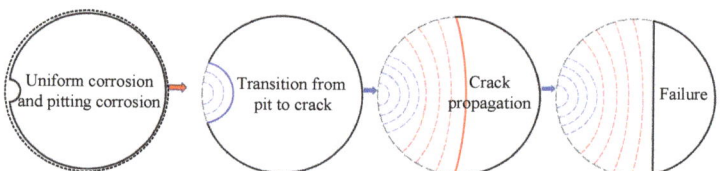

Figure 8. Degradation of suspender steel wires.

3.2. Uniform Corrosion and Pitting Corrosion

The corrosion of steel wire includes uniform corrosion and pitting corrosion. Uniform corrosion describes the degree of average corrosion of the steel wire surface, which directly causes the reduction in the diameter of the steel wire, and the extent of diameter reduction is assumed to stay unchanged along the steel wire length [25]. The steel wire parameters adopted in this study are shown in Table 2. The surface of high-strength steel wire is usually protected by a coating for corrosion resistance. In the prototype bridge, the suspender consists of Galfan-coated steel wires. The Galfan coating should not be less than 300 g/m^2 due to the specification of bridge designation [26]. According to the survey of relevant cable manufacturers, the coating quality is usually controlled within 350 g/m^2. Thus, the depth of a Zn-Al alloy coating can be calculated according to its density (6.58 g/cm^3) and ranges from about 29 μm to 34 μm.

Table 2. Parameters of steel wire samples.

	Nominal Diameter (mm)	Tensile Strength (MPa)	Yield Strength (MPa)	Modulus of Elasticity (GPa)	Coating Quality (g/cm^2)	Coating Depth (μm)
Galfan steel wire	5.25	1926	1775	2.08×10^5	337	31.05

The uniform corrosion of high-strength steel wire undergoes a two-stage corrosion process; that is, the corrosion of the coating and the corrosion of the steel wire substrate. The corrosion rate can be described as Equation (1).

$$a_u(t) = \begin{cases} d_c(t) & t \leq t_c \\ d_s(t) + d_c(t) & t \geq t_c \end{cases} \quad (1)$$

where $a_u(t)$ is the depth of uniform corrosion, d_c is the corrosion depth of the zinc–aluminum alloy coating, d_s is the corrosion depth of the steel wire, t is corrosion time, and t_c is the time when the coating is totally corroded.

According to the preliminary work of the research team, the corrosion process of Galfan steel wire is measured by an accelerated corrosion test and can be simulated by parabola distribution as Equation (2) [25]. The corrosion rate decreases gradually. The oxidation products of aluminum in the coating form a passive film, which slows down the corrosion rate.

$$a_u(t) = 0.04431t - 0.000014t^2 \quad (2)$$

where t is corrosion time.

In service conditions, the corrosion rate of Galfan coating is significantly different because of the exposure environment. As is well known, field exposure tests are difficult to conduct due to the high cost of time, and it is also difficult to find exactly matched field exposure test results. Thus, the time conversion scale was determined by the field exposure test of Galfan coating by Aoki and Katayama, in which a hot-dipped Galfan-coated steel plate with a 25 μm coating was investigated [27,28]. Assuming that the influence of coating thickness and surface shape on the corrosion rate is negligible, and the test results are applicable to the conversion time scale, it is suggested that 1 h of the accelerated corrosion test corresponds to 0.033~0.052 years in a rural environment, 0.018~0.024 years in an industrial environment, 0.019~0.028 years in a marine environment, and 0.014~0.022 years in a severe marine environment.

When the metal material surface has a passive or protective film, the pitting pit on the substrate surface appears after the protective layer is consumed, greatly affecting the characteristics of the steel wire. Pitting corrosion occurs randomly, accompanied by uniform corrosion [29]. Given the stress concentration effect, pitting corrosion is the site where steel wire fatigue fracture may occur. The pitting pit with the largest depth determines the working state of the steel wire; thus, the pitting pit with the largest depth is the key analysis point in corrosion fatigue analysis. Pitting pit depth can be calculated by uniform corrosion depth and pitting coefficient.

$$\wedge(t) = a_p(t)/a_u(t) \quad (3)$$

where $a_p(t)$ and $a_u(t)$ denote the depth of pitting corrosion and uniform corrosion.

The distribution of the maximum pitting coefficient conforms $\wedge(t)$ to the Gumbel distribution [7], which can be expressed as

$$F(\wedge(t)) = exp\{-exp[-\frac{(\wedge(t) - \beta_0)}{\alpha_0}]\} \quad (4)$$

where $F(\wedge(t))$ is the cumulative probability density function; $\wedge(t)$ is the maximum pitting coefficient; and α_0 and β_0 are the distribution parameters.

Then, the distribution parameter of any wires with different lengths and diameters can be calculated by Equation (5):

$$\beta_k = \beta_0 + \frac{1}{\alpha_0} \ln\left(\frac{A_k}{A_0}\right), \alpha_k = \alpha_0 \quad (5)$$

where A_k is the surface area of the analysis target, and A_0 is the surface area of the wire with a 125 mm length and 8 mm diameter.

3.3. Corrosion Fatigue Crack

Stress concentration happens due to the shape characteristics of the corrosion pit. As the depth of the corrosion pit increases, a crack will occur when the stress intensity reaches a critical value. The transition process from pitting to cracking can be determined by two methods: (1) the growth rate of the fatigue crack exceeding that of the corrosion pit and (2) the stress intensity factor of the corrosion pit reaching the critical threshold of fatigue crack propagation. This study adopts the former method. The steel wire crack dominates when the development speed of the pitting pit depth exceeds that of the crack.

Corrosion cracks expand until failure under the stress cycle caused by an operating live load. The Forman formula is used to analyze the growth rate of a metal corrosion fatigue crack, as shown in Equation (6).

$$\frac{da}{dN} = C(\Delta K)^m / [K_c(1-R) - \Delta K] \quad (6)$$

where $\frac{da}{dN}$ is the growth rate of the crack, a is the depth of the crack, C and m are the parameters of the Paris criterion [30], K_c is the fracture toughness of the material, ΔK is the stress intensity factor range, and R is the stress ratio of alternating load.

The stress intensity factor ΔK is given by Forman, as follows:

$$\Delta K = F_a\left(\frac{a}{b}\right) \Delta \sigma_a \sqrt{\pi a} + F_b\left(\frac{a}{b}\right) \Delta \sigma_b \sqrt{\pi a} \quad (7)$$

where $F_a\left(\frac{a}{b}\right)$ denotes a coefficient related to axial stress, $F_b\left(\frac{a}{b}\right)$ denotes a coefficient related to bending stress, a is the crack depth, b is the diameter of the steel wire, $\Delta \sigma_a$ is the equivalent axial stress amplitude, and $\Delta \sigma_b$ is the equivalent axial stress amplitude. $F\left(\frac{a}{b}\right)$ is calculated by Equation (8) [31].

$$\begin{cases} F_a\left(\frac{a}{b}\right) = 0.92 \cdot \frac{2}{\pi} \cdot \sqrt{\frac{2b}{\pi a} \cdot \tan\frac{\pi a}{2b}} \cdot \frac{0.752 + 1.286\left(\frac{a}{b}\right) + 0.37\left(1 - \sin\frac{\pi a}{2b}\right)^3}{\cos\frac{\pi a}{2b}} \\ F_b\left(\frac{a}{b}\right) = 0.92 \cdot \frac{2}{\pi} \cdot \sqrt{\frac{2b}{\pi a} \cdot \tan\frac{\pi a}{2b}} \cdot \frac{0.923 + 0.199\left(1 - \sin\frac{\pi a}{2b}\right)^4}{\cos\frac{\pi a}{2b}} \end{cases} \quad (8)$$

where a is the crack depth, and b is the diameter of the steel wire.

To consider the effect of daily traffic flow on the structure comprehensively, a crack depth development model is established on the basis of daily traffic flow operation according to the traffic load investigation, as shown in Equation (9).

$$\begin{cases} a_i = \Delta a + a_{i-1} \\ \Delta a = C \sum e_j N_j (\Delta K_j)^m / [K_c(1-R) - \Delta K] \end{cases} \quad (9)$$

where a_i is the depth of the crack at time i; Δa is the increment of the crack; e_j is the operating time of traffic flow with different intensities; $\sum e_j = 24$ h; and ΔK_j and N_j are the stress intensity factor range and the number of cycles, respectively. Mayrbaurl pointed out that the critical relative crack depth conforms to the lognormal distribution with an average value of 0.390 and a coefficient of variation of 0.414. Based on the test, the maximum critical relative depth is 0.5, which is used as the judgment standard for steel wire failure.

4. Traffic-Induced Stress Responses of Suspenders

4.1. Vehicle Bridge System

Vehicles can be classified into different types according to axle distance, axle number, vehicle load, etc. Vehicle subsystems are commonly simplified as a car body, wheels, a shock mitigation system, and a damping system. The corresponding dynamic models are established on the basis of the hypothesis that the mass of the damper and spring components are ignored. For example, a three-axle vehicle is shown in Figure 9 [32]. The longitudinal vibration of the vehicle is neglected for its few effects on the bridge; thus, the longitudinal degree of freedom is ignored in the analysis [33]. Thus, five degrees of freedom (vertical, horizontal, head nodding, side rolling, and head shaking) are considered for the integral vehicle. The vehicle dynamic models are also classified into five types, and the corresponding dynamic models are constructed.

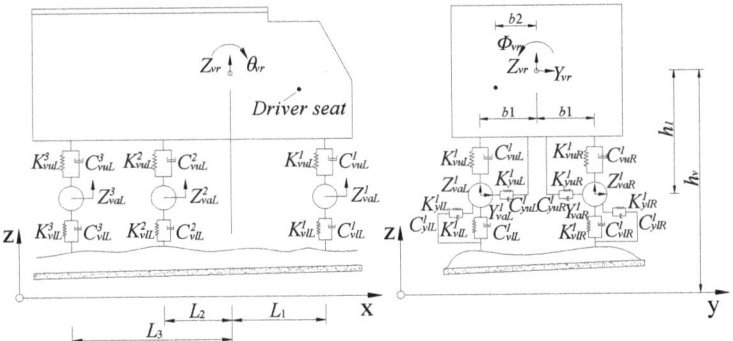

Figure 9. Dynamic model of a three-axle vehicle.

Vehicle wheels always keep contact with the deck; the bridge deformation caused by an external load leads to the vibration response of the vehicle and bridge subsystems; the dynamic response is influenced by the overall total mass matrix, damping matrix, and the overall stiffness matrix of the subsystem; and the road surface roughness is the main excitation source. Therefore, the interaction force between the vehicle and bridge system is a function of the vehicle–bridge system's motion state and road roughness, which can be analyzed in the established vehicle–bridge analysis system [34]. The road surface roughness is described by a power spectral density function, which can be generated through Fourier inversion [35].

$$\begin{cases} F_v = F_{vi}\left(Z_v, \dot{Z}_v, \ddot{Z}_v, Z_b, \dot{Z}_b, \ddot{Z}_b, i\right) \\ F_b = F_{bi}\left(Z_v, \dot{Z}_v, \ddot{Z}_v, Z_b, \dot{Z}_b, \ddot{Z}_b, i\right) \end{cases} \quad (10)$$

where Z_v denotes vehicle displacement, Z_b denotes bridge displacement, and i denotes road surface roughness.

4.2. Traffic Load Simulation

The vehicle load data monitored in a region is used to further evaluate the degradation process of suspenders under traffic load. The data were collected from the traffic load of a long-span bridge for one month by a weigh-in-motion (WIM) system. Figure 10 shows the hourly traffic volume results of the traffic flow, which are divided into five levels based on the range of traffic volume, including level 1 (<300 passenger car unit (pcu)/h), level 2 (300~600 pcu/h), level 3 (600~900 pcu/h), level 4 (900~1050 pcu/h) and level 5 (>1050 pcu/h). The error bar of the hourly traffic volume proves that the traffic volume is relatively stable. Although the standard deviation of peak traffic volume is larger than the trough period, the overall distribution is consistent, which does not affect the division of traffic intensity. The time proportions are 0.25, 0.21, 0.165, 0.21, and 0.165, respectively. The established random traffic flow simulation method is used to generate traffic flow loads of different strengths for loading [34], so as to obtain the impact of traffic flow level on the stress response of the suspenders.

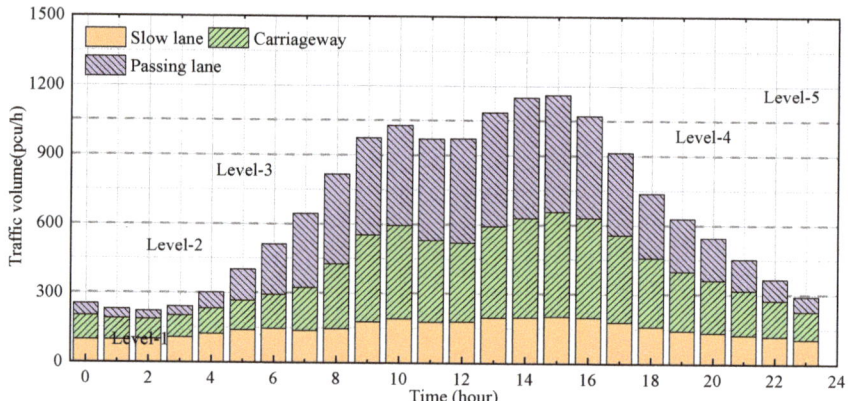

Figure 10. Average traffic volume from WIM data.

5. Numerical Analysis

5.1. Analysis Conditions

As a tensioned component, flexible central buckles cannot support the vibration response of the midspan main beam or cable as the rigid central buckles, but they can still affect the overall response of the structure by changing the fastening force. The connection system formed by the central cable and the suspenders changes the distribution of the force of the suspender near the midspan under traffic load. Traffic load is the main inducement of the bridge vibration response. To study the improvement effect of central buckles on bridge vibration, this study analyzes the response of bridges under conditions such as no central buckle and settled flexible central buckles. The detailed analysis conditions are shown in Table 3.

Table 3. Analysis conditions of the FE model.

Condition	Description	Schematic
N-C	No central buckle	
F-C-1	Single flexible central buckle	
F-C-2	Double flexible central buckles	
F-C-3	Three flexible central buckles	

5.2. Evaluation Process

The fatigue life of suspender wires can be predicted by the corrosion fatigue theory, and the detailed prediction process is shown in Figure 11. Considering the traffic density variation, the contribution of the corresponding stress cycle times to crack depth is calculated on the basis of the hourly occupancy rate of different traffic flows in one year, and the change law of crack depth and crack development rate with time is obtained. The process includes the following steps: (1) analyzing the characteristics of traffic flow parameters and generating random traffic flow samples on the basis of the WIM data; (2) taking traffic flow into the vehicle bridge coupling analysis system and obtaining the time history results of suspender stress; (3) simulating the uniform corrosion process of steel wire and generating random samples of pitting corrosion; (4) calculating crack propagation by integrating vehicle flow effects of different levels until failure.

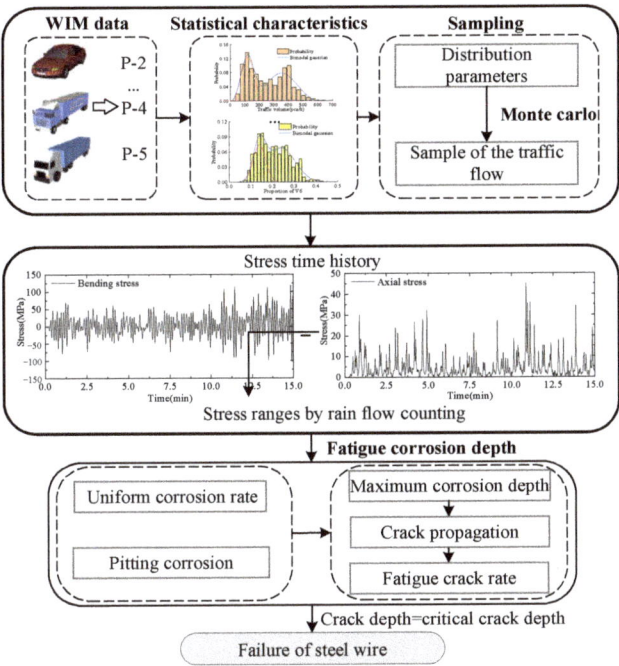

Figure 11. Simulation process of wire life.

5.3. Result Discussion

The dynamic test under truck load in Figure 6 is used to analyze the dynamic response of the bridge structure with or without the central buckle. First, the suspender response of the running test is shown in Figure 12 to analyze the difference between suspenders. The suspender bears axial stress and bending stress due to the relative movement between the main cable and the stiffening beam. The bending stress of the suspender cannot be directly obtained by the LINK10 element. Thus, the Wyatt theoretical formula is introduced to calculate bending stress according to computed axial stress and the angle caused by relative movement between the main cable and the stiffening girder [36]. The Wyatt theoretical formula can only be applied to an object that is a round wire, which is not suitable for a set of strands such as the prototype bridge. Kondoh proposed that the bending stress in this kind of suspender at the joint was assumed to be 60% of the theoretical Wyatt formula based on the experimental results as Equation (11) [37]

$$\sigma_b = 1.2 \tan\theta \cdot \sqrt{\sigma_a E} \qquad (11)$$

where σ_a is axial stress, E is the elasticity modulus of steel wire, and θ is the angle caused by the relative movement between the main cable and the stiffening girder.

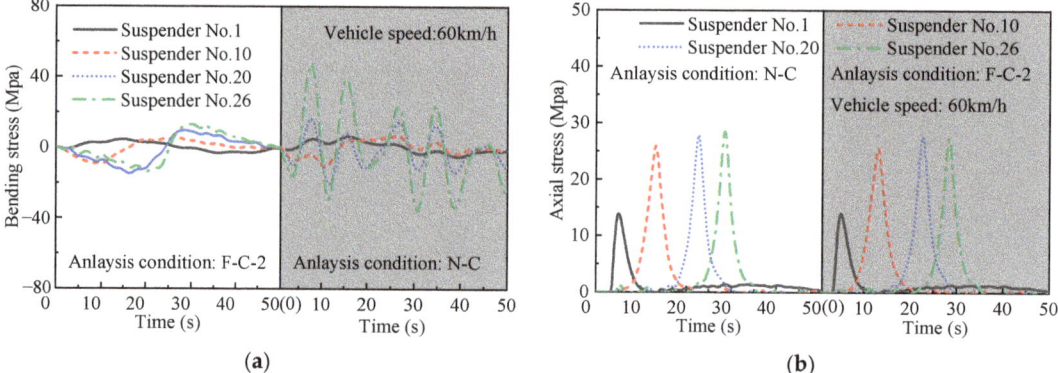

Figure 12. Comparison of suspender stress. (**a**) Bending stress; (**b**) axial stress.

Only the bending stress and axial stress of partial suspenders are shown due to layout constraints. The variation of the axial stress of short suspenders is small, whereas the length of short suspenders near the midspan is too small to release stress; the bending stress is greater than long suspenders. The influence of bending stress cannot be neglected in the analysis of suspender degradation. The settlement of flexible buckles considerably reduces the bending stress of suspenders but has minimal effect on the axial stress. The bending stress of short suspenders near suspender no. 26 (midspan) slightly decreases, whereas the long suspenders are almost unaffected. Thus, short suspenders nos. 21–26 are selected as analysis objects.

The traffic load is divided into different levels according to traffic density and then used for loading to calculate the structural dynamic response under traffic conditions. Figure 13 shows the suspender stress under the traffic flow at level 5. The time history of axial stress is consistent for different conditions, and the bending stress presents a remarkable difference in that the peak values are greatly reduced.

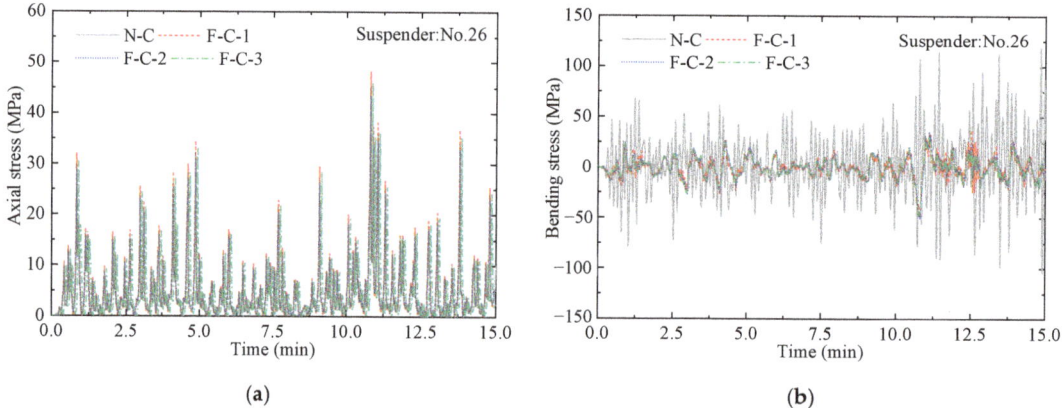

Figure 13. Time history of suspender stress (level 5). (**a**) Axial stress; (**b**) bending stress.

Figure 14 shows the peak values of bending stress and axial stress. The suspender stress response under different traffic flows is different, and the axial stress is slightly influenced by different traffic flows, whereas the bending stress is greatly reduced by flexible central buckles. The settlement of flexible central buckles has a certain influence on the axial stress and bending stress of short suspenders. The axial stress peak values vary under different conditions. F-C-1 has an improvement effect on suspender no. 26, whereas F-C-2 has an improvement effect on suspender no. 25; these results are related to the position of buckles. The settlement of buckles shares the axial stress of the suspender between inclined cables. In terms of bending stress, the flexible central buckle can remarkably reduce the peak value of short suspenders, but the weakening effect is not significantly improved with the increase in the number of buckles.

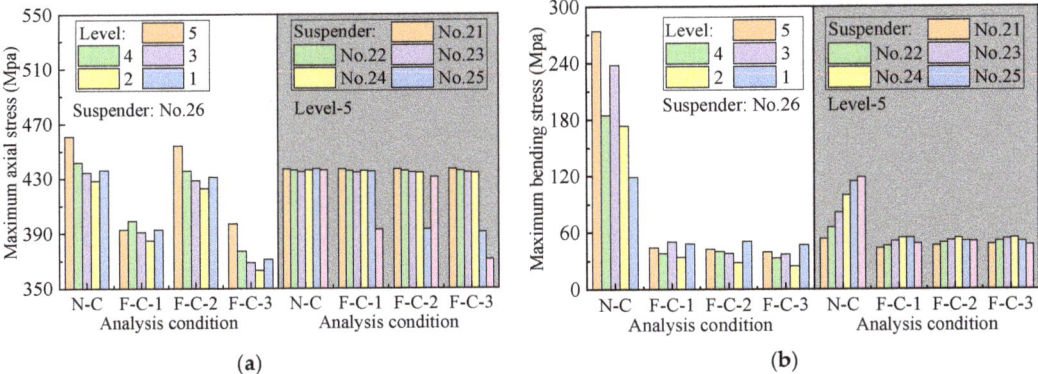

Figure 14. Comparison of peak values of stress. (a) Axial stress; (b) bending stress.

The generation of pitting corrosion is a random process, and the maximum pitting depth directly affects the generation of cracks and fatigue life. In order to reflect the difference in steel wire life, the corrosion fatigue degradation of steel wire under different working conditions was simulated. A total of 150 samples for each analysis condition were sampled based on randomly generated maximum pitting coefficients, and then the transition from pitting corrosion to cracking and the crack development were calculated on the basis of the proposed predicting process. Figure 15 shows the crack development of the steel wire samples of suspender nos. 21–26. Under the N-C condition, the average crack development in the steel wire samples of suspender nos. 24–26 is remarkably faster than that of suspender nos. 21–23, satisfying the service requirements. Although the corrosion resistance of the aluminum alloy steel wire is better than that of the galvanized steel wire, once the steel wire coating is consumed, the crack growth rate caused by the substrate pitting pit mainly depends on the stress response of the suspender. The bending stress of the short suspender under the N-C condition is larger, but the steel wire fatigue life remains lower than the design life of the bridge. Figure 16 shows the crack speed of the wires of suspender no. 26; the settlement of flexible central buckles substantially improves their fatigue life, and the improvement effect is similar to that of the rigid central buckle [17]. However, the increase in the number of buckles does not considerably weaken the crack growth rate. The fatigue lives were fitted, and the results are shown in Figure 17; all of them obey a normal distribution. The mean values of the fatigue lives of suspender nos. 24, 25, and 26 have small differences because the length of these suspenders is close. The 5% fractiles are taken as the characteristic service life, in which the service life of the steel wire has a 95% assurance rate. The service life of the short suspender is 20.04, 21.02, and 24.18, respectively. All of them are lower than expected, but the service life can be significantly improved by increasing the length of the steel wire; that is, reducing the bending stress.

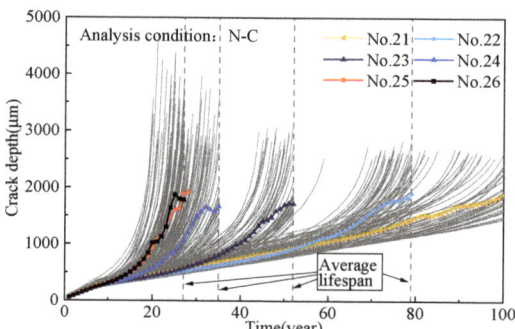

Figure 15. Development of average crack depth.

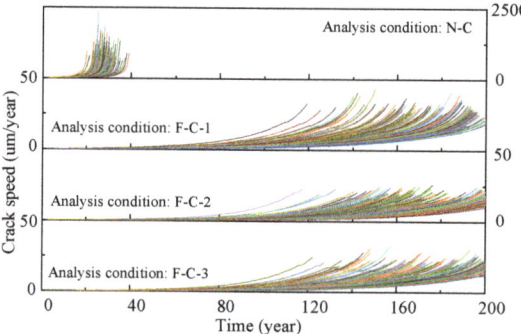

Figure 16. Crack speed of wire samples (suspender No. 26).

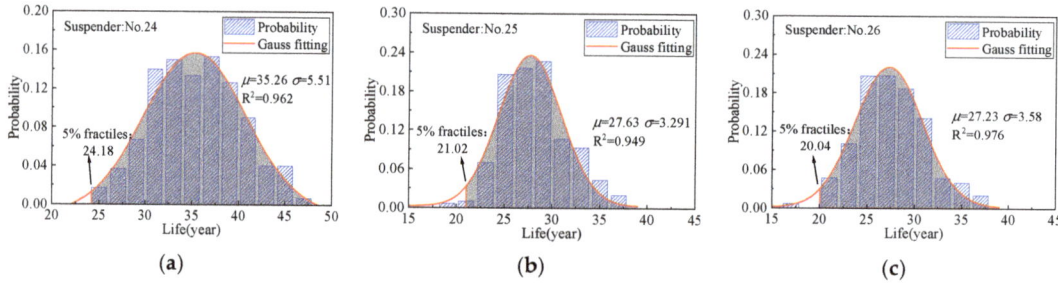

Figure 17. Distribution of steel wire life under the N-C condition. (**a**) Suspender no. 24; (**b**) suspender no. 25; (**c**) suspender no. 26.

Table 4 shows the comparison of the equivalent stress amplitude and fatigue life of the steel wires of suspender nos. 24–26. With consideration of the responses under traffic flow of five levels, the fatigue life of the steel wire after setting buckles meets the service requirements, but the increase in the number of buckles has hardly improved the life of the suspender steel wire. The 95% confidence interval results of fatigue life are shown in Table 4. When the number of samples is sufficient, the confidence interval length is small. The length of the confidence intervals of N-C steel wire is less than 1.5 years, and those of F-C are all less than 8 years. The average life of steel wire tends to be stable, thus the sampling results are proven to be reliable. The setting of buckles can considerably improve the extreme value of bending stress and the equivalent stress replication. When two buckles are settled, the extreme value and the equivalent stress amplitude of the steel wire tend to be stable, and increasing the number of buckles is unnecessary. The equivalent stress amplitude of axial stress is unaffected, and the suspender stress of long-span suspension bridges is determined by the dead load.

Table 4. Fatigue life of suspender wire under different analysis conditions.

Suspender Number (Length)	Analysis Condition	Maximum Bending Stress (MPa)	Equivalent Bending Stress (MPa)	Equivalent Axial Stress (MPa)	Fatigue Life (Year)		Confidence Intervals (95% CI)
					μ	σ	
24 (2.75 m)	N-C	208.31	37.30	10.23	35.2	5.51	(34.3,36.1)
	F-C-1	70.52	14.19	9.60	179.3	19.31	(176.2,182.4)
	F-C-2	54.14	11.38	9.51	178.5	19.54	(175.4,181.6)
	F-C-3	54.67	12.46	9.45	181.3	19.59	(178.2,184.4)
25 (2.45 m)	N-C	238.91	44.32	9.63	27.6	3.29	(27.1,28.1)
	F-C-1	77.16	15.68	10.12	177.8	19.32	(174.7,180.9)
	F-C-2	50.81	11.62	10.33	177.9	20.39	(174.6,181.2)
	F-C-3	50.79	11.81	9.38	181.1	20.10	(177.9,184.3)
26 (2.41 m)	N-C	274.10	46.66	10.03	27.2	3.58	(26.6,27.8)
	F-C-1	72.17	14.38	10.17	174.2	19.49	(171.1,177.3)
	F-C-2	50.68	11.93	8.98	178.1	21.35	(174.7,181.5)
	F-C-3	46.85	12.07	9.30	179.4	20.60	(176.1,182.7)

6. Conclusions

The influence of a flexible central buckle on suspension bridge vibration was remarkable, but the control effect on short suspenders is still unknown. This study established the corrosion fatigue degradation model of high-strength steel wire based on traffic composition and explored the influence of flexible central buckles on the corrosion fatigue life of suspenders under traffic flow. To improve the consideration of traffic flow, the WIM data were processed according to traffic density and used to analyze the suspender response under traffic flow of different densities. The fatigue life of short suspenders without buckles and with different numbers of buckles was analyzed based on monitoring traffic data. The following conclusions were drawn:

1. The intensity of traffic flow greatly influences the stress response of suspenders. The bending stress of short suspenders is considerably greater than that of long suspenders. The setting of flexible central buckles can effectively reduce the peak value of bending stress, but when the number of central buckles exceeds two, the increase in number does not remarkably weaken the bending stress. In addition, the buckles can share the axial stress of the suspender between inclined cables, and the weakening effect is affected by the setting position.

2. According to numerical analysis results, the fatigue life of short suspender wires under traffic load is remarkably lower than that of the other suspenders due to large bending stress (about 27–35 years). The setting of buckles can effectively reduce the equivalent bending stress amplitude, but the equivalent axial stress amplitude does not remarkably decrease. The improved stress state of the short suspenders considerably extends the fatigue life of the steel wires under traffic flow (about 174–179 years); by contrast, the increase in the number of buckles has a minimal effect on steel wire life and extreme stress values.

The dynamic motion of the bridges is complex for diverse loads. Moreover, the fatigue behavior of short suspenders and the vibration control effect are influenced by other loads, such as wind, earthquakes, and other special conditions. The optimal design of flexible central buckles should be studied further.

Author Contributions: Conceptualization, Y.Z. and X.G.; methodology, Y.Z.; software, X.G.; validation, X.G. and B.S.; formal analysis, X.G.; investigation, B.S. and Y.S.; resources, Y.S.; data curation, X.L.; writing—original draft preparation, Y.Z.; writing—review and editing, X.G.; visualization, B.S.; supervision, X.L.; project administration, Y.Z.; funding acquisition, Y.Z. All authors have read and agreed to the published version of the manuscript.

Funding: This research was funded by the Natural Science Basic Research Program of Shaanxi (Program No. 2022JQ-336); the open fund of Shaanxi Provincial Key Laboratory (Chang'an University) of Highway Bridges and Tunnels (Program No. 300102212509); and the Foundation of Xi'an University of Technology (Grant no. 256082109).

Data Availability Statement: Some or all data, models, or codes that support the findings of this study are available from the corresponding author upon reasonable request.

Conflicts of Interest: The authors declare no conflict of interest.

References

1. Guo, T.; Liu, J.; Zhang, Y.F.; Pan, S. Displacement monitoring and analysis of expansion joints of long-span steel bridges with viscous dampers. *J. Bridge Eng.* **2015**, *20*, 4014099. [CrossRef]
2. Yuan, A.M.; Yang, T.; Xia, Y.F.; Qian, L.; Dong, L.; Jin, X. Replacement Technology of Long Suspenders of Runyang Suspension Bridge. *China J. Highw. Transp.* **2021**, *34*, 289–297.
3. Liu, X.; Zhang, W.; Sun, P.; Liu, M. Time-Dependent Seismic Fragility of Typical Concrete Girder Bridges under Chloride-Induced Corrosion. *Materials* **2022**, *15*, 5020. [CrossRef]
4. Liu, M. Effect of uniform corrosion on mechanical behavior of E690 high-strength steel lattice corrugated panel in marine environment: A finite element analysis. *Mater. Res. Express* **2021**, *8*, 066510. [CrossRef]
5. Roffey, P. The Fracture Mechanisms of Main Cable Wires from the Forth Road Suspension. *Eng. Fail. Anal.* **2013**, *31*, 430–441. [CrossRef]
6. Qiao, Y.; Liao, C.Q.; Sun, C.Z. Evaluation of corrosion fatigue life for corroded wire for cable-supported bridge. *J. Civ. Environ. Eng.* **2017**, *39*, 115–121. (In Chinese)
7. Valor, A.; Caleyo, F.; Alfonso, L. Stochastic Modeling of Pitting Corrosion: A New Model for Initiation and Growth of Multiple Corrosion Pits. *Corros. Sci.* **2007**, *49*, 559–579. [CrossRef]
8. Nakamura, S.I.; Suzumura, K. Hydrogen Embrittlement and Corrosion Fatigue of Corroded Bridge Wires. *J. Constr. Steel Res.* **2009**, *65*, 269–277. [CrossRef]
9. Suzumura, K.; Nakamura, S.I. Environmental Factors Affecting Corrosion of Galvanized Steel Wires. *J. Mater. Civ. Eng.* **2004**, *16*, 1–7. [CrossRef]
10. Misaelides, P.; Hatzidimitriou, A.; Noli, F.; Pogrebnjak, A.D.; Tyurinc, Y.N.; Kosionidisd, S. Preparation, characterization, and corrosion behavior of protective coatings on stainless steel samples deposited by plasma detonation techniques. *Surf. Coat. Technol.* **2004**, *180*, 290–296. [CrossRef]
11. Liu, X.D.; Han, W.S.; Yuan, Y.G.; Chen, X.; Xie, Q. Corrosion fatigue assessment and reliability analysis of short suspender of suspension bridge depending on refined traffic and wind load condition. *Eng. Struct.* **2021**, *234*, 111950. [CrossRef]
12. Liu, Z.; Guo, T.; Pan, S.; Liu, J. Forensic investigation on cracking in hanger-to-girder connections of long-span suspension bridges. *IABSE Symp. Rep.* **2017**, *27*, 344–352. [CrossRef]
13. Viola, J.M.; Syed, S.; Clenance, J. The New Tacoma Narrows Suspension Bridge: Construction Support and Engineering. In Proceedings of the 2005 Structures Congress and the 2005 Forensic Engineering Symposium, New York, NY, USA, 20–24 April 2005.
14. Hu, T.F.; Hua, X.G.; Zhang, W.W.; Xian, Q.S. Influence of central buckles on the modal characteristics of long-span suspension bridge. *J. Highw. Transp. Res. Dev.* **2016**, *10*, 72–77. [CrossRef]
15. Li, G.L.; Su, Q.K.; Gao, W.B.; Han, W.S. The influence of the central buckle on the dynamic characteristics of the suspension bridge and the on-board excitation response of the short suspension cable. *China J. Highw. Transp.* **2021**, *34*, 174–186.
16. Wang, H.; Li, A.Q.; Guo, T. Accurate stress analysis on rigid central buckle of long-span suspension bridges based on submodel method. *Sci. China Ser. E* **2009**, *52*, 1019–1026. [CrossRef]
17. Wang, H.; Li, A. Influence of central buckle on wind-induced buffeting response of long-span suspension bridges. *China Civ. Eng. J.* **2009**, *42*, 78–84.
18. Liu, Z.X.; Guo, T.; Huang, L.Y.; Pan, Z. Fatigue Life Evaluation on Short Suspenders of Long-Span Suspension Bridge with Central Clamps. *J. Bridge Eng.* **2017**, *22*, 04017074. [CrossRef]
19. Wang, L.H.; Sun, Z.H.; Cui, J.F.; Hu, S.C. Effects of Central Buckle on End Displacement of Suspension Bridges under Vehicle Excitation. *J. Hunan Univ.* **2019**, *46*, 18–24.
20. Han, W.S.; Liu, X.D.; Gao, G.Z.; Xie, Q. Site-Specific Extra-Heavy Truck Load Characteristics and Bridge Safety Assessment. *J. Aerosp. Eng.* **2018**, *31*, 04018098. [CrossRef]
21. Gao, C.; Liu, Y.J.; Zhang, J.G.; Yao, Z.G. Traffic Flow Simulation of Multi-Lane Highway Bridge Based on Monte Carlo Method. *J. Chongqing Jiaotong Univ.* **2011**, *30*, 1375–1378+1434.
22. Caprani, C.C.; Obrien, E.J.; Mclachlan, G.J. Characteristic Traffic Load Effects from A Mixture Of Loading Events On Short To Medium Span Bridges. *Struct. Saf.* **2008**, *30*, 394–404. [CrossRef]
23. Bernard, E.; O"Brien, E. Monte Carlo Simulation of Extreme Traffic Loading on Short and Medium Span Bridges. *Struct. Infrastruct. Eng.* **2013**, *9*, 1267–1282.
24. Xu, X.; Qiang, S.Z. Influence of Central Buckle on Dynamic Behavior and Seismic Response of Long-span Suspension Bridge. *J. China Railw. Soc.* **2010**, *32*, 84–91.

25. Yuan, Y.G.; Liu, X.D.; Pu, G.N.; Wang, T.; Guo, Q. Corrosion Features and Time-Dependent Corrosion Model of Galfan Coating of High Strength Steel Wires. *Constr. Build. Mater.* **2021**, *313*, 125534. [CrossRef]
26. *JT/T 1104-2016*; Hot-Dip Zinc-Aluminium Coated Steel Wires for Bridge. Ministry of Transport of the People's Republic of China: Beijing, China, 2016.
27. Aoki, T.; Kittaka, T. Results of 10-year atmospheric corrosion testing of hot dip Zn-5mass% Al alloy coated sheet steel. In *Society IaS*; Galvatech: Chicago, IL, USA, 1995.
28. Katayama, H.; Kuroda, S. Long-term atmospheric corrosion properties of thermally sprayed Zn, Al and Zn–Al coatings exposed in a coastal area. *Corros. Sci.* **2013**, *76*, 35–41. [CrossRef]
29. Liu, M. Finite element analysis of pitting corrosion on mechanical behavior of E690 steel panel. *Anti-Corros. Methods Mater.* **2022**, *28*, 7527–7536. [CrossRef]
30. Toribio, J.; Matos, J.C.; González, B. Micro- and macro-approach to the fatigue crack growth in progressively drawn pearlitic steels at different R-ratios. *Int. J. Fatigue* **2009**, *31*, 2014–2021. [CrossRef]
31. Mayrbaurl, R.M.; Camo, S. Cracking and fracture of suspension bridge wire. *J. Bridge Eng.* **2001**, *6*, 645–650. [CrossRef]
32. Han, W.S.; Ma, L.; Wang, B. Refinement Analysis and Dynamic Visualization of Traffic-Bridge Coupling Vibration System. *China J. Highw. Transp.* **2013**, *26*, 78–87.
33. Wang, T.L.; Shahawy, M.; Huang, D.Z. Dynamic Response of Highway Trucks Due To Road Surface Roughness. *Comput. Struct.* **1993**, *49*, 1055–1067. [CrossRef]
34. Zhao, Y.; Huang, P.M.; Long, G.X.; Sun, Y. Influence of Fluid Viscous Damper on the Dynamic Response of Suspension Bridge under Random Traffic Load. *Adv. Civ. Eng.* **2020**, *2020*, 1857378. [CrossRef]
35. *ISO-8608*; Mechanical Vibration-Road Surface Profiles-Reporting of Measured Data. International Organization for Standardization: Geneva, Switzerland, 1995.
36. Wyatt, T.A. Secondary Stress in Parallel Wire Suspension Cables. *J. Struct. Div.* **1960**, *86*, 37–59. [CrossRef]
37. Kondoh, M.; Okuda, M.; Kawaguchi, K.; Yamazaki, T. Design Method of A Hanger System For Long-Span Suspension Bridge. *J. Bridge Eng.* **2001**, *6*, 176–182. [CrossRef]

Disclaimer/Publisher's Note: The statements, opinions and data contained in all publications are solely those of the individual author(s) and contributor(s) and not of MDPI and/or the editor(s). MDPI and/or the editor(s) disclaim responsibility for any injury to people or property resulting from any ideas, methods, instructions or products referred to in the content.

Article

Corrosion Behavior of High-Strength C71500 Copper-Nickel Alloy in Simulated Seawater with High Concentration of Sulfide

Xin Gao [1,2,*,†] and Ming Liu [3,*,†]

1. Research and Development Department, Beijing Med-Zenith Medical Scientific Corporation Limited, Beijing 101316, China
2. Key Laboratory of Biomechanics and Mechanobiology, Ministry of Education, Beijing Advanced Innovation Center for Biomedical Engineering, School of Biological Science and Medical Engineering, Beihang University, Beijing 100083, China
3. Center for Advancing Materials Performance from the Nanoscale (CAMP-Nano), State Key Laboratory for Mechanical Behavior of Materials, Xi'an Jiaotong University, Xi'an 710049, China

* Correspondence: 15901462422@163.com (X.G.); liuming0313@xjtu.edu.cn (M.L.)
† These authors contributed equally to this work.

Abstract: The corrosion behavior of high-strength C71500 copper-nickel alloy in high concentrations of sulfide-polluted seawater was studied by potentiodynamic polarization measurements, electrochemical impedance spectroscopy (EIS), immersion testing, and combined with SEM, EDS, XPS, and XRD surface analysis methods. The results showed that the C71500 alloy shows activation polarization during the entire corrosion process, the corrosion rate is much higher (0.15 mm/a) at the initial stage of immersion, and the appearance of diffusion limitation by corrosion product formation was in line with the appearance of a Warburg element in the EIS fitting after 24 h of immersion. As the corrosion process progressed, the formed dark-brown corrosion product film had a certain protective effect preventing the alloy from corrosion, and the corrosion rate gradually decreased. After 168 h of immersion, the corrosion rate stabilized at about 0.09 mm/a. The alloy was uniformly corroded, and the corrosion products were mainly composed of Cu_2S, CuS, $Cu_2(OH)_3Cl$, Mn_2O_3, Mn_2O, MnS_2, FeO(OH), etc. The content of Cu_2S gradually increased with the extension of immersion time. The addition of S^{2-} caused a large amount of dissolution of Fe and Ni, and prevented the simultaneous formation of a more protective Cu_2O film, which promoted the corrosion process to some extent.

Keywords: C71500 copper-nickel alloy; sulfide; immersion test; corrosion behavior; seawater environment

Citation: Gao, X.; Liu, M. Corrosion Behavior of High-Strength C71500 Copper-Nickel Alloy in Simulated Seawater with High Concentration of Sulfide. *Materials* 2022, *15*, 8513. https://doi.org/10.3390/ma15238513

Academic Editors: Jose M. Bastidas and Amir Mostafaei

Received: 28 October 2022
Accepted: 27 November 2022
Published: 29 November 2022

Publisher's Note: MDPI stays neutral with regard to jurisdictional claims in published maps and institutional affiliations.

Copyright: © 2022 by the authors. Licensee MDPI, Basel, Switzerland. This article is an open access article distributed under the terms and conditions of the Creative Commons Attribution (CC BY) license (https://creativecommons.org/licenses/by/4.0/).

1. Introduction

Copper-nickel alloys are widely used in marine environments because of their good corrosion resistance, machinability, high thermal conductivity, and electrical conductivity, as well as their moderate biological scaling resistance. They can be utilized in many marine engineering structures, such as seawater desalination, seawater reverse irrigation generators, marine ship power systems, marine ship power generation systems, etc. [1,2]. The corrosion rate of copper-nickel alloy pipes in polluted seawater is much faster than that in unpolluted areas; corrosion resistance can be lost in the presence of sulfide ions and other sulfur-containing substances [3,4]. The sulfide in polluted seawater in coastal areas can originate from industrial waste discharge, biological and bacterial reproduction processes in seawater (algae, marine organisms or microorganisms, bacteria that reduce sulfide), and atmospheric sulfide emissions in coastal areas, which may all also cause seawater pollution [5,6].

Copper and its alloys have been widely studied in seawater environments. Kong et al. [7] studied the effect of sulfide concentrations on copper corrosion behavior in anaerobic chloride solution, and found that the corrosion resistance of copper decreased with the increase of sulfide concentration and sulfide addition. Chen et al. [8] demonstrated that the migration of Cu occurred between the surface of the net anode and cathode by using a pre-corroded and fresh Cu electrode in NaCl solution containing sulfide. Rao and Kumar [6,9] studied the corrosion inhibition behavior of copper-nickel alloy in simulated seawater and synthetic seawater containing 10 ppm sulfide, which confirmed that 5-(3-Aminophenyl) tetrazole shows good corrosion inhibition. Radovanović et al. [10] studied the protective effect of the non-toxic compound 2-amino- 5-ethyl- 1,3,4-thiadiazole (AETDA) on copper in acidic chloride solution, and the results showed that the stability of the protective layer mainly depends on the concentration of the inhibitor and the potential value of the protective film. Nady et al. [11] studied the electrochemical properties of Cu and Cu-10Al-10Ni in sulfide ion containing 3.5 wt.% NaCl solution, and found that the addition of Ni into the Cu_2O barrier film could enhance the corrosion resistance of the alloy. Jandaghi et al. [12] conducted in-depth research on the microstructure evolution and corrosion resistance of aluminum/copper joints manufactured by explosive welding processes and revealed the cause of corrosion.

Our group [13–17] has previously studied the grain boundary engineering treatment on the mechanical behavior of C71500 copper-nickel alloy. Using the process of "multi pass deformation and grain boundary treatment + single pass deformation recovery" could obtain a large number of low coincidence site lattice grain boundary structures. In order to improve the corrosion resistance of the alloy, the harmful elements such as C, S, and O are reduced to less than 5 ppm by vacuum refining technology; the strength of the material reaches 430 MPa, and the elongation rate reaches 45%. Additionally, we found that this alloy has good corrosion resistance in a 3.5 wt.% NaCl simulated seawater environment [18]. In the actual use process, it is found that the copper-nickel alloy cooling equipment stably runs during the process of navigation, but after the ship berths for a long time, the corrosion damage of cooling equipment often occurs. It is found that the S^{2-} content of flowing seawater is not high, but in the non-flowing area of the pipeline, the S^{2-} content of seawater reaches 0.5 wt.%; the corrosion behavior is obviously different from previous research results [19,20]. Under such conditions, stress corrosion cracking easily occurs. In the marine environment, a large number of sulfur ions exist due to the discharge of industrial wastes, the reproduction process of organisms and bacteria in seawater, and the discharge of atmospheric sulfide in coastal areas. The corrosion behavior of seawater polluted by sulfur ions is significantly different from previous studies [21]. Based on this, the corrosion behavior of the C71500 Cu-Ni alloy in high concentration S^{2-} polluted seawater was systematically studied.

2. Experimental Procedure

2.1. Specimen and Solution

The material used in this study was C71500 copper-nickel alloy manufactured according to ASTM-B224. The chemical composition (wt.%) was: Ni = 30.54, Mn = 0.93, and Fe = 0.80, the content of C, S, and O was strictly controlled, and the residual was Cu. After casting, forging, hot piercing, cold rolling and heat treatment, the material was fabricated into a pipe with the dimensions of ⌀ 60 × 5.03 mm. In order to ensure the original structure of the material, all samples were directly cut from the alloy pipe. The as-revived microstructure of the alloy was a single-phase, and a large number of low coincidence site lattice (CSL) grain boundaries could be obtained after thermomechanical treatment [13–17].

The sample for the immersion test was cut into four equal parts in the radial direction, at a length of 35 mm. The electrochemical test specimen was cut with a dimension of 10 × 10 mm toward the pipe wall thickness; the outer surface of the pipe was used as the testing surface, and the remaining part was sealed with epoxy resin. The testing arc surface of the pipe was ground by silicon carbide paper down to 1200#. The outer diameter, wall thickness, and length of the sample were measured by a micrometer and the surface area of the sample was drawn using SolidWorks software. Then, the sample was ultrasonically degreased in acetone for 10 min, and put into a dish to dry for 24 h before testing.

A solution of 3.5% NaCl + 0.5% Na$_2$S (wt.%) was chosen to simulate a high concentration sulfide-containing seawater environment with a pH 8.20. All experiments were carried out in a constant temperature water bath at around 35 ± 2 °C.

2.2. Immersion Test

The immersion test was performed according to ASTM G31-2012a standard. The exposure time was set to 6 durations of 24, 72, 120, 168, 336, and 672 h. Five samples were used for each duration; three were used for weight loss testing, and the other two were used for surface analysis. After each period, the corrosion products of weight loss samples were removed by HCl solution with a volume ratio (HCl/H$_2$O) of 2: 1. The weight of samples before and after immersion testing were measured by an electronic balance (METTLER TOLEDO ME204/02) with an accuracy of 0.0001 g. The corrosion rate can be obtained by Equation (1):

$$v = \frac{8.76(m_0 - m_1)}{S_0 t d} \tag{1}$$

where: v is the corrosion rate, mm/a; m_0 is the mass of the sample before corrosion, g; m_1 is the mass of the sample after removing corrosion products, g; S_0 is the area of the tested surface, m^2; t is the corrosion test cycle, h; and d is the density of the alloy, g/cm^3.

2.3. Electrochemical Measurement

A CS2350H/CORRTEST electrochemical workstation with a three electrode system was used to carry out the electrochemical analysis, the working electrode was an arc sample, the saturated calomel electrode (SCE) was used as a reference electrode and a platinum mesh electrode with a surface area of 8 cm^2 was the auxiliary electrode. The open circuit potential (OCP) was tested for 30 min before other electrochemical features were measured. The frequency range (100 kHz–0.01 Hz) of electrochemical impedance spectrum (EIS) was selected and the AC amplitude signal was 10 mV. Potentiodynamic polarization was performed from cathodic 0.3 V (OCP) to anodic 0.8 V (vs. OCP) with a scan rate of 1 mV/s.

In order to study the electrochemical corrosion characteristics of the alloy at different immersion times, we also tested the samples immersed in the corrosion solution for different periods, as for the immersion test procedure.

2.4. Surface Analysis

A Zeiss Gemini 300 scanning electron microscope (SEM) was applied for surface observation, and the relevant corrosion product was analyzed by EDS.

The corroded product compositions were analyzed by X-ray diffractometer (Rigaku SmartLab 9 kW) in the range of 20°–100° and a step length of 0.02°, and Jade 6.5 software was used to analyze the test results.

The corrosion products were quantificationally analyzed by XPS diffractometer (Thermo Fisher Scientific Escalab 250Xi). The scanning range, interval, and pitch was 0–1200 eV, 1 eV, and 0.1 eV, respectively, with a spot diameter of 400 µm. Peak fitting and analysis was performed using Advantage software and the binding energy calibration standard was the C1s spectral line (284.8 eV).

3. Results and Discussion

3.1. Weight Loss Experiment

Figure 1 shows the C71500 alloy average corrosion rate variation with different corrosion times. As can be seen from Figure 1, the maximum corrosion rate of the alloy is 0.15 mm/a at the initial stage of immersion. With the extension of immersion time, the average corrosion rate of the alloy rapidly decreases. When the corrosion time reaches 168 h, the average corrosion rate of the C71500 alloy tends to be stable at about 0.09 mm/a. This change may be related to the formation and densification of the protective corrosion product film on the surface of the alloy. With the formation and dissolution of the protective film gradually balanced, the corrosion rate of the alloy after long-term immersion tends to be stable.

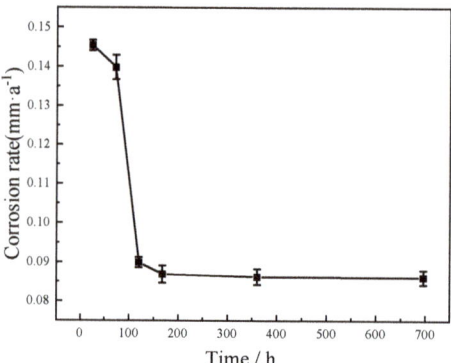

Figure 1. Corrosion rate of C71500 alloy after immersion in simulated polluted marine environment at room temperature for different times (24, 72, 120, 168, 336, and 672 h).

3.2. Potentiodynamic Polarization Measurements

Figure 2 shows the potentiodynamic polarization curves of the alloy under different corrosion conditions. It can be seen from Figure 2a that the electrochemical response of C71500 alloy in 3.5 wt.% NaCl solution is significantly different by adding S^{2-}. The corrosion potential of C71500 alloy shifts to the negative direction and the anodic curve is no longer smooth. It can be seen from Figure 2b that the polarization curves of C71500 alloy in simulated polluted marine solution at immersion for various times shift to the left, i.e., lower current densities, and the corrosion rate decreases with immersion time. It should be noted that one or more current density decrease inflection points could be observed before the peak current density is reached after adding 0.5 wt.% Na_2S. With increasing immersion time, the anodic polarization curves of the alloy change: at the early stage of corrosion, two secondary anodic peaks appear at about −525 mV and −404 mV (SCE) [5,22]. When the immersion time increases to 336 h and 672 h, the peak of −525 mV (SCE) disappears, and only a −404 mV (SCE) peak can be observed. The Tafel extrapolation method was applied to fit the parameters of the polarization curves at various immersion times. It can be seen from Table 1 that, at the initial stage of immersion, the corrosion current density (i_{corr}) is much higher, which is about 96.6 µA/cm²; i_{corr} decreases and tends to stabilize for longer immersion times, which is consistent with the corrosion weight loss results presented in Figure 1.

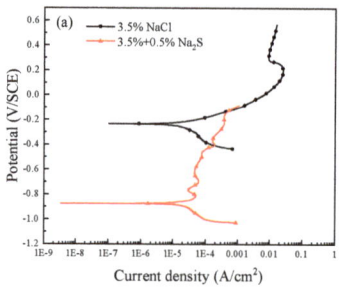 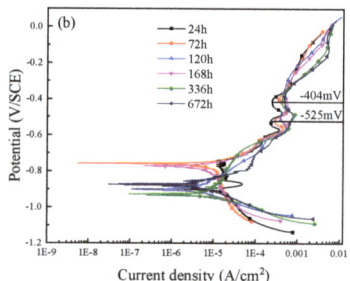

Figure 2. Polarization curves of C71500 alloy under different corrosion conditions: (**a**) immersion in 3.5 wt.% NaCl solution for 30 min with and without sulfide; (**b**) immersion in 3.5 wt.% NaCl solution containing sulfide for different times.

Table 1. Characteristic parameters of polarization curve of alloy fitted for different corrosion times. (Residual is the polarization curve fitting residual error).

Time (h)	β_a (mV/dec)	β_c (mV/dec)	i_{corr} (μA/cm^2)	E_{corr} (mV)
0.5	337 ± 12	−541 ± 16	27.4 ± 10	−877 ± 25
24	15 ± 7	−18 ± 5	2.72 ± 0.5	−899 ± 8
72	33 ± 5	−41 ± 4	1.7 ± 0.3	−758 ± 7
120	54 ± 4	−30 ± 3	1.3 ± 0.2	−902 ± 8
168	40 ± 5	−23 ± 3	1.2 ± 0.2	−773 ± 6
336	42 ± 3	−23 ± 2	0.8 ± 0.1	−929 ± 7
672	39 ± 3	−42 ± 2	0.6 ± 0.1	−873 ± 5

3.3. Electrochemical Impedance Spectroscopy

The Nyquist and Bode plots of C71500 alloy immersed in simulated polluted marine solution for different periods are shown in Figure 3. It can be seen from the Nyquist diagram that only a small capacitive arc could be observed after immersion for 30 min, which indicates that the corrosion rate is much higher at the initial stage of immersion [23]. Warburg impedance could be seen in the low frequency parts of the Nyquist plots after immersion for 24 and 72 h, which indicates that the speed of corrosion process is faster and the substances involved in the reaction are quickly consumed, and that the characteristic of the diffusion control process may exist [4]; after 120 h, the diffusion impedance disappears and the capacitive reactance arc radius further increases, which indicates the corrosion rate is greatly reduced. The modulus of impedance |Z| in Figure 3b in the low frequency region gradually increases with exposure time and tends to be stable at immersion for 120 h. Therefore, the corrosion process can be mainly divided into three stages: the first stage (0–72 h) includes the formation of a corrosion product layer; the second stage (72–120 h) is the transition stage; the third stage (120–672 h) is the stable corrosion product film. The appearance of Warburg impedance reflects the growth process of the corrosion product film and its related mass transfer limitations. With the increase in immersion time, the arc radius of the capacitive gradually increases (72–672 h), indicating increasing corrosion resistance, and then tends to stabilize at even longer immersion times, which indicates that the formed corrosion product film is in a dynamic equilibrium process of growth/dissolution, thus reducing the corrosion rate.

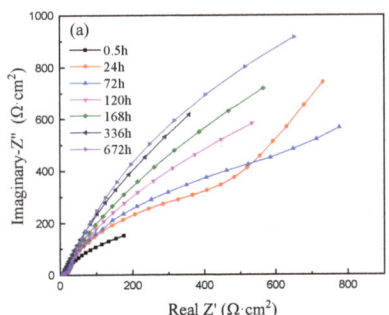
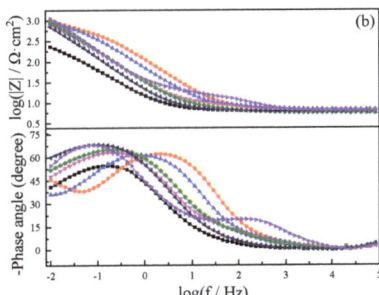

Figure 3. EIS of C71500 alloy in simulated polluted marine solution. (**a**) Nyquist; (**b**) Bode.

Based on prior research [24–26], the equivalent circuits (Figure 4) were used to fit the EIS of C71500 alloy after immersion in simulated polluted marine solution for various times; where R_s R_t, R_f, CPE$_1$, CPE$_2$, and W is the solution resistance, charge transfer resistance, corrosion product film resistance, constant phase element of the corrosion product film

(Q_{dl}), the constant phase element (Q_f) related to charge transfer, and finite length diffusion Warburg impedance, Z_W [27] respectively. The Q_{dl} impedance can be obtained by [27]:

$$\frac{1}{Z} = Q_{dl}(j\omega)^n \quad (2)$$

where j, ω, and n are the imaginary number ($j^2 = -1$), the angular frequency ($\omega = 2\,pf$) and the exponent of Q_{dl}, respectively. When $n = 1$, it relates to resistance; $n = -1$ relates to an inductance; and $n = 0.5$ is equal to Warburg impedance [28]. The CPE is often used as a pure capacitance to eliminate the dispersion effect caused by electrode surface roughness and for other reasons in an analog equivalent circuit [18].

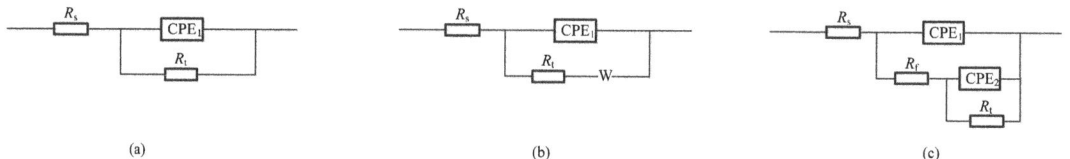

(a)   (b)   (c)

Figure 4. Equivalent circuits of C71500 alloy in simulated polluted marine solution: (**a**) 0.5 h; (**b**) 24 h, 72 h; and (**c**) 120 h.

The equivalent circuit parameters fitted by ZsimpWin software are shown in Table 2. With increasing immersion time, R_t gradually increases and reaches the maximum value of 6619 $\Omega \cdot cm^2$ after immersion for 672 h, indicating the corrosion rate gradually decreases with immersion time. Q_{dl} also increases with immersion time, which indicates the gradual growth of a corrosion product film. R_f reaches the maximum value at about 120 h, then decreases and increases again, which indicates the dynamic equilibrium of the formed corrosion product film. Q_f is the double layer capacitance associated with corrosion products, which indicates that the corrosion product film may be changed with immersion times.

Table 2. Electrochemical parameters of EIS obtained by equivalent circuit fitting after different immersion times.

Time, h	R_s, $\Omega \cdot cm^2$	R_t, $\Omega \cdot cm^2$	n_1	Q_{dl} Y_0 $\Omega^{-1} \cdot cm^{-2} \cdot s^n$	R_f, $\Omega \cdot cm^2$	n_2	Q_f Y_0 $\Omega^{-1} \cdot cm^{-2} \cdot s^n$	W $10^{-4} \cdot \Omega^{-1} \cdot cm^{-2} \cdot s^{1/2}$
0.5	6.11 ± 1.81	1817 ± 20	0.73 ± 0.10	459 ± 7	–	–	–	–
24	6.96 ± 0.55	2213 ± 12	0.82 ± 0.04	485 ± 6	–	–	–	43.4 ± 2.4
72	7.17 ± 0.51	3454 ± 7	0.77 ± 0.04	908 ± 6	–	–	–	76.4 ± 1.9
120	6.36 ± 0.26	5084 ± 8	0.69 ± 0.03	9 ± 2	3069 ± 15	0.89 ± 0.03	2068 ± 18	–
168	5.53 ± 0.21	5134 ± 6	0.72 ± 0.02	7 ± 1	2617 ± 12	0.89 ± 0.03	2752 ± 11	–
336	5.98 ± 0.16	5950 ± 6	0.78 ± 0.01	9 ± 1	2658 ± 8	0.90 ± 0.02	3164 ± 9	–
672	6.17 ± 0.12	6619 ± 4	0.71 ± 0.01	15 ± 1	2686 ± 7	0.84 ± 0.01	3844 ± 10	–

3.4. Morphology Analysis

3.4.1. Macroscopic Morphology

Figure 5 shows the macro morphology of C71500 alloy after immersion in simulated polluted marine solution for various times. It can be seen that the alloy mainly shows uniform corrosion. After immersion for 24 h, the sample surface is covered with a dark brown corrosion products layer, and the corrosion degree gradually increases with immersion time. The corrosion products fall off and deposit at the bottom of the solution with immersion time, thus resulting in different corrosion morphologies at local areas of the sample.

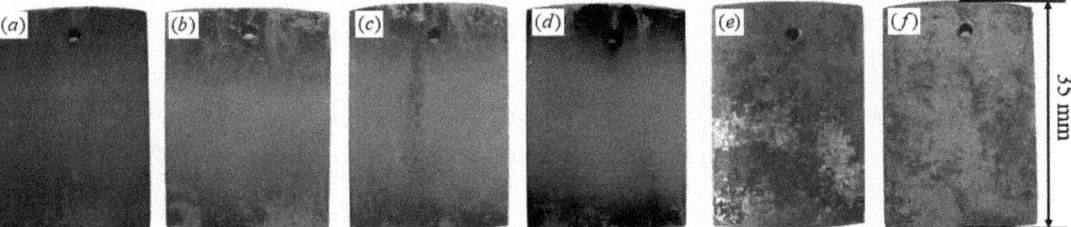

Figure 5. Macroscopic morphology of sample after immersion for different times: (**a**) 24 h; (**b**) 72 h; (**c**) 120 h; (**d**) 168 h; (**e**) 336 h; (**f**) 672 h.

3.4.2. SEM Analysis

In order to further analyze the evolution of corrosion products on C71500 alloy at high lateral resolution, the corrosion product morphology after different immersion times were analyzed by SEM. In the initial 24 h of immersion (Figure 6a), less than 1 μm granular products are formed on the surface, and the gap between particles is larger. The particle size gradually increases with immersion time, and flat sheet corrosion product films could be locally detected, the surface of the specimen is not completely covered by the corrosion product film, and the film is in its lateral growth stage [29]. After immersion for 72 h, it can be seen that the surface of the alloy is completely covered by the granular corrosion product layer, and the size of the particles gradually increases from less than 1 to 3 μm; there are also some scattered corrosion films between the corrosion products. The particles continue to increase when the immersion time reaches 120 h, and the thickness of the corrosion product film also further increases. Small spheres are difficult to form by precipitation from supersaturated solutions [30]. After 168 h of immersion, cracks can be seen in the corrosion product film, and the cracks show grain boundary related patterns, which indicates that there may be local intergranular corrosion in C71500 alloy; the initial corrosion product film also shows as irregular (see Figure 6m). After 336 h of immersion, a large amount of corrosion products adhere to the surface of the sample, and the formed film is loose and porous. After immersion for 672 h, the formed corrosion product film has gradually become dense. It can be seen that the corrosion morphology of the alloy has greatly changed with various immersion times. The SEM morphologies shown in Figure 6d–i indicate the shape of spherical corrosion particles at the beginning of immersion, and the edges and corners become more and more clear with immersion time.

The corrosion product EDS analysis of C71500 alloy after different immersion times is shown in Table 3, and the surface scan energy spectrum analysis is depicted in Figure 7. It can be seen that the content of Fe and Ni in the corrosion products decreases with immersion time, which indicates that the alloy is subject to Fe and Ni removal [1]; the content of Cu in the corrosion products increases. The content of S fluctuates; it obviously increases in the early stage, indicating that S may participate in the corrosion reaction and could be incorporated in the corrosion product on the substrate surface.

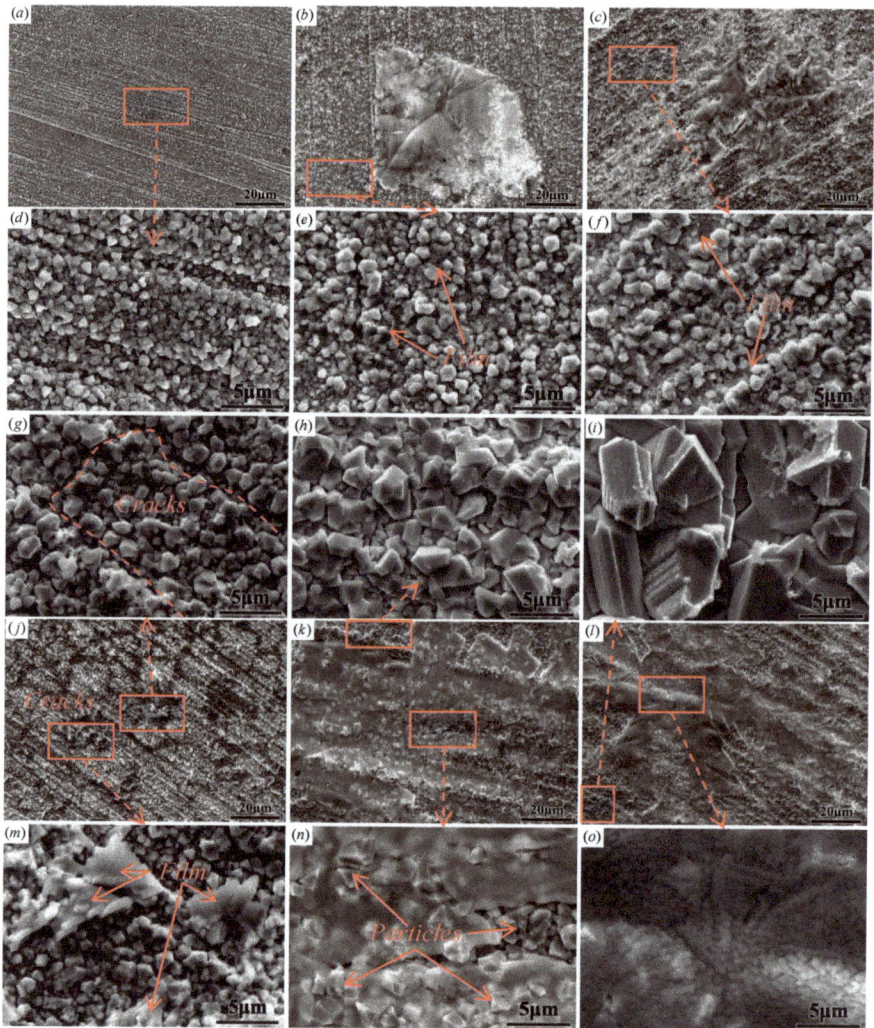

Figure 6. SEM morphology of the corrosion product layer of the alloy after different immersion times: (**a,d**) 24 h; (**b,e**) 72 h; (**c,f**) 120 h; (**g,j,m**) 168 h; (**h,k,n**) 336 h; (**i,l,o**) 672 h. The arrows and red boxes represent the original positions and relationships of the enlarged image.

Table 3. Corrosion products EDS analysis of C71500 alloy after immersion for different times (wt.%).

Exposure Time (h)	O	S	Mn	Fe	Ni	Cu
24	0.4	6.2	0.8	1.7	30.8	60.1
72	2.6	18.1	1.3	0.6	5.5	71.9
120	4.7	13.0	1.1	0.5	6.9	73.9
168	0.8	17.9	0.3	0.3	6.4	74.4
336	1.3	23.5	0.6	0.2	1.0	73.4
672	0.3	18.4	1.6	0.2	0.6	78.8

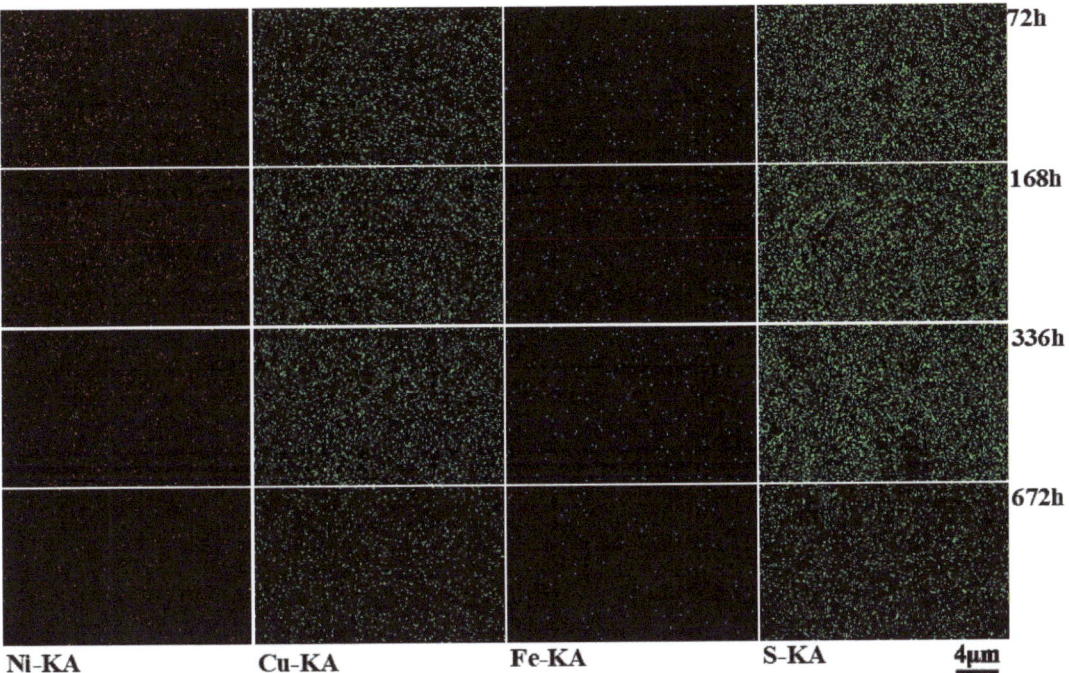

Figure 7. Scanning energy spectrum analysis of C71500 alloy surface corrosion products after different immersion times (72, 168, 336, and 672 h).

3.5. XPS Analysis

The corrosion product film composition of samples after immersion for 72, 168, and 336 h was analyzed by XPS (Figure 8). It can be seen that the corrosion product film is mainly composed of Cu, S, Cl, Mn, O, and Na. After 72 h of immersion, a small amount of Fe-based corrosion products are detected on the surface of the sample, while it disappears after longer immersion times, and a small amount of Ni-based corrosion products are seen after 336 h of immersion.

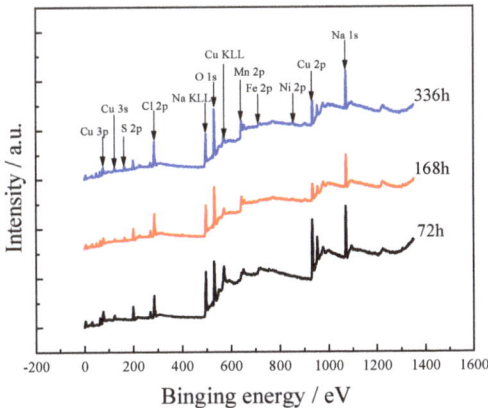

Figure 8. XPS spectrum of C71500 alloy corroded samples after various immersion times (72, 168, and 336 h).

3.5.1. Cu Spectrum

In Table 2, the p peak for all tested samples are composed of the Cu 2p1/2 (952 eV) and Cu 2p3/2 (932.2 eV) peaks (Figure 9a). Except for a slight fluctuation in the early stage of immersion (72 h), no obvious S polarization and satellite peaks were detected. Since Cu and Cu_2O compounds do not have the excitation peak of the Cu 2p3/2 spectrum, it can be preliminarily concluded that the superficial corrosion product films do not contain Cu^{2+}. It shows that Cu is mainly composed of Cu^+, except for a small amount of Cu^{2+} that may be formed in the early stages of immersion. The Auger spectrum of Cu was analyzed to further verify the valence state of Cu (Figure 9b). It shows that the Cu Auger spectrum is between 916.4 and 916.8 eV, again establishing that the corrosion product is composed of Cu^+ [31,32].

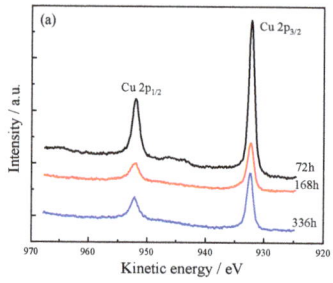 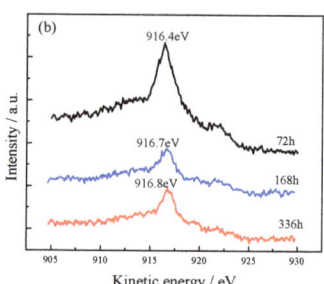

Figure 9. Cu spectrum of alloy after different corrosion times (72, 168, and 336 h). (**a**) Cu 2p; (**b**) Cu CLM.

The Cu 2p3/2 spectra of the corrosion product after various immersion times are shown in Figure 10. The binding spectrum was analyzed in high-resolution conditions (929–939 eV). The corrosion products are mainly comprised CuS, Cu_2S, and Cu_2O. The binding energy peak of Cu was 160,000 CPS in the early stages of immersion, indicating more Cu could be observed in the corrosion film. After immersion for 168 h, the binding energy peak of Cu 2p3/2 stabilized at 75,000–85,000 CPS. Only Cu_2S and CuS could be detected in the 72-h immersed sample, indicating that a protective Cu_2O layer had not been formed. The Cu_2O film was considered to be a uniform and dense corrosion product film that adhered well to the substrate and promoted passivity of the copper-nickel alloy, thereby improving the corrosion resistance of the alloy [33]. When the immersion time reached 168 h, Cu_2O could be observed. In this work, the CuS component gradually decreased and the Cu_2O component gradually stabilized with immersion time (see Table 4).

Table 4. The corrosion product phase analysis of Cu 2p3/2 spectrum after different immersion times.

Valence	Exposure Time (h)	Proposed Compounds	Binding Energy (eV)	Intensity Area	Atomic (%)
Cu 2p3/2	72	Cu_2S	932.3	115,144	71.4
		CuS	932.1	46,123	28.6
	168	Cu_2S	932.3	27,119	44.3
		CuS	932.2	12,969	21.2
		Cu_2O	932.9	21,124	34.5
	336	Cu_2S	932.1	41,307	62.6
		CuS	932.0	8467	12.8
		Cu_2O	932.6	16,175	24.5

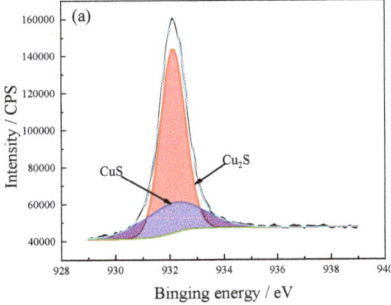
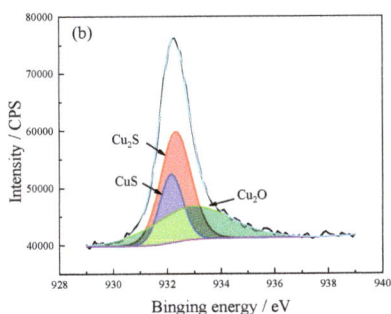
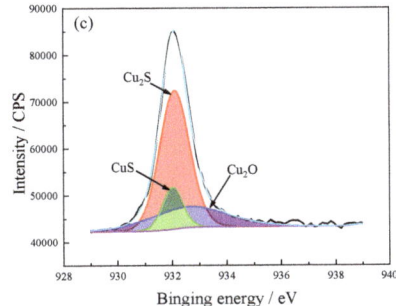

Figure 10. Cu 2p3/2 spectrum of C71500 alloy corrosion product layer after different immersion times. (**a**) 72 h; (**b**) 168 h; (**c**) 336 h. The black line is the original result and the color line is the fitting result.

3.5.2. Mn Spectrum

The binding spectrum of Mn 3s under high-resolution conditions (72.5–97.5 eV) is shown in Figure 11a. No small satellite peaks could be detected (72 h), while the energy differences between the two satellite peaks (168 and 336 h) were 6.0 and 6.1 eV, respectively, indicating that the Mn^{2+} was the main product in the corrosion product film.

The Mn 2p spectra after different immersion times are shown in Figure 11b–d. It can be seen that there is no obvious difference between the peak values. The composition of the corrosion product after 72-h immersion is mainly Mn, MnO, and $MnCl_2$, among which the content of Mn is the least. With the extension of the immersion time, MnS and high Mn-oxides (MnO_2, Mn_2O_3) are observed. For prolonged immersion times, the high Mn-oxide products disappear and the components become MnS, MnO, and $MnCl_2$. The Ni, Mn, Fe, and other alloying elements are oxidized into high-oxides at the metal/film interface, and then migrate at the corrosion product membrane/solution interface. High-price ions increase the ionic and electronic resistance of the Cu_2O, thereby improving the corrosion resistance of Cu alloys [34]. This also means that corrosion-resistant Cu alloys can be improved by forming high-oxide ions (see Table 5).

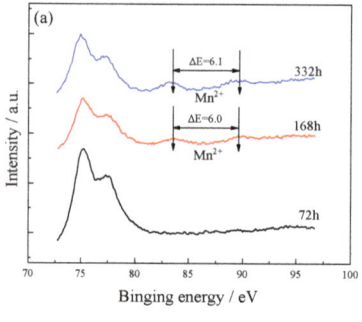
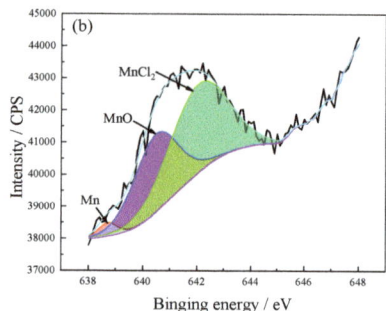
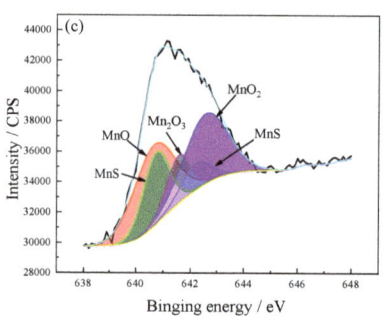
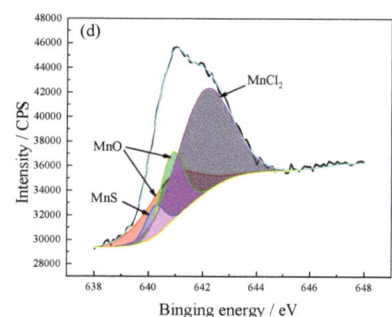

Figure 11. The corrosion product layer of Mn spectra of C71500 alloy after different immersion times. (**a**) Mn 3s; (**b**) 72 h; (**c**) 168 h; (**d**) 336 h. The black line is the original result and the color line is the fitting result.

Table 5. Mn 2p spectrum phase analysis of corrosion product after different immersion times.

Valence	Exposure Time (h)	Proposed Compounds	Binding Energy (eV)	Intensity Area	Atomic (%)
Mn 2p3/2	72	Mn	638.7	367	2.8
		MnO	640.5	5061	39.0
		MnO	640.7	10,260	32.4
		MnS	640.7	5713	18.1
		MnS	642.2	2526	8.0
	168	MnO	640.8	9187	23.9
		$MnCl_2$	642.0	7557	58.2
		MnO_2	642.6	9584	30.3
		MnS	640.2	2266	5.9
	336	MnO	640.9	6228	16.2
		MnO	640.8	9187	23.9
		$MnCl_2$	642.1	20,745	54.0

3.5.3. O Spectrum

The O 1s spectrum of the C71500 alloy corrosion product after different immersion times is shown in Figure 12. The oxide after the initial immersion is mainly composed of Cu_2O. With increasing immersion time, Ni_2O_3 is observed in the corrosion product film. Ni atoms could be oxidized in the form of NiO or Ni_2O_3, and doped in the Cu_2O film interacting with O_2 (see Table 6).

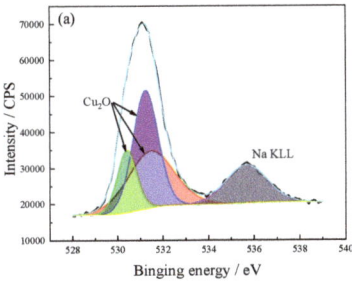
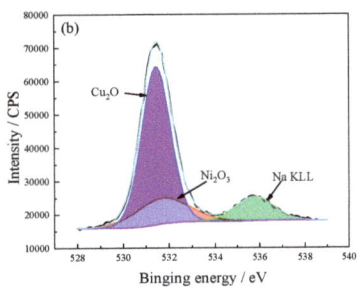
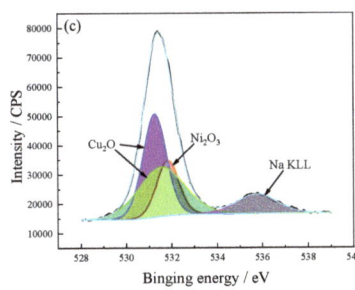

Figure 12. O 1s spectrum of the C71500 alloy surface corrosion product layer after different immersion times. (**a**) 72 h; (**b**) 168 h; (**c**) 336 h. The black line is the original result and the color line is the fitting result.

Table 6. Corrosion product phase analysis of O 1s spectrum after different immersion times.

Valence	Exposure Time (h)	Proposed Compounds	Binding Energy (eV)	Intensity Area	Atomic (%)
O 1s	72	Cu_2O	531.5	41,584	31.5
		Cu_2O	531.2	47,032	35.6
		Cu_2O	530.4	18,914	14.3
		Na KLL	535.7	24,612	18.7
	168	Cu_2O	531.4	75,394	66.1
		Ni_2O_3	531.7	22,763	20.0
		Na KLL	535.8	15,830	13.9
	336	Cu_2O	531.2	46,810	36.5
		Cu_2O	531.5	43,086	33.6
		Ni_2O_3	531.8	23,414	18.3
		Na KLL	535.8	14,872	11.6

3.5.4. S Spectrum

The corrosion product layer S 2p spectra after different immersion times are shown in Figure 13. The composition of the corrosion products is not stable at the early stages of immersion, the peak spectrum of the S 2p after 72 h immersion is as low as 6400 CPS, and many saw-tooth patterns are observed on the waveform. The corrosion products are mainly Cu_2S, FeS_2, CuS, and some of the interference peaks of Na_2S and $Na(SO_3)_2$. When the immersion time reaches 168 h, the waveform of the S 2p spectrum smoothens; FeS components are detected at an immersion time of 336 h (see Table 7).

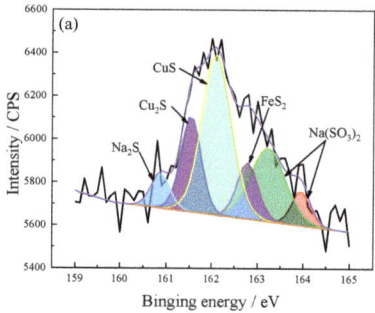
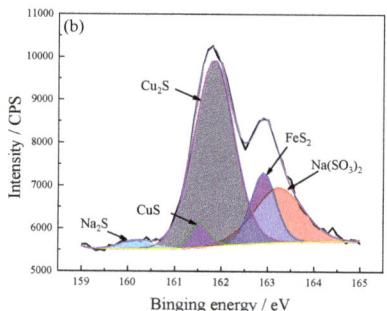
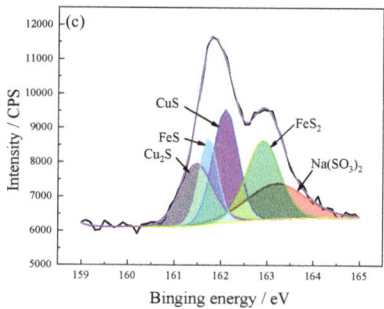

Figure 13. The corrosion product S 2p spectra after different immersion times. (**a**) 72 h; (**b**) 168 h; (**c**) 336 h. The black line is the original result and the color line is the fitting result.

Table 7. The corrosion product phase analysis of S 2p spectrum after different corrosion times.

Valence	Exposure Time (h)	Proposed Compounds	Binding Energy (eV)	Intensity Area	Atomic (%)
S 2p	72	Cu_2S	161.5	254	16.8
		CuS	162.1	620	40.8
		FeS_2	162.8	150	9.9
		Na_2S	160.9	84	5.5
		$Na(SO_3)_2$	163.2	327	21.5
		$Na(SO_3)_2$	164.0	83	5.5
	168	Cu_2S	161.9	4654	58.4
		CuS	161.6	300	3.8
		FeS_2	162.9	1132	14.2
		Na_2S	160.2	174	2.2
		$Na(SO_3)_2$	163.2	1713	21.5
	336	Cu_2S	161.7	1310	14.6
		CuS	162.1	2226	24.8
		FeS_2	162.9	2131	23.8
		FeS	161.5	1619	18.1
		$Na(SO_3)_2$	163.2	1673	18.7

3.6. XRD Analysis

Figure 14 shows the XRD pattern of the C71500 alloy surface corrosion product layer after different immersion times. The observed corrosion products are mainly $Cu_2(OH)_3Cl$, CuS, Cu_2S, Mn_2O_3, Mn_2O, MnS_2, and FeO(OH). Among them, the peak distribution of Cu_2S is the most obvious, and increases with immersion time. After 72 h of immersion, the corrosion product is mainly $Cu_2(OH)_3Cl$ and FeO(OH); the peak of $Cu_2(OH)_3Cl$ gradually disappears and a large amount of Cu_2S is detected at an immersion time of 168h, the accompanying peaks of Mn$_2$O, Mn$_2$O$_3$, and MnS$_2$ could also be observed; the accompanying peaks of Cu$_2$(OH)$_3$Cl and Cu$_2$S in the corrosion products significantly increase when the immersion time reaches 336 h.

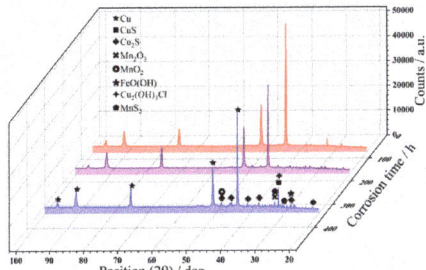

Figure 14. XRD analysis of C71500 alloy surface corrosion product after 72-h, 168-h, and 332-h corrosion.

4. Corrosion Mechanism
4.1. Cu Reaction Mechanism

Cu is ionized through electron transfer with Cl$^-$, resulting in CuCl$_{ads}$ [35].

$$Cu + Cl^- \rightarrow CuCl_{ads} + e^- \quad (3)$$

It is easy to dissolve and form CuCl$_2^-$ [27,35]:

$$CuCl_{ads} + Cl^- \rightarrow CuCl_2^- \quad (4)$$

In the neutral solution, the cathode reaction can be determined as follows:

$$O_2 + 2H_2O + 4e^- \rightarrow 4OH^- \quad (5)$$

The C71500 alloy polarization behavior in chloride containing solution mainly dissolves Cu in soluble CuCl$_2^-$ according to the process described above [36,37]. In the case of high chloride concentration, CuCl$_3^{2-}$ and CuCl$_4^{3-}$ [35,38] may also be formed. Subsequently, in the environment of OH$^-$, Cu$_2$O is generated through the following reaction:

$$2CuCl_2^- + 2OH^- \rightarrow Cu_2O + H_2O + 4Cl^- \quad (6)$$

It is also possible to hydrolyze the resulting soluble ion complex to form a corrosion-resistant Cu$_2$O layer [36].

$$2CuCl_2^- + H_2O \rightarrow Cu_2O + 4Cl^- + 2H^+ \quad (7)$$

Under dissolved oxygen conditions, as immersion time increases, the corrosion product is CuO [39]. Stable Cu$_2$O can be oxidized to CuO through the following series of reactions [40]:

$$Cu_2O + O_2 + H_2O \rightarrow 2CuO + H_2O_2 \quad (8)$$

$$Cu_2O + H_2O_2 \rightarrow 2CuO + H_2O \quad (9)$$

In the XRD analysis, Cu$_2$(OH)$_3$Cl could be detected, and then Cu(OH)$_2$ and Cu$_2$(OH)$_3$Cl are formed by the reaction of Cu$_2$O and H$_2$O [40]:

$$Cu_2O + 3H_2O \rightarrow 2Cu(OH)_2 + 2H^+ + 2e^- \quad (10)$$

$$Cu_2O + 2H_2O + Cl^- \rightarrow Cu_2(OH)_3Cl + H^+ \quad (11)$$

In addition, local acidification of corrosion products could induce CuO to dissolve and generate $Cu_2(OH)_3Cl$ [41]:

$$2CuO + Cl^- + H^+ + H_2O \rightarrow Cu_2(OH)_3Cl \tag{12}$$

The corrosion behavior of C71500 alloy is related to the stability of the formed corrosion product layer. When exposed to an aerated water solution, the copper surface undergoes an electrochemical transformation to form an oxide film. A double $CuCl_{ads}/Cu_2O$ layer will be formed in the initial stage, and then a pure Cu_2O layer will eventually be formed [42]. The protectiveness of the film will be increased with an extension of immersion time, and can also be significantly improved due to the presence of Ni and Fe elements [37].

The above analysis is the first immersion corrosion reaction stage of the C71500 alloy, which is the same as the corrosion process without adding Na_2S. The specific reaction process is shown in Figure 15a.

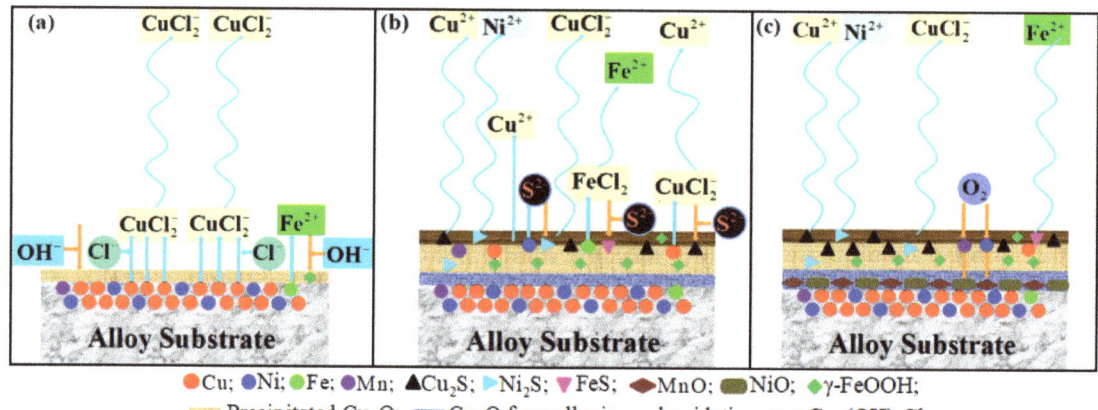

Figure 15. Schematic diagram of the formation process of corrosion product film on C71500 copper-nickel alloy in heavily polluted seawater environment: (a) the initial stage of corrosion, copper dissolves to form $CuCl_2^-$ and Cu_2O precipitation layer is formed; (b) the middle stage of corrosion, unprotected $Cu_2(OH)_3Cl$ is formed, Cu_2O film starts to grow endogenous, the Cu_2O precipitation layer thickens, S^{2+} leads to the formation of Cu_2S, Ni_2S, and Fe_2S, and the accelerated dissolution of Fe and Ni; (c) the later stage of corrosion, diffused oxygen forms MnO and NiO in the Cu_2O precipitation layer, and the corrosion performance of the alloy increases.

After adding sulfides, sulfur can be hydrogenated to produce HS^- [43]:

$$S^{2-} + H_2O \leftrightarrow HS^- + OH^- \tag{13}$$

As can be seen from the anodic polarization curves (Figure 2b), a secondary anode peak could be observed. In all studied electrolytes, the current increases as the potential reaches a definite limit value, which remains unchanged within a large potential range of several hundreds of millivolts. The presence of sulfide ions increases the current limit, which may be caused by the deterioration of the protective film or the anodic oxidation of sulfide ions [44]:

$$HS^-_{aq} \leftrightarrow S_s + H^+ + 2e^- \tag{14}$$

$$3HS^-_{aq} \rightarrow S_3^{2-} + 3H^+ + 4e^- \tag{15}$$

$$2HS^-_{aq} + 2H_2O \rightarrow S_2O_3^{2-} + 8H^+ + 8e^- \tag{16}$$

Since a thin porous and non-protective Cu_2S film is formed on the alloy surface, the corrosion rate and corrosion current will increase, thereby catalyzing the corrosion reaction and preventing the formation of a protective oxide layer [1,31]. The formation of Cu_2S interferes with the protective oxide film and reduces its corrosion resistance [5]. The higher corrosion rate is due to the highly defective Cu_2O layer containing Cu_2S, which could lead to the rapid exchange of ions and electrons; Cu_2S is not as protective as Cu_2O [45].

The presence of sulfide ions increased the corrosion process of C71500 alloys in the studied electrolytes. The increase in corrosion rate is due to the formed film (porous Cu_2S) which catalyzes the corrosion reaction and prevents the formation of a more protective corrosion product layer [46]. The presence of sulfide ions could promote the corrosion of Cu to form adsorbed sulfides. These adsorbed species catalyze the anode dissolution reaction through the following conditions [46]:

$$Cu_s + HS^- \rightarrow Cu(HS^-)_{ads} \tag{17}$$

Then, Cu may undergo anodic dissolution:

$$Cu(HS^-) \rightarrow Cu(HS)_s + e^- \tag{18}$$

Followed by the dissociation and reorganization process:

$$Cu(HS)_s \rightarrow Cu^+{}_{aq} + HS^-{}_{aq} \tag{19}$$

$$2Cu^+ + HS^-{}_{aq} + OH^-{}_{aq} \rightarrow Cu_2S_s + H_2O \tag{20}$$

The overall reaction equation can be written as:

$$4Cu + 2S^{2-} + O_2 + 2H_2O \leftrightarrow 2Cu_2S + 4OH^- \tag{21}$$

The formation of hydroxide is proceeded through the following reaction:

$$Cu^+{}_{aq} + OH^-{}_{aq} \rightarrow Cu(OH)_s \tag{22}$$

In parallel with these reactions, Equation (7) could also form Cu_2O. However, the presence of Cu_2S greatly hinders the protectiveness of the Cu_2O film. Therefore, the corrosion resistance of the film can be greatly reduced by increasing the concentration of sulfide ion.

Under low HS^- and high Cl^- concentration conditions, soluble chloride complexes $CuCl_2^-$ formed by reacting with surface intermediates $Cu(HS^-)_{ads}$ can compete with the film formation reaction [23]. This will lead to a large amount of Cu dissolution. When the HS^- is low on Cu alloy surface, Cu^+ could be transferred to the Cu_2S/solution interface by dissolving $CuCl_2^-$ (Equation (23)), where the concentration of HS^- is much higher, and Cu_2S would be formed by Equation (24) which will dominate at the film/electrolyte interface:

$$Cu(HS^-)_{ads} + 2Cl^- \rightarrow CuCl_2^- + HS^- \tag{23}$$

$$2CuCl_2^- + HS^- \rightarrow Cu_2S + 4Cl^- + H^+ \tag{24}$$

As can be seen from XPS results of samples at the early stages of immersion, the corrosion products in the outer layer are mainly Cu_2S, Cu_2O, and CuS, while the intermediate layer analyzed by XRD is mainly $Cu_2(OH)_3Cl$. From the analysis of the spectra of O 1s and S 2p in XPS, it can be seen that the content of Cu_2O is basically unchanged and the content of Cu_2S gradually increases as the reaction progresses, indicating that Equation (24) is the main form of reaction at this time. A schematic diagram of the whole reaction process is shown in Figure 15b.

4.2. Fe Reaction Mechanism

In the early stages of corrosion product formation, the dissolution (oxidation) kinetics of iron in seawater is much faster than that of copper [1]:

$$Fe \rightarrow Fe^{2+} + 2e^- \tag{25}$$

Only a small part of the dissolved Cu can precipitate as Cu_2O using Equation (6), and most of them dissolve in the solution due to hydrodynamic conditions. In contrast, most of the Fe-oxides could convert to γ-FeOOH through the following reactions [1]:

$$Fe^{2+} + 2OH^- \rightarrow Fe(OH)_2 \tag{26}$$

$$Fe(OH)_2 \rightarrow FeO + H_2O \tag{27}$$

$$FeO + H_2O \rightarrow \gamma - FeOOH + H^+ + e^- \tag{28}$$

The total formation reaction of γ-FeOOH in the copper-nickel alloy corrosion product film in seawater can be described as:

$$Fe + 2H_2O \rightarrow \gamma - FeOOH + 3H^+ + 3e^- \tag{29}$$

Therefore, although Fe is a trace element in the alloy, it can gradually accumulate and enrich in the form of γ-FeOOH in the corrosion product film. The content of γ-FeOOH in the corrosion product film can be used as an indicator to reflect the corrosion rate of copper-nickel alloys in seawater, because iron is oxidized and unevenly precipitated during the corrosion (Cu dissolution) process.

However, the function of γ-FeOOH in the corrosion product film of copper-nickel alloy is still controversial. Campbell et al. [41] reported that γ-FeOOH simultaneously precipitated with Cu_2O at the early stage of corrosion to form a complete protective film, and γ-FeOOH also has an important effect on the cathode polarization reaction due to its high resistivity. Zanoni et al. [47] pointed out that the surface film of C71640 copper-nickel alloy is mainly composed of γ-FeOOH in seawater, the corrosion product does not adhere or protect under such low E_{corr}, resulting in the active corrosion state of the alloy. Vreeland et al. [48] attributed the inhibitory effect of ferrous sulfate ($FeSO_4$) on copper-nickel pipes to the deposition of a brown γ-FeOOH layer.

In the presence of a large amount of Na_2S, NaCl and H_2O first react to form acidic substances [49]:

$$2NaCl + H_2O \rightarrow NaOH + 2HCl \tag{30}$$

Then, Fe reacts with acid:

$$2HCl + Fe \rightarrow FeCl_2 + H_2 \tag{31}$$

$$FeCl_2 + H_2S \rightarrow FeS + 2HCl \tag{32}$$

The total formation reaction of FeS in copper-nickel alloy in polluted seawater can be described as [45,46]:

$$Fe + H_2S \rightarrow FeS + H_2 \tag{33}$$

From the analysis of XPS, it can be seen that Equation (29) is the main reaction in the initial stage of immersion, and the corrosion product is mainly composed of γ-FeOOH. The content of H_2S in the solution gradually increases with the extension of corrosion time, and thus Equation (33) becomes the main corrosion process; significant amounts of FeS could be observed, and the high-priced FeS_2 also could be detected. The dissolution rate of Fe increases with the progress of the reaction, leading to a gradual decrease of Fe content in the EDS analysis. The specific schematic diagram of Fe dissolution process is illustrated in Figure 15b.

4.3. Ni Reaction Mechanism

The enriched Ni elements in the outer and the middle layer of the film are mainly in metallic form, and a small amount of Ni^{2+} could incorporate into $Cu_2(OH)_3Cl/Cu_2O$ and replace Cu^{2+}/Cu^+ [50]. The corrosion rate of Ni is approximately two orders of magnitude lower than that of Cu [51]. The content of Ni incorporated in the Cu_2O lattice accordingly increases when the metallic Ni is enriched at the alloy and film interface. The amorphous or microcrystalline NiO are transited by Cu_2O in certain local areas, where the doped Ni concentration reaches a certain value, and then, the NiO phase grows through appropriate conditions. The production of $Ni(OH)_2$ may be due to the NiO hydrolysis reaction caused by the dissolution of the outer layer and the penetration of seawater at certain locations with the increase of immersion time, this reaction process is explained in the formation process of the NiO inner surface layer in Figure 15c.

The presence of sulfide ions promotes the corrosion of nickel in a way similar to that of copper, and forms adsorbed sulfides on the alloy surface. These adsorbed species catalyze the anode dissolution reaction according to the following conditions [45]:

$$Ni_s + HS^- \rightarrow Ni(HS^-)_{ads} \tag{34}$$

Ni may undergo anodic dissolution:

$$Ni(HS^-) \rightarrow Ni(HS)_s + e^- \tag{35}$$

Followed by the dissociation and reorganization process:

$$Ni(HS)_s \rightarrow Ni^+_{aq} + HS^-_{aq} \tag{36}$$

$$2Ni^+ + HS^-_{aq} + OH^-_{aq} \rightarrow Ni_2S_s + H_2O \tag{37}$$

The whole reaction is:

$$4Ni + 2S^{2-} + O_2 + 2H_2O \leftrightarrow 2Ni_2S + 4OH^- \tag{38}$$

The formation of hydroxide proceeds through the following reactions:

$$Ni^+_{aq} + OH^-_{aq} \rightarrow Ni(OH)_s \tag{39}$$

Although Ni-sulfide is not found in the corrosion products, the Ni content in the corrosion products significantly decreases with immersion time. This process is verified by the results of the EDS analysis.

5. Conclusions

(1) The initial corrosion rate of C71500 copper-nickel alloy in the polluted marine environment is much higher. As the immersion time increases, the corrosion rate greatly decreases, and finally stabilizes in the range of 0.09 mm/a.

(2) The C71500 alloy is in activated corrosion state in the polluted marine environment. Diffusion resistance can be observed at the early stage where a certain protective corrosion product film is formed and tends to stabilize with immersion time, thus reducing the corrosion rate.

(3) With the extension of immersion time, the corrosion product of C71500 alloy begins to clump and adhere to the surface of the sample and becomes dense with sharp edges and corners, and the corrosion products change from spherical to polygonal.

(4) The corrosion product film is mainly Cu_2S in the early stage of immersion, and Cu_2O in the later stage; the content of Mn^{2+} is relatively stable; the oxides are mainly Cu_2O and Ni_2O_3; the sulfide is less present in the early immersion stage and gradually increases in the later stage. It is mainly composed of Cu_2S, CuS, and FeS_2, and a stable $Cu_2(OH)_3Cl$ layer is formed.

(5) The corrosion mechanism of C71500 alloy in sulfide-polluted seawater is similar to that in conventional seawater in the early stages of immersion; however, Cu_2S gradually replaces the reaction process of Cu_2O with immersion time, thus accelerating the corrosion. At the same time, with the dissolution reaction of increased S^{2-}, Ni, and Fe on the surface of the alloy, the corrosion resistance of the alloy further reduces.

Author Contributions: Conceptualization, M.L.; Methodology, X.G.; Data curation, X.G. and M.L.; Writing—original draft, X.G.; Writing—review & editing, M.L.; Supervision, X.G. and M.L. All authors have read and agreed to the published version of the manuscript.

Funding: The financial supports from the Chinese National Natural Science Foundation (No. 51801149), Postdoctoral research Fund subsidy of Shunyi District, Beijing and the International Postdoctoral Exchange Fellowship program 2019 by the Office of China Postdoctoral Council (No. 20190086).

Institutional Review Board Statement: Not applicable.

Informed Consent Statement: Not applicable.

Data Availability Statement: No data were used to support this study.

Conflicts of Interest: The authors declare no conflict of interest.

References

1. Ma, A.L.; Jiang, S.L.; Zheng, Y.G.; Ke, W. Corrosion product film formed on the 90/10 copper-nickel tube in natural seawater: Composition/structure and formation mechanism. *Corros. Sci.* **2015**, *91*, 245–261. [CrossRef]
2. Wei, M.; Yang, B.; Liu, Y.; Wang, X.; Yao, J.; Gao, L. Research progress and prospect on erosion-corrosion of Cu-Ni alloy pipe in seawater. *J. Chin. Soc. Corros. Prot.* **2016**, *36*, 513–521.
3. Liu, M. Effect of uniform corrosion on mechanical behavior of E690 high-strength steel lattice corrugated panel in marine environment: A finite element analysis. *Mater. Res. Express* **2021**, *8*, 066510. [CrossRef]
4. Song, Y.; Wang, H.; En-Hou, F.; Wang, G.; Huang, H. Corrosion Behavior of Cupronickel Alloy in Simulated Seawater in the Presence of Sulfate-Reducing Bacteria. *Acta Metall. Sin. (Engl. Lett.)* **2017**, *30*, 1201–1209. [CrossRef]
5. Sayed, S.; Ashour, E.; Youssef, G. Effect of sulfide ions on the corrosion behaviour of Al–brass and Cu10Ni alloys in salt water. *Mater. Chem. Phys.* **2003**, *78*, 825–834. [CrossRef]
6. Kumar, K.C.; Rao, B.V.A. Mitigation of microbially influenced corrosion of Cu–Ni (90/10) alloy in a seawater environment. *Res. Chem. Intermed.* **2016**, *42*, 5807–5823. [CrossRef]
7. Liu, M.; Li, J.; Zhang, Y.X.; Xue, Y.N. Recent Advances in Corrosion Research of Biomedical NiTi Shape Memory Alloy. *Rare Met. Mater. Eng.* **2021**, *50*, 4165–4173.
8. Chen, J.; Qin, Z.; Martino, T.; Guo, M.; Shoesmith, D. Copper transport and sulphide sequestration during copper corrosion in anaerobic aqueous sulphide solutions. *Corros. Sci.* **2017**, *131*, 245–251. [CrossRef]
9. Rao, B.V.A.; Kumar, K.C. 5-(3-Aminophenyl)tetrazole—A new corrosion inhibitor for Cu–Ni (90/10) alloy in seawater and sulphide containing seawater. *Arab. J. Chem.* **2017**, *10*, S2245–S2259.
10. Radovanović, M.B.; Antonijević, M.M. Protection of copper surface in acidic chloride solution by non-toxic thiadiazole derivative. *J. Adhes. Sci. Technol.* **2016**, *31*, 369–387. [CrossRef]
11. Nady, H.; El-Rabiei, M.; Samy, M. Corrosion behavior and electrochemical properties of carbon steel, commercial pure titanium, copper and copper–aluminum–nickel alloy in 3.5% sodium chloride containing sulfide ions. *Egypt. J. Pet.* **2017**, *26*, 79–94. [CrossRef]
12. Jandaghi, M.R.; Saboori, A.; Khalaj, G.; Shiran, M.K.G. Microstructural Evolutions and its Impact on the Corrosion Behaviour of Explosively Welded Al/Cu Bimetal. *Metals* **2020**, *10*, 634. [CrossRef]
13. Gao, X.; Wu, H.; Liu, M.; Zhou, X.; Zhang, Y.-X. Analysis of the Influence of Sulfur on the Hot Tensile Fracture of C71500 Cu-Ni Alloy. *J. Mater. Eng. Perform.* **2021**, *30*, 312–319. [CrossRef]
14. Gao, X.; Wu, H.-B.; Liu, M.; Zhang, Y.-X.; Zhou, X.-D. Dynamic Recovery and Recrystallization Behaviors of C71500 Copper-Nickel Alloy Under Hot Deformation. *J. Mater. Eng. Perform.* **2020**, *29*, 7678–7692. [CrossRef]
15. Gao, X.; Wu, H.-B.; Liu, M.; Zhang, Y.-X. Effect of annealing time on grain boundary characteristics of C71500 cupronickel alloy tubes with different deformation. *Mater. Charact.* **2020**, *169*, 110603. [CrossRef]
16. Gao, X.; Wu, H.; Liu, M.; Zhang, Y.; Gao, F.; Sun, H. Processing Map of C71500 Copper-nickel Alloy and Application in Production Practice. *J. Wuhan Univ. Technol. Sci. Ed.* **2020**, *35*, 1104–1115. [CrossRef]
17. Gao, X.; Wu, H.; Tang, D.; Li, D.; Liu, M.; Zhou, X. Six Different Mathematical Models to Predict the Hot Deformation Behavior of C71500 Cupronickel Alloy. *Rare Met. Mater. Eng.* **2020**, *49*, 4129–4141.
18. Gao, X.; Wu, H.; Liu, M.; Zhang, Y.; Gao, F. Corrosion Behavior of High Strength C71500 Cu-Ni Alloy Pipe in Simulated High Sulfide Polluted Seawater at Different Temperatures. *Int. J. Electrochem. Sci.* **2021**, *16*, 1–12. [CrossRef]

19. Melchers, R.E. Long-term immersion corrosion of steels in seawaters with elevated nutrient concentration. *Corros. Sci.* **2014**, *81*, 110–116. [CrossRef]
20. Nishimoto, M.; Muto, I.; Sugawara, Y.; Hara, N. Role of Cerium Ions for Improving Pitting Corrosion Resistance of Sulfide Inclusions in Stainless Steels. *ECS Meet. Abstr.* **2017**, 698–699. [CrossRef]
21. Kiahosseini, S.R.; Baygi, S.J.M.; Khalaj, G.; Khoshakhlagh, A.; Samadipour, R. A Study on Structural, Corrosion, and Sensitization Behavior of Ultrafine and Coarse Grain 316 Stainless Steel Processed by Multiaxial Forging and Heat Treatment. *J. Mater. Eng. Perform.* **2017**, *27*, 271–281. [CrossRef]
22. Ezuber, H.M.; al Shater, A. Influence of environmental parameters on the corrosion behavior of 90/10 cupronickel tubes in 3.5% NaCl. *Desalination Water Treat.* **2016**, *57*, 1–10. [CrossRef]
23. Hu, S.; Liu, R.; Liu, L.; Cui, Y.; Oguzie, E.E.; Wang, F. Effect of hydrostatic pressure on the galvanic corrosion of 90/10 Cu-Ni alloy coupled to Ti6Al4V alloy. *Corros. Sci.* **2020**, *163*, 108242. [CrossRef]
24. Liu, M.; Cheng, X.; Li, X.; Yue, P.; Li, J. Corrosion Behavior and Durability of Low-Alloy Steel Rebars in Marine Environment. *J. Mater. Eng. Perform.* **2016**, *25*, 4967–4979. [CrossRef]
25. Liu, M.; Cheng, X.; Li, X.; Zhou, C.; Tan, H. Effect of carbonation on the electrochemical behavior of corrosion resistance low alloy steel rebars in cement extract solution. *Constr. Build. Mater.* **2017**, *130*, 193–201. [CrossRef]
26. Liu, M. Finite element analysis of pitting corrosion on mechanical behavior of E690 steel panel. *Anti-Corrosion Methods Mater.* **2022**, *69*, 351–361. [CrossRef]
27. Xiao, Z.; Li, Z.; Zhu, A.; Zhao, Y.; Chen, J.; Zhu, Y. Surface characterization and corrosion behavior of a novel gold-imitation copper alloy with high tarnish resistance in salt spray environment. *Corros. Sci.* **2013**, *76*, 42–51. [CrossRef]
28. Nam, N.D. Role of Zinc in Enhancing the Corrosion Resistance of Mg-5Ca Alloys. *J. Electrochem. Soc.* **2015**, *163*, C76–C84. [CrossRef]
29. Liu, T.; Chen, H.; Zhang, W.; Lou, W. Accelerated Corrosion Behavior of B10 Cu-Ni Alloy in Seawater. *J. Mater. Eng.* **2017**, *45*, 31–37.
30. Patil, A.P.; Tupkary, R.H. Corrosion resistance of new copper alloy containing 29Zn, 10Ni and up to 5Mn vis-a-vis Cu-10Ni in sulphide polluted synthetic seawater. *Trans. Indian Inst. Met.* **2009**, *62*, 71–79. [CrossRef]
31. Deroubaix, G.; Marcus, P. X-ray photoelectron spectroscopy analysis of copper and zinc oxides and sulphides. *Surf. Interface Anal.* **2010**, *18*, 39–46. [CrossRef]
32. Galtayries, A.; Bonnelle, J.-P. XPS and ISS studies on the interaction of H2S with polycrystalline Cu, Cu_2O and CuO surfaces. *Surf. Interface Anal.* **2010**, *23*, 171–179. [CrossRef]
33. Song, Q.N.; Xu, N.; Bao, Y.F.; Jiang, Y.F.; Qiao, Y.X. Corrosion Behavior of Cu40Zn in Sulfide-Polluted 3.5% NaCl Solution. *J. Mater. Eng. Perform.* **2017**, *26*, 1–9. [CrossRef]
34. Patil, A.P.; Tupkary, R.H. Development of single-phased copper alloy for seawater applications: As a cost-effective substitute for Cu-10Ni alloy. *Trans. Indian Inst. Met.* **2008**, *61*, 13–25. [CrossRef]
35. Liu, M.; Li, J. In-Situ Raman Characterization of Initial Corrosion Behavior of Copper in Neutral 3.5% (wt.) NaCl Solution. *Materials* **2019**, *12*, 2164. [CrossRef]
36. Ferreira, J.P.; Rodrigues, J.A.; Da Fonseca, I.T.E. Copper corrosion in buffered and non-buffered synthetic seawater: A comparative study. *J. Solid State Electrochem.* **2004**, *8*, 260–271. [CrossRef]
37. Kear, G.; Barker, B.D.; Stokes, K.; Walsh, F.C. Electrochemical Corrosion Behaviour of 90–10 Cu–Ni Alloy in Chloride-Based Electrolytes. *J. Appl. Electrochem.* **2004**, *34*, 659–669. [CrossRef]
38. Hong, S.; Chen, W.; Zhang, Y.; Qun, H.; Ming, L.; Nian, L.; Li, B. Investigation of the inhibition effect of trithiocyanuric acid on corrosion of copper in 3.0 wt.% NaCl. *Corros. Sci.* **2013**, *66*, 308–314. [CrossRef]
39. Bonora, P.L.; Rossi, S.; Benedetti, L.; Draghetti, M. Improved Sacrificial Anode for the Protection of Off-Shore Structures. *Dev. Mar. Corros.* **1998**, 155–162. [CrossRef]
40. de Sanchez, S.R.; Berlouis, L.E.; Schiffrin, D.J. Difference reflectance spectroscopy of anodic films on copper and copper base alloys. *J. Electroanal. Chem. Interfacial Electrochem.* **1991**, *307*, 73–86. [CrossRef]
41. Campbell, S.A.; Radford, G.J.W.; Tuck, C.D.S.; Barker, B.D. Corrosion and Galvanic Compatibility Studies of a High-Strength Copper-Nickel Alloy. *Corrosion* **2002**, *58*, 57–71. [CrossRef]
42. Bech-Nielsen, G.; Jaskula, M.; Chorkendorff, I.; Larsen, J. The initial behaviour of freshly etched copper in moderately acid, aerated chloride solutions. *Electrochim. Acta* **2003**, *47*, 4279–4290. [CrossRef]
43. Al-Hajji, J.; Reda, M. The corrosion of copper-nickel alloys in sulfide-polluted seawater: The effect of sulfide concentration. *Corros. Sci.* **1993**, *34*, 163–177. [CrossRef]
44. Alkharafi, F.M.; Nazeer, A.A.; Abdullah, R.M.; Galal, A. Effect of Sulfide-Containing Solutions on the Corrosion of Cu-Ni Alloys; Electrochemical and Surface Studies. In Proceedings of the ECS & SMEQ Joint International Meeting, Cancun, Mexico, 5–9 October 2014; Volume 64, p. 121.
45. Al-Kharafi, F.M.; Abdel-Nazeer, A.; Abdullah, R.M.; Galal, A. Effect of Sulfide-Containing Solutions on the Corrosion of Cu-Ni Alloys; Electrochemical and Surface Studies. *ECS Trans.* **2015**, *64*, 121–134. [CrossRef]
46. Vazquez, M.; De Sanchez, S.R. Influence of sulphide ions on the cathodic behaviour of copper in 0.1 M borax solution. *J. Appl. Electrochem.* **1998**, *28*, 1383–1388. [CrossRef]
47. Zanoni, R.; Gusmano, G.; Montesperelli, G.; Traversa, E. X-ray Photoelectron Spectroscopy Investigation of Corrosion Behavior of ASTM C71640 Copper-Nickel Alloy in Seawater. *Corrosion* **1992**, *48*, 404–410. [CrossRef]

48. Burleigh, T.D.; Waldeck, D.H. Effect of Alloying on the Resistance of Cu-10% Ni Alloys to Seawater Impingement. *Corrosion* **1999**, *55*, 800–804. [CrossRef]
49. Wang, X.H.; Wei, Y.; Shao, C.Y.; Shi, Y.J.; Xue, W.; Zhu, J.F. Study on Corrosion Behavior of the 304 Stainless Steel in the Heavy Oil with High Salt, High Sulfur and High Acid Value. *Appl. Mech. Mater.* **2013**, *252*, 271–275. [CrossRef]
50. Frost, R.L. Raman spectroscopy of selected copper minerals of significance in corrosion. *Spectrochim. Acta Part A Mol. Biomol. Spectrosc.* **2003**, *59*, 1195–1204. [CrossRef]
51. Efird, K.D. Potential-pH Diagrams for 90-10 and 70-30 Cu-Ni in Sea Water. *Corrosion* **2013**, *31*, 77–83. [CrossRef]

Article

Formation of Nanoscale Al₂O₃ Protective Layer by Preheating Treatment for Improving Corrosion Resistance of Dilute Fe-Al Alloys

Chenglong Li [1], Katharina Freiberg [2], Yuntong Tang [1], Stephanie Lippmann [2] and Yongfu Zhu [1,*]

1. Key Laboratory of Automobile Materials, Ministry of Education, School of Materials Science and Engineering, Jilin University, Changchun 130022, China
2. Otto Schott Institute of Materials Research, Friedrich Schiller University, 07743 Jena, Germany
* Correspondence: yfzhu@jlu.edu.cn

Abstract: In this work, an attempt was made to improve the corrosion resistance of dilute Fe-Al alloys (1.0 mass% Al) by preheating treatment at 1073 K in H_2 atmosphere. In comparison with pure Fe and unpreheated Fe-Al alloys, the resistance to oxidation at 673 K in pure O_2 and to electrochemical corrosion in 5 wt.% NaCl solution is significantly improved for preheated Fe-Al alloys. This improvement is attributed to the formation of a 20 nm thin, but dense Al_2O_3 protective layer on the surface of preheated Fe-Al alloys.

Keywords: corrosion resistance; oxidation; heat treatment; Fe-Al alloy; ferroalloy

1. Introduction

As one of the most widely used structural materials, ferroalloys are widely utilized in vehicle engineering, the petrochemical industry, machinery manufacturing, aerospace and marine engineering [1]. In modern society, high-temperature oxidation in air and electrochemical corrosion in wet air or chemical solution are the most common ways to cause ferroalloy failure, leading to economic losses and potential security risks [2]. Consequently, the high-temperature oxidation resistance [3,4] and corrosion resistance [5–8] of ferroalloys must be considered.

For ferroalloys, the commonly used anticorrosion methods include alloying, surface coating and electroplating. Although all mentioned methods improve the oxidation resistance and the electrochemical corrosion resistance of ferroalloys, they require additional process steps that are costly, and all methods are known to cause pollution. Numerous reports on tons of pollutants caused by alloying [9,10], electroplating [11–13] and protective painting [14,15] that cause serious damage to the environment are available. Due to the lack of a better alternative (so far), most commonly, the resistance to the oxidation and electrochemical corrosion of Fe is increased by adding high ratios of Cr and Ni [5,6,16–18] which are expensive, and which also have a negative impact on the plasticity of ferroalloys [19].

In fact, Al as an alloying element has been reported to have a positive effect on the oxidation resistance and corrosion resistance of the metallic alloys including Fe [20–27]. In comparison with Cr and Ni, Al is cheap and widely available, which makes it an interesting candidate for industrial use. However, the amount of Al that can be added to Fe is limited. Contents higher than 20 wt.% seriously reduce the plasticity of ferroalloys [19]. In recent years, to address this issue, a low alloying technique combined with a preheating treatment in reducing the atmosphere has been developed for Cu-Al alloys with the additions of <3 mass% Al [20,28,29]. The increase of the oxidation resistance was attributed to the formation of a surface Al_2O_3 protective layer that prevents the outward diffusion of Cu as well as the inward diffusion of O. At the same time, the matrix is covered by an Al_2O_3

protective layer, and the matrix does not directly contact air or an aqueous solution, which greatly improves the corrosion resistance of the Cu-based alloy.

In this work, the low alloying technique combined with the preheating treatment is applied to Fe to investigate its effect on the corrosion resistance of ferroalloys. It is expected that an Al_2O_3 protective layer can be formed on the surface of preheated Fe-Al alloys to improve its corrosion resistance by preventing the outward diffusion of Fe and inward diffusion of O. The low Fe-Al alloy is prepared by adding 1.0 mass% Al into Fe, and the alloy is preheated in a high purity H_2 atmosphere and subsequently oxidized in a pure oxygen atmosphere. The resistance of the preheated Fe-Al alloy to oxidation at 673 K in pure O_2 and to electrochemical corrosion in 5 wt.% NaCl solution will be measured. Correspondingly, the surface morphology, structure and the composition of the protective layer formed on the preheated Fe-Al alloy will also be characterized.

2. Materials and Experimental Procedures

Fe-Al alloys with 1.0 mass% Al are made from pure Fe (99.999 mass%) and Al (99.9999 mass%) by repeated melting in a vacuum electric arc furnace (DHL-300, SKY Technology Development CO., Shenyang, China) with a nonconsumable electrode under a protective atmosphere (99.999% Ar). The composition is chosen below the solubility limit of Al in Fe [30]. That is, there is no intermetallic compound in the sample to ensure its processability. The sample ingots were cut and cold rolled into thin plates with a thickness of 0.5 mm, and then subjected to mechanical and electrolytic polishing, where the electrolytic polishing current is 0.5 A and the reaction time is 1 min. After that, the samples were preheated in a high purity H_2 atmosphere (99.9999%) at 101,325 Pa and 1073 K for 1440 min, referred to as P-FeAl. The samples used for the oxidation experiment were stamped into a round piece with a diameter of 4 mm, and the sample used for the electrochemical corrosion experiment was cut into a rectangular piece with a length of 10 mm and a width of 20 mm. During this process, the surface of the sample is covered with high-quality sulfuric acid paper to prevent external pollution and damage to the sample.

The mass gain of the preheated sample and those control samples with pure Fe and unpreheated Fe-Al alloys (uP-FeAl) during oxidation was measured using a thermogravimetric method (METTLER 1100LF, Mettler Toledo, Switzerland). When a sample is heated to the required test temperature in the reaction gas, the increase in mass is recorded. A scanning electron microscope (SEM, JEOL, Tokyo, Japan) and Energy Dispersive X-ray Spectroscopy (EDS, JEOL, Tokyo, Japan) were used to analyze the surface. The detection of surface elements was carried out by X-ray photoelectron spectroscopy (XPS, ESCALAB 250Xi, Thermo Fisher Scientific Inc., Waltham, MA USA). Transmission electron microscopy (TEM, Jeol-NEOARM200F, Jeol, Tokyo, Japan) analysis operating at 200 kV was performed to observe the microstructure of the Al_2O_3 layer on the preheated Fe-Al alloy.

The corrosion resistance was assessed with an electrochemical impedance spectroscopy (EIS) and potentiodynamic polarization measurement operated by a CHI660E electrochemical workstation (SUMAT, Beijing, China). The sample was installed in an electrochemical cell and exposed to solutions of 5 wt.% NaCl for 30 min. EIS spectral were then recorded at the open circuit potential (OCP), and measured with a three-electrode configuration, consisting of the sample as the working electrode (surface area 1 cm^2), a platinum gauze counter electrode and a saturated calomel electrode (SCE) as the reference electrode in a Faraday cage. The EIS experiments were conducted in the frequency range of 10^5 Hz to 10^{-2} Hz at OCP by applying 10 mV sinusoidal amplitude. Potentiodynamic polarizations were performed after 30 min exposure to solutions of 5 wt.% NaCl. The samples were polarized using a scanning rate of 10 mV/s and a scanning range of -0.3 V (vs. OCP) ~ 0 V. The solution was open to air and not stirred during the measurement. All experiments were performed at room temperature.

3. Results and Discussion

To see the effects of the preheating treatment on the oxidation resistance of low Fe-Al alloys, the weight gain curves of pure Fe, uP-FeAl and P-FeAl recorded during oxidation at 673 K in 0.1 MPa O_2 are given in Figure 1. The weight gain of pure Fe increases the fastest as the oxidation proceeds. In comparison, the weight gain of uP-FeAl is significantly lowered, but still visibly increases, meaning that uP-FeAl is also oxidized but slower than pure Fe. Interestingly, the weight gain of P-FeAl is the lowest and increases the slowest, showing that the oxidation resistance of Fe-Al alloys is significantly improved after the preheating treatment.

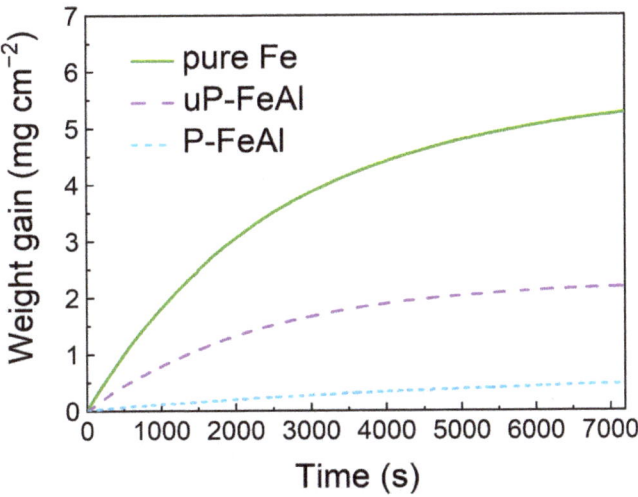

Figure 1. Weight gain curves of pure Fe, uP-FeAl and P-FeAl oxidized at 673 K in 0.1 MPa O_2 for 120 min showing the superior oxidation resistance of preheated Fe-Al alloys.

To further evaluate the oxidation resistance of low Fe-Al alloys after preheating, the SEM images of the surface morphology of pure Fe, uP-FeAl and P-FeAl oxidized at 673 K for 2880 min in 0.1 MPa O_2 are shown in Figure 2. The micrographs of oxidized pure Fe in Figure 2a,b and uP-FeAl in Figure 2c,d show significantly rough surfaces in comparison with that of P-FeAl. Oxide whiskers grown during the oxidation fully cover the alloy surface of pure Fe and uP-FeAl. The whiskers are composed of Fe_2O_3 [17]. The surface of P-FeAl in Figure 2e,f is strikingly smooth without oxide grains or whiskers. The oxide morphology images support the previous results of the superior antioxidant ability of P-FeAl from the weight gain curve in Figure 1.

To evaluate the corrosion resistance of one metal or alloy, one efficient way is to provide its electrochemical corrosion diagram, where a high polarization impedance and low corrosion current mean excellent corrosion resistance [31–37]. In this work, the electrochemical corrosion diagram of pure Fe, uP-FeAl and P-FeAl measured in 5 wt.% NaCl solution is given in Figure 3, with the electrochemical impedance spectroscopy (EIS) diagram in (a), the potentiodynamic polarization curves in (b), and the bode and phase angle plots in (c). In Figure 3a, the EIS curve of P-FeAl has the largest diameter of the semicircle, indicating that it has the highest resistance among the three. With reference to previous studies [38], a two time-constant equivalent circuit model inserted in Figure 3a is constructed to illustrate the electrochemical impedance of a sample that is exposed to the solution of 5 wt.% NaCl. In this model, R_s represents the solution resistance between the reference electrode and working electrode, the first time-constant represents the resistance of the Al_2O_3 protective layer (R_c) and its capacitance (Q_c), while the second one describes the electrochemical processes (corrosion) at the substrate in terms of the charge transfer resistance (R_{ct}) and the double layer capacitance (Q_{dl}). Using such an equivalent circuit model, the fitting result gives that the charge transfer resistance of P-FeAl (R_{ct} = 4948 Ω) is significantly higher than those of pure Fe (R_{ct} = 788 Ω) and uP-FeAl (R_{ct} = 1906 Ω). In Figure 3b, the potentiodynamic polarization curve can be adopted to obtain the corrosion current by the extrapolation method. Due to this, one sees that the corrosion potential (E_{corr}) of P-FeAl (E_{corr} = −0.87 V) is lower than that of uP-FeAl (E_{corr} = −0.81 V) and pure Fe (E_{corr} = −0.86 V), and the corrosion current of P-FeAl ($I_{P\text{-}FeAl}$ = 9.8 µA cm^{-2}) is significantly lower than those of pure Fe ($I_{pure\ Fe}$ = 35.9 µA cm^{-2}) and uP-FeAl ($I_{uP\text{-}FeAl}$ = 79.6 µA cm^{-2}). Figure 3c shows the Bode and phase angle plots of P-FeAl, pure Fe and uP-FeAl. The phase angle plots show two time-constants; one is for the Al_2O_3 protective layer (high frequency range 10^3–10^5 Hz), the other is for the electrochemical activity at the matrix (middle frequency range 10^{-1}–10^2 Hz). As for the Bode plot, it can provide the polarization resistance (R_p) from the difference in the real impedance at a lower and higher frequency [34]. In light of this, P-FeAl has the highest impedance modulus value after 30 min exposure to the 5 wt.% NaCl solution at a low frequency region. Interestingly, the R_p of P-FeAl (R_p = 6656 Ω) is considerably higher than those of pure Fe (R_p = 1157 Ω) and uP-FeAl (R_p = 2786 Ω). All these results show that P-FeAl possesses a high corrosion resistance.

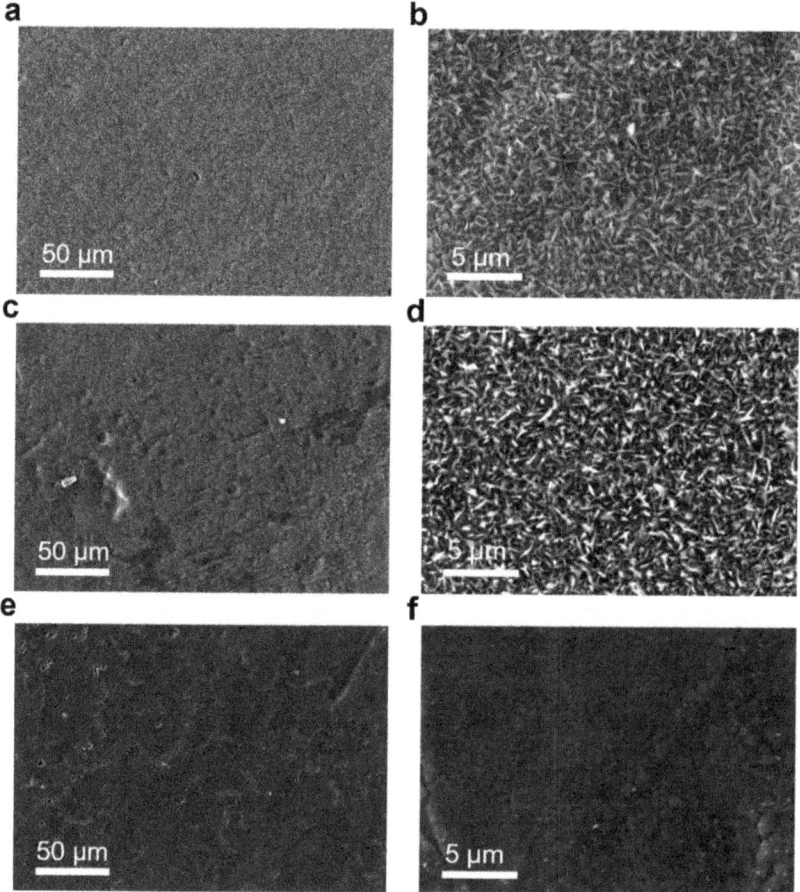

Figure 2. SEM images of surface (oxide) morphologies of pure Fe, uP-FeAl and P-FeAl after oxidation in hydrogen with residual oxygen content of 0.1 MPa at 673 K for 2880 min. (**a**,**b**) pure Fe and (**c**,**d**) uP-FeAl with oxide flakes and whiskers; (**e**,**f**) P-FeAl covered by a uniform and dense oxide layer.

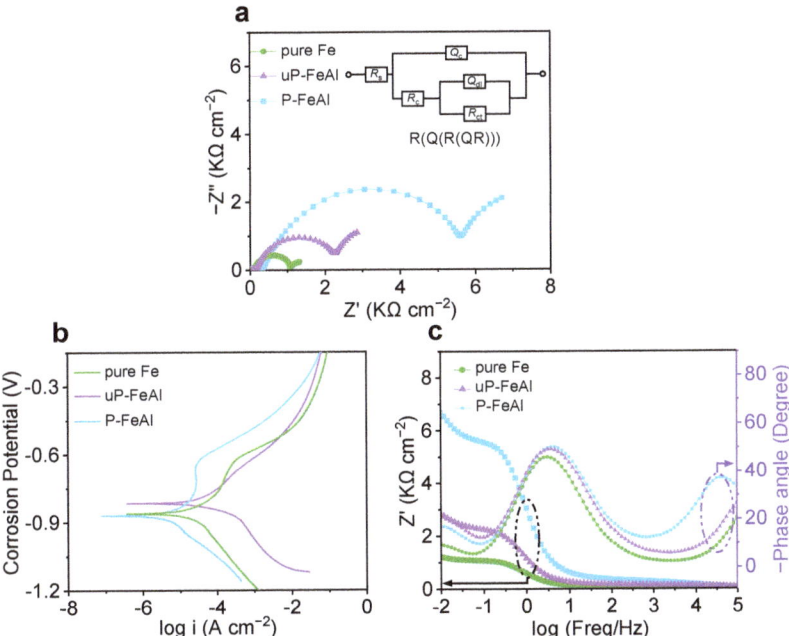

Figure 3. Electrochemical corrosion curves of pure Fe, uP-FeAl and P-FeAl tested in 5 wt.% NaCl solution, with (**a**) EIS diagram at E_{corr}, (**b**) Potentiodynamic polarization curves and (**c**) Bode and phase angle plots. Inset of (**a**) shows equivalent circuit model for EIS in (**a**) with R_s: solution resistance, R_c: film resistance, Q_c: film capacitance, Q_{dl}: double layer capacitance, R_{ct}: charge transfer resistance.

In Figures 1–3, the excellent corrosion resistance of P-FeAl alloys should be attributed to the preheating treatment prior to the oxidation or corrosion experiments, and a surface protective layer that be formed might be responsible for the improvement. This will be further investigated in the following.

Figure 4 shows the EDS concentration depth profiles of Fe, Al and O measured on P-FeAl. In the vicinity of the surface, the signals of Al and O are strong, but that of Fe is negligibly weak. As the distance increases, the signals of Al and O decline deeply to a low level, especially for Al, while that of Fe increases quickly up to the maximum level. The depth profiles suggest the accumulation of Al and O in the vicinity of the surface, showing that a thin protective Al_2O_3 layer about 20 nm thick is formed on the surface of FeAl during the preheating treatment. The aluminum oxide layer formation is originated from the outward segregation of Al to the surface vicinity. Note that the composition of the aluminum oxide layer is not exactly Al_2O_3 but higher in Al content. In addition to this, the profiles show further that oxygen is solved in the Fe-Al alloy, but the level is not in accordance with any stochiometric iron oxide.

To characterize the composition of the surface protective layer, Figure 5 gives the XPS pattern of the P-FeAl surface. The O 1s spectrum displayed three characteristic peaks of metal–oxygen bonds (529.7 eV for O1), defect sites with a low oxygen coordination (531.2 eV for O2), and hydroxyl groups (532.8 eV for O3), consistent with that reported in the literature [39]. A further characteristic peak appears near 74.3 eV, which comes from Al 2p, also consistent with those reported results [39,40]. These measurements suggest that an Al_2O_3 is formed on the surface of P-FeAl, which should be responsible for the improvement in the oxidation resistance and corrosion resistance of P-FeAl.

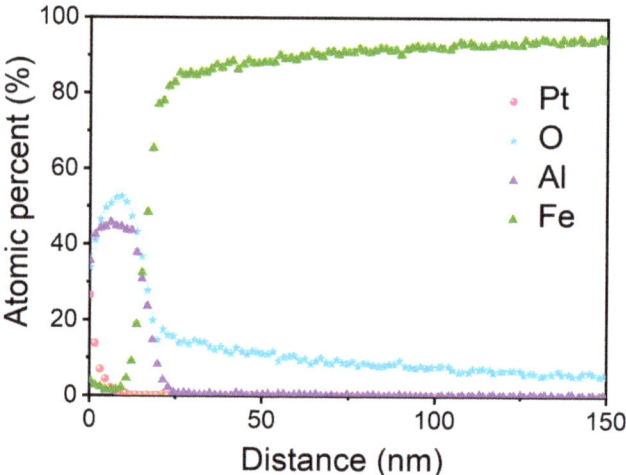

Figure 4. EDS concentration depth profiles of Fe, Al and O measured on P-FeAl.

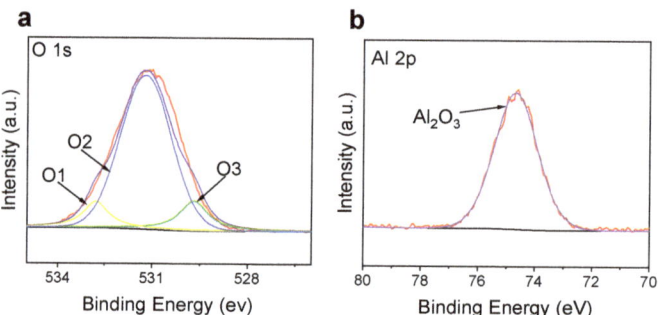

Figure 5. The XPS pattern of the P-FeAl surface confirm the formation of Al_2O_3 oxide layer with the characteristic peak positions for (**a**) O 1s and (**b**) Al 2p.

To observe the microstructure of the surface protective layer, a TEM cross section of P-FeAl with EDS mapping of Al, O and Fe are given in Figure 6. In the brightfield image of the surface region shown in Figure 6a, a dense aluminum oxide layer formed on the surface is visible. The mean thickness is 20 ± 5 nm. Figure 6b shows an enlarged view of (a), where the Al_2O_3 protective layer can be clearly observed. The Fast Fourier Transform (FFT) inset of the crystalline aluminum oxide layer in Figure 6b was indexed, using the singlecrystal® 4 software, as Al_2O_3 orthorhombic (SG 33) structure in $[0\bar{3}1]$ orientation. Undoubtedly, this Al_2O_3 surface layer is the key to the improvement of the corrosion resistance of P-FeAl. Figure 6d–g exhibit the EDS mapping of Fe, Al, O and Pt over the cross section of P-FeAl as (c). The accumulation of Al and O can be found in the vicinity of the surface of P-FeAl, corresponding to Figure 6. The elemental distributions confirm the formation of a thin protective Al_2O_3 layer on the surface of P-FeAl during the preheating treatment. It should be noted here that, in Figure 6d–g, the signals of Al and O can also be observed at the interior place of the Fe-Al base a little further away from the surface, suggesting the internal oxidation of Al in Fe during the preheating process. The internal oxidation is attributed to the solution of O in Fe as shown in Figure 4. Unfortunately, this will lead to a decrease in the actual concentration of Al in Fe, limiting further growth of the surface Al_2O_3 layer.

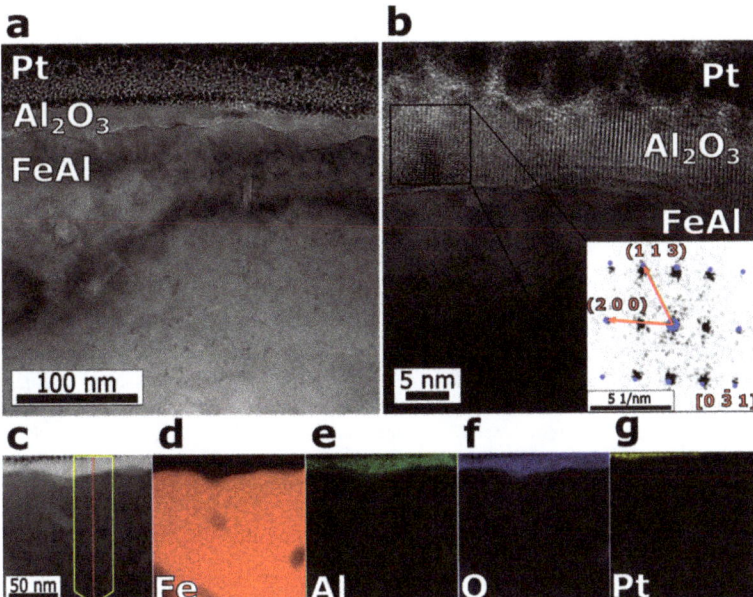

Figure 6. (**a**) BF TEM image of P-FeAl cross section. (**b**) BF TEM image of P-FeAl cross section, where the inset gives FFT image Al_2O_3 phase with overlay fit of single crystal®. Fitted orientation $[0\bar{3}1]$ of orthorhombic Al_2O_3 (blue). (**c**) BF TEM image of P-FeAl cross section with corresponding EDS image of (**d**) Fe, (**e**) Al, (**f**) O and (**g**) Pt.

4. Conclusions

In summary, P-FeAl is prepared with dilute Fe-Al alloys (1.0 mass% Al) as a precursor through a preheating treatment in an H_2 atmosphere. After the preheating treatment, P-FeAl shows excellent oxidation resistance. Compared to pure Fe and uP-FeAl, the mass gain of P-FeAl during the oxidation in 0.1 MPa O_2 at 673 K for 2880 min is much lower, where almost no oxides can be observed on its surface. In the electrochemical corrosion test, P-FeAl also showed excellent electrochemical corrosion resistance with R_p = 6656 Ω and $I_{P\text{-FeAl}}$ = 9.8 µA cm^{-2}, which is significantly better than pure Fe (R_p = 1157 Ω and $I_{pure\ Fe}$ = 35.9 µA cm^{-2}) and uP-FeAl (R_p = 2786 Ω and $I_{uP\text{-FeAl}}$ = 79.6 µA cm^{-2}). These improvements are attributed to an Al_2O_3 protective layer about 20 nm thick, which is self-formed on the surface due to the reaction of Al outward diffusion from the inner part of the FeAl base with an O_2 remnant in the annealing atmosphere during the preheating treatment. Such a thin protective Al_2O_3 layer can prevent the diffusion of atoms or ions through it during the corrosion process.

Author Contributions: Conceptualization, Y.Z. and C.L.; methodology, Y.Z.; software, C.L. and K.F.; validation, Y.Z., K.F. and C.L.; formal analysis, Y.Z., S.L. and C.L.; investigation, C.L.; resources, Y.Z.; data curation, C.L., K.F. and Y.T.; writing—original draft preparation, C.L. and K.F.; writing—review and editing, Y.Z., S.L. and C.L.; visualization, Y.Z. and C.L.; supervision, Y.Z. and S.L.; project administration, Y.Z. and S.L.; funding acquisition, Y.Z. and S.L. All authors have read and agreed to the published version of the manuscript.

Funding: This research was funded by the National Natural Science Foundation of China (Nos. 51871107, 52130101 and 51631004), the Natural Science Foundation of Jilin Province (No. 20200201019JC), the Top-notch Young Talent Program of China (W02070051), the Chang Jiang Scholar Program of China (Q2016064), Key scientific and technological research and development projects of Jilin Provincial Department of Science and Technology (20220201113GX), the Fundamental Research Funds for the Central Universities, the Program for Innovative Research Team (in Science and Technology) in University of Jilin Province, and the Program for JLU Science and Technology Innovative Research Team (2017TD-09). S. Lippmann acknowledges financial support of the German Research Foundation (Deutsche Forschungsgemeinschaft DFG) under Grant Nos. INST 275/3911.

Institutional Review Board Statement: Not applicable.

Informed Consent Statement: Not applicable.

Data Availability Statement: Not applicable.

Acknowledgments: S. Lippmann thanks C. Ronning, FSU Jena for granting access to the FIB facility. The authors thank S.Y. Liu, X. Li and Z.Y. Yang for experimental work.

Conflicts of Interest: The authors declare no conflict of interest.

References

1. Qiao, Y.X.; Wang, X.Y.; Yang, L.L.; Wang, X.J.; Chen, J.; Wang, Z.B.; Zhou, H.L.; Zou, J.S.; Wang, F.H. Effect of aging treatment on microstructure and corrosion behavior of a Fe-18Cr-15Mn-0.66N stainless steel. *J. Mater. Sci. Technol.* **2022**, *107*, 197–206. [CrossRef]
2. Liu, M.; Li, J.; Zhang, Y.X.; Xue, Y.N. Recent Advances in Corrosion Research of Biomedical NiTi Shape Memory Alloy. *Rare Metal Mat. Eng.* **2021**, *50*, 4165–4173.
3. Raman, R.K.S. Mechanical Alloying of Elemental Powders into Nanocrystalline (NC) Fe-Cr Alloys: Remarkable Oxidation Resistance of NC Alloys. *Metals* **2021**, *11*, 695. [CrossRef]
4. Backman, D.G.; Williams, J.C. Advanced materials for aircraft engine applications. *Science* **1992**, *255*, 1082–1087. [CrossRef] [PubMed]
5. Aye, K.K.; Nguyen, T.D.; Zhang, J.; Young, D.J. Effect of silicon on corrosion of Fe-20Cr and Fe-20Cr-20Ni alloys in wet CO_2 with and without HCl at 650 °C. *Corros. Sci.* **2021**, *179*, 109096. [CrossRef]
6. Liu, M. Finite element analysis of pitting corrosion on mechanical behavior of E690 steel panel. *Anti-Corros. Method M.* **2022**, *69*, 351–361. [CrossRef]
7. Gong, Y.; Young, D.J.; Kontis, P.; Chiu, Y.L.; Larsson, H.; Shin, A.; Pearson, J.M.; Moody, M.P.; Reed, R.C. On the breakaway oxidation of Fe9Cr1Mo steel in high pressure CO_2. *Acta Mater.* **2017**, *130*, 361–374. [CrossRef]
8. Liu, M. Effect of uniform corrosion on mechanical behavior of E690 high-strength steel lattice corrugated panel in marine environment: A finite element analysis. *Mater. Res. Express* **2021**, *8*, 066510. [CrossRef]
9. Wang, F.; Shu, Y. Influence of Cr content on the corrosion of Fe-Cr alloys: The synergistic effect of NaCl and water vapor. *Oxid. Met.* **2003**, *59*, 201–214. [CrossRef]
10. Jung, K.; Ahn, S.; Kim, Y.; Oh, S.; Ryu, W.H.; Kwon, H. Alloy design employing high Cr concentrations for Mo-free stainless steels with enhanced corrosion resistance. *Corros Sci.* **2018**, *140*, 61–72. [CrossRef]
11. Huang, C.A.; Chang, J.H.; Chen, C.Y.; Liao, K.Y.; Mayer, J. Microstructure and electrochemical corrosion behavior of Cr-Ni-Fe alloy deposits electroplated in the presence of trivalent Cr ions. *Thin Solid Films* **2013**, *544*, 69–73. [CrossRef]
12. Choi, Y.I.; Shin, E.S.; Kuroda, K.; Okido, M.; Park, C.J. Improved surface morphology and corrosion resistance for galvannealed coatings by pre-electroplating iron. *Corros. Sci.* **2012**, *58*, 152–158. [CrossRef]
13. Hu, W.B.; Cai, H.L.; Yang, M.H.; Tong, X.L.; Zhou, C.M.; Chen, W. Fe-C-coated fibre Bragg grating sensor for steel corrosion monitoring. *Corros Sci.* **2011**, *53*, 1933–1938. [CrossRef]
14. Kim, J.H.; Lee, M.H. A Study on Cavitation Erosion and Corrosion Behavior of Al-, Zn-, Cu-, and Fe-Based Coatings Prepared by Arc Spraying. *J. Therm. Spray Technol.* **2010**, *19*, 1224–1230. [CrossRef]
15. Jeong, Y.S.; Kainuma, S.; Ahn, J.H. Structural response of orthotropic bridge deck depending on the corroded deck surface. *Constr. Build Mater.* **2013**, *43*, 87–97. [CrossRef]
16. Yu, C.; Nguyen, T.D.; Zhang, J.; Young, D.J. Sulfur Effect on Corrosion Behavior of Fe-20Cr-(Mn, Si) and Fe-20Ni-20Cr-(Mn, Si) in CO_2-H_2O at 650 °C. *J. Electrochem. Soc.* **2015**, *163*, C106–C115. [CrossRef]
17. Rao, V.S. The influence of temperature on the oxidation behaviour of Fe_3Al-$Fe_3AlC_{0.69}$ and $FeAl$-$Fe_3AlC_{0.69}$ intermetallics. *Intermetallics* **2003**, *11*, 713–719. [CrossRef]
18. Wei, W.; Geng, S.J.; Chen, G.; Wang, F.H. Growth mechanism of surface scales on Ni-Fe-Cr alloys at 960 degrees C in air. *Corros. Sci.* **2020**, *173*, 108737. [CrossRef]
19. Zhang, C.-H.; Huang, S.; Shen, J.; Chen, N.-X. Structural and mechanical properties of Fe–Al compounds: An atomistic study by EAM simulation. *Intermetallics* **2014**, *52*, 86–91. [CrossRef]
20. Hong, S.H.; Zhu, Y.F.; Mimura, K.; Isshiki, M. Role of Al_2O_3 layer in oxidation resistance of Cu-Al dilute alloys pre-annealed in H_2 atmospheres. *Corros. Sci.* **2006**, *48*, 3692–3702. [CrossRef]
21. Novak, P.; Nova, K. Oxidation Behavior of Fe-Al, Fe-Si and Fe-Al-Si Intermetallics. *Materials* **2019**, *12*, 1748. [CrossRef] [PubMed]
22. Jang, P.; Shin, S.; Jung, C.S.; Kim, K.H.; Seomoon, K. Fabrication of Fe-Al nanoparticles by selective oxidation of Fe-Al thin films. *Nanoscale Res. Lett.* **2013**, *8*, 1–6. [CrossRef] [PubMed]

23. Li, H.; Zhang, J.Q.; Young, D.J. Oxidation of Fe-Si, Fe-Al and Fe-Si-Al alloys in CO_2-H_2O gas at 800 degrees C. *Corros. Sci.* **2012**, *54*, 127–138. [CrossRef]
24. Xu, Y.F.; Jeurgens, L.P.H.; Bo, H.; Lin, L.C.; Zhu, S.L.; Huang, Y.; Liu, Y.C.; Qiao, J.W.; Wang, Z.M. On the competition between synchronous oxidation and preferential oxidation in Cu-Zr-Al metallic glasses. *Corros. Sci.* **2020**, *177*, 108996. [CrossRef]
25. Pang, X.J.; Li, S.S.; Qin, L.; Pei, Y.L.; Gong, S.K. Effect of trace Ce on high-temperature oxidation behavior of an Al-Si-coated Ni-based single crystal superalloz. *J. IronSteel Res. Int.* **2019**, *26*, 78–83.
26. Wang, X.H.; Li, F.Z.; Chen, J.X.; Zhou, Y.C. Insights into high temperature oxidation of Al_2O_3-forming Ti_3AlC_2. *Corros Sci.* **2012**, *58*, 95–103. [CrossRef]
27. Lu, S.D.; Li, X.X.; Liang, X.Y.; Shao, W.T.; Yang, W.; Chen, J. Effect of Al content on the oxidation behavior of refractory high-entropy alloy $AlMo_{0.5}NbTa_{0.5}TiZr$ at elevated temperatures. *Int. J. Refract Met. H.* **2022**, *105*, 105812. [CrossRef]
28. Zhu, Y.F.; Mimura, K.; Isshiki, M. Oxidation mechanism of copper at 623–1073 K. *Mater. Trans.* **2002**, *43*, 2173–2176. [CrossRef]
29. Ogbuji, L.U. The oxidation behavior of an ODS copper alloy Cu-Al_2O_3. *Oxid. Met.* **2004**, *62*, 141–151. [CrossRef]
30. Liu, Y.X.; Yin, F.C.; Hu, J.X.; Li, Z.; Cheng, S.H. Phase equilibria of Al-Fe-Sn ternary system. *T Nonferr. Metal. Soc.* **2018**, *28*, 282–289. [CrossRef]
31. Li, R.; Wang, S.; Zhou, D.; Pu, J.; Yu, M.; Guo, W. A new insight into the NaCl-induced hot corrosion mechanism of TiN coatings at 500 °C. *Corros. Sci.* **2020**, *174*, 108794. [CrossRef]
32. Dafali, A.; Hammouti, B.; Mokhlisse, R.; Kertit, S. Substituted uracils as corrosion inhibitors for copper in 3% NaCl solution. *Corros. Sci.* **2003**, *45*, 1619–1630. [CrossRef]
33. Ozcan, M.; Dehri, I.; Erbil, M. Organic sulphur-containing compounds as corrosion inhibitors for mild steel in acidic media: Correlation between inhibition efficiency and chemical structure. *Appl. Surf. Sci.* **2004**, *236*, 155–164. [CrossRef]
34. Saha, S.K.; Banerjee, P. Introduction of newly synthesized Schiff base molecules as efficient corrosion inhibitors for mild steel in 1 M HCl medium: An experimental, density functional theory and molecular dynamics simulation study. *Mater. Chem. Front.* **2018**, *2*, 1674–1691. [CrossRef]
35. Solmaz, R.; Kardas, G.; Culha, M.; Yazici, B.; Erbil, M. Investigation of adsorption and inhibitive effect of 2-mercaptothiazoline on corrosion of mild steel in hydrochloric acid media. *Electrochim. Acta* **2008**, *53*, 5941–5952. [CrossRef]
36. Saha, S.K.; Dutta, A.; Ghosh, P.; Sukul, D.; Banerjee, P. Novel Schiff-base molecules as efficient corrosion inhibitors for mild steel surface in 1 M HCl medium: Experimental and theoretical approach, *Phys. Chem. Chem. Phys.* **2016**, *18*, 17898–17911. [CrossRef]
37. Visser, P.; Terryn, H.; Mol, J.M.C. On the importance of irreversibility of corrosion inhibitors for active coating protection of AA2024-T3. *Corros. Sci.* **2018**, *140*, 272–285. [CrossRef]
38. Xu, J.; Zhou, C.; Chen, Z.; Wang, Y.; Jiang, S. Corrosion behaviors of $(Cr,Fe)_3Si/Cr_{13}Fe_5Si_2$ composite coating under condition of synergistic effects of electrochemical corrosion and mechanical erosion. *J. Alloys Compd.* **2010**, *496*, 429–432. [CrossRef]
39. Zhu, X.X.; Sun, M.Y.; Zhao, R.; Li, Y.Q.; Zhang, B.; Zhang, Y.L.; Lang, X.Y.; Zhu, Y.F.; Jiang, Q. 3D hierarchical self-supported NiO/Co_3O_4@C/CoS_2 nanocomposites as electrode materials for high-performance supercapacitors. *Nanoscale Adv.* **2020**, *2*, 2785–2791. [CrossRef]
40. van den Brand, J.; Sloof, W.G.; Terryn, H.; de Wit, J.H.W. Correlation between hydroxyl fraction and O/Al atomic ratio as determined from XPS spectra of aluminium oxide layers. *Surf. Interface Anal.* **2004**, *36*, 81–88. [CrossRef]

Article

Effect of Air Storage on Stress Corrosion Cracking of ZK60 Alloy Induced by Preliminary Immersion in NaCl-Based Corrosion Solution

Evgeniy Merson [1,*], Vitaliy Poluyanov [1], Pavel Myagkikh [1], Dmitri Merson [1] and Alexei Vinogradov [2]

1. Institute of Advanced Technologies, Togliatti State University, 445020 Togliatti, Russia
2. Department of Mechanical and Industrial Engineering, Norwegian University of Science and Technology, 4791 Trondheim, Norway
* Correspondence: mersoned@gmail.com

Abstract: The preliminary exposure of Mg alloys to corrosion solutions can cause their embrittlement. The phenomenon is referred to as pre-exposure stress corrosion cracking (PESCC). It has been reported that relatively long storage in air after pre-exposure to the corrosion solution is capable of eliminating PESCC. This effect was attributed to the egress of diffusible hydrogen that accumulated in the metal during pre-exposure. However, recent findings challenged this viewpoint and suggested that the corrosion solution retained within the side surface layer of corrosion products could be responsible for PESCC. The present study is aimed at the clarification of the role of hydrogen and the corrosion solution sealed within the corrosion products in the "healing" effect caused by post-exposure storage in air. Using the slow strain rate tensile (SSRT) testing in air and detailed fractographic analysis of the ZK60 specimens subjected to the liquid corrosion followed by storage in air, we found that PESCC was gradually reduced and finally suppressed with the increasing time and temperature of air storage. The complete elimination of PESCC accompanied by recovery of elongation to failure from 20% to 38% was achieved after 24 h of air storage at 150–200 °C. It is established that the characteristic PESCC zone on the fracture surface is composed of two regions, of which the first is always covered by the crust of corrosion products, whereas the second one is free of corrosion products and is characterised by quasi-brittle morphology. It is argued that the corrosion solution and hydrogen stored within the corrosion product layer are responsible for the formation of these two zones, respectively.

Keywords: air storage; corrosion products; fractography; Mg alloys; pre-exposure; stress corrosion cracking

Citation: Merson, E.; Poluyanov, V.; Myagkikh, P.; Merson, D.; Vinogradov, A. Effect of Air Storage on Stress Corrosion Cracking of ZK60 Alloy Induced by Preliminary Immersion in NaCl-Based Corrosion Solution. *Materials* **2022**, *15*, 7862. https://doi.org/10.3390/ma15217862

Academic Editor: Ming Liu

Received: 15 October 2022
Accepted: 3 November 2022
Published: 7 November 2022

Publisher's Note: MDPI stays neutral with regard to jurisdictional claims in published maps and institutional affiliations.

Copyright: © 2022 by the authors. Licensee MDPI, Basel, Switzerland. This article is an open access article distributed under the terms and conditions of the Creative Commons Attribution (CC BY) license (https://creativecommons.org/licenses/by/4.0/).

1. Introduction

Magnesium alloys being tensile tested in air after preliminary exposure to corrosive media suffer from a significant drop in mechanical properties, which is accompanied by a change in the fracture mode from ductile to brittle [1–5]. This phenomenon is known as pre-exposure stress corrosion cracking (PESCC) [6], also referred to as pre-exposure embrittlement [7,8]. Since the specific conditions inducing PESCC are less common in practice than those responsible for true stress corrosion cracking (SCC), significantly less attention has been paid to the former phenomenon. Nevertheless, understanding the mechanisms governing PESCC is of great importance because it sheds light on the nature of SCC, which is still under debate. For example, the widespread belief that hydrogen plays a key role in the mechanism of SCC [9–12] is primarily based on the observed similarity between the features of PESCC and hydrogen embrittlement (HE) in other metallic materials [13]. It was established that the degree of PESCC in many Mg alloys increases with decreasing strain rate [6,14] and increasing time of pre-exposure in various corrosion solutions [7,8,15–20]. Longer pre-exposure was found to cause more significant corrosion damage [18–20] and a larger brittle-like area on the peripheral part of the fracture surface [7,16,20], accompanied by an increase in the hydrogen concentration in the test specimens [14,20]. It was suggested [5,7,8,14,16–18] that hydrogen, being an inevitable by-product of a corrosion reaction, is absorbed by the metal in the form of atoms having a high diffusion rate that is sufficient to cause hydrogen-assisted cracking through the well-known HE mechanisms, including hydrogen-enhanced decohesion, hydrogen-enhanced localised plasticity, or delayed hydride cracking. The details of these mechanisms can be found elsewhere [21,22]. What should be stressed here is that all of them require the presence of a sufficiently high concentration of diffusible hydrogen in the metal matrix. A recent study has challenged this belief and shown that the role of diffusible hydrogen in PESCC can be substantially less important than that of the corrosion product layer deposited on the side surface of the metal during pre-exposure [6,20,23]. In particular, it was reported that PESCC of the alloys ZK60 and AZ31 preliminarily immersed in the NaCl-based corrosion solutions was completely eliminated by the chemical removal of the corrosion product layer, provided that irreversible corrosion damage was not too high [20,23]. Furthermore, a negligible concentration of diffusible hydrogen was found in the base metal of these alloys after the removal of corrosion products [20,23–25]. It was, therefore, suggested that the corrosion product layer may serve as a sealant for embrittling agents, such as hydrogen or retained corrosion solution, which facilitate the propagation of the brittle cracks under external loading [6,20,23]. The presence of the corrosion solution inside this layer is indicated by the crust of corrosion products, which was observed on the peripheral part of the fracture surface of the alloy ZK60 embrittled due to PESCC [6,16,20,23]. Originally, this crust was attributed to corrosion damage, which is induced during pre-exposure [16,23]. However, it was demonstrated recently that these corrosion products were formed not during the pre-exposure process but exactly during post-exposure mechanical testing in air [6]. Two independent experimental observations corroborated this finding. Firstly, the size of the area covered by corrosion products on the fracture surface decreased with an increasing strain rate. Secondly, many secondary cracks with corrosion products inside were found in the cross-section of the pre-exposed specimens after tensile testing, while none were observed right after pre-exposure. Obviously, this would not be the case if the cracks were present with the corresponding region of the fracture surface covered with the crust of corrosion products in the specimens before tensile testing. Thus, it was concluded that there must be an aggressive medium retaining inside the corrosion product layer on the side surface to form the corrosion products on the fracture surface. Later on, the presence of the corrosion solution within the corrosion product layer was additionally evidenced by the evolution of extra hydrogen generated by the corrosion reaction inside the specimens after pre-exposure to corrosive media [26].

The suggestion that the corrosion product layer containing the corrosion solution inside can be responsible for PESCC of Mg alloys prompted us to review this topic with new experimental findings relevant to PESCC, which has been exclusively associated with the effect of hydrogen so far. In particular, it was reported that vacuum annealing [14] or long exposure to dry air [5,7] after immersion in corrosion solution suppresses PESCC of Mg alloys to a certain extent. Chakrapani and Pugh [14] showed that vacuum annealing at 385 °C for 4 h results in a slight recovery of both the elongation to failure and ultimate tensile strength of the Mg-7.5%Al alloy embrittled by pre-exposure to the NaCl-K_2CrO_4 solution. The observed recovery of mechanical properties was attributed to the reduction in the hydrogen concentration in the specimens, as was assessed by the gas analyser after annealing. Moreover, the positive strain rate sensitivity of mechanical properties, which is a characteristic feature of HE in many alloys [27–29], was eliminated by annealing. This observation was considered as an additional argument blaming hydrogen for being the root cause of PESCC. Similar partial recovery of mechanical properties was documented after storing the pre-exposed specimens of pure Mg [5] and AZ31B [7] alloy in a desiccator. This was also explained by the removal of hydrogen from the specimens. Nevertheless, it was found recently that the complete elimination of PESCC in ZK60 can be achieved by the removal of the side surface corrosion product layer [6]. Concurrently, the strain rate sensitivity of ductility changed from positive to negative. Bearing in mind that, besides hydrogen, the corrosion solution can be contained within the corrosion product layer, one can plausibly suppose that the suppression of PESCC by air or vacuum exposure at room or elevated temperatures could be attributed not only to the extraction of hydrogen but also to the evaporation of water from the corrosion products. Thus, the primary aim of the present study was to clarify the significance and the role of post-exposure storage in air in the suppression of PESCC of the Mg alloy ZK60. To be more specific, we endeavour to establish relationships, or at least correlations, between mechanical properties, hydrogen concentration, and fractographic characteristics, including the area of fracture surface covered by corrosion products. To this end, the slow-strain rate tensile (SSRT) testing coupled with thermal desorption analysis of hydrogen and detailed quantitative fractographic and side surface examination were utilised. The ZK60 alloy was chosen as a typical representative of a class of wrought Mg alloys suffering from high susceptibility to PESCC. The grain microstructure and distribution of primary and secondary phases in the commercial hot-extruded alloy used in this work were documented in detail in the previous reports [24,25,30,31] by the present and other investigators; interested readers are encouraged to review the above-cited publications.

2. Materials and Methods

The hot-extruded commercial alloy ZK60 was used in the present investigation in the as-received state. The chemical composition, microstructure, and the manufacturing process of the alloy were the same as reported in the previous studies [6,20,23,24]. The microstructure of the alloy is represented by α-Mg fine grains with a 3 μm average diameter and numerous secondary phase particles [24]. The cylindrical threaded specimens for SSRT testing, as well as the samples for the gas analysis, were machined from round bars of 25 mm diameter along the extrusion direction. The samples for the gas analysis and the gauge parts of the specimens for tensile testing had the same geometry and dimensions of 30 mm length and 6 mm diameter. All specimens and samples were annealed in vacuum at 250 °C for 2 h to relieve residual stresses.

Before SSRT testing, the gauge part of the specimens was placed into the plexiglass cell, which was then filled with the 4% NaCl (NevaReactiv, St. Petersburg, Russia) + 4% $K_2Cr_2O_7$ (ChimProm, Novocheboksarsk, Russia) corrosive solution. After 1.5 h of pre-exposure in this solution at an open-circuit potential without external stress, the specimens were extracted from the cell, cleaned with ethanol, and dried with compressed air. The pre-exposure was optionally followed by removal of corrosion products, which was executed by submerging the specimens in the standard 20% CrO_3 (ChimProm, Novocheboksarsk, Russia) + 1% $AgNO_3$ (NevaReactiv, St. Petersburg, Russia) aqueous solution for 1 min, followed by rinsing the specimens with ethanol (NevaReactiv, St. Petersburg, Russia) and drying by compressed air. Several pre-exposed specimens (with and without corrosion products) were SSRT tested within two minutes after pre-exposure, while others, including the reference ones, which did not undergo pre-exposure, were subjected to storage in ambient air for 24 h at temperatures ranging from 24 to 200 °C. The uniaxial SSRT testing of all specimens was performed in air at 24 °C with the 5×10^{-6} s^{-1} nominal strain rate (0.01 mm/min traverse velocity) using the AG-X Plus (Shimadzu, Kyoto, Japan) screw-driven frame. The samples for the gas analysis underwent the same treatments as their counterparts for the SSRT testing.

The fracture and side surfaces of the SSRT tested specimens were investigated by a scanning electron microscope (SEM) (SIGMA, Zeiss, Jena, Germany) equipped with an energy dispersive X-ray (EDS) detector (EDAX, Mahwah, NJ, USA). The quantitative geometrical measurements of the specific areas on the fracture surfaces were conducted either by the ImageJ (v1.52u, NIH, Bethesda, MD, USA) freeware or by the built-in SmartSEM (v05.04.05.00, Zeiss, Jena, Germany) software.

The thermal desorption of hydrogen from the pre-exposed samples was investigated by the G8 Galileo (Bruker, Kalkar, Germany) gas analyser via the hot extraction method in N_2 (99.999%, LindeGas Rus, Samara, Russia) carrier gas flux. The concentration of hydrogen in the equipment was measured by the difference in the thermal conductivity of the pure carrier gas and the mixture of carrier gas with the extracted hydrogen. The contaminating admixtures such as CO_2 and water were removed from the gas mixture by the system of filters. The pre-exposed samples were placed into the quartz tube of the gas analyser within 2 min after pre-exposure or after 24 h of air storage at a specific temperature. The procedure of the thermal desorption analysis (TDA) included (i) heating the sample from room temperature of 24 °C to 450 °C with a constant heating rate of 38 °C/min, followed by (ii) 15 min of baking at 450 °C and (iii) cooling for 10 min. The additional details of the gas analysis procedure have been thoroughly described elsewhere [24].

3. Results and Discussion

3.1. Mechanical Properties

The mechanical testing showed that the specimens which were SSRT tested in air within a few minutes after pre-exposure to the corrosion solution demonstrated a substantial loss in strength and ductility compared to the reference specimens (Figures 1 and 2). This is typical behaviour for ZK60 embrittled by PESCC. In harmony with our previous reports [6,20,23], the removal of corrosion products resulted in complete recovery of ductility and partial recovery of strength in all specimens tested. The apparent irreversible loss of strength, which remains after the removal of corrosion products, is not related to embrittlement but rather is due to the reduction in the specimens' cross-section caused by irreversible corrosion damage [20]. It was found that similarly to the removal of corrosion products, post-exposure air storage inhibits PESCC. The efficiency of this recovery process, however, depends on the time and temperature of the storage of specimens in air. As can be seen in Figures 1 and 2, both ultimate tensile strength (UTS) and elongation to failure (EF) of the pre-exposed specimens after air storage at 24 °C for 24 h are significantly higher than those after 2 min of air storage at the same temperature. Nevertheless, even after 24 h of storage at room temperature, the pre-exposed specimens show roughly 50% loss of EF and approximately 25 MPa smaller UTS in comparison with their reference counterparts. It was found that the recovery of mechanical properties can be substantially enhanced by storage of the specimens in air at elevated temperatures. Figure 2a demonstrates that the increased temperature of air storage results in the increase in EF. The improvement in the EF value is non-monotonic and is most pronounced in the temperature range from 50 to 100 °C. After storage in air at 100 °C, the ductility loss of the pre-exposed specimens does not exceed 20%. However, at a chosen post-exposure storage time, the complete recovery of ductility can be achieved only at 150–200 °C. After such treatment, the measured EF values were the same as those of the reference specimens and the specimens with the removed corrosion product layer (referred to hereafter as CP-free specimens).

The effect of storage temperature on the strength of the pre-exposed specimens is more complex than on ductility. It can be seen from Figure 2b that UTS of the specimens monotonically grows when the storage temperature increases from 25 to 80 °C, while the further temperature increase up to 200 °C results in the non-monotonic decrease in UTS with a short plateau around 120–150 °C. To clarify whether the observed complex behaviour of strength and ductility is associated with temperature-induced microstructural changes or if it is attributed specifically to PESCC, the effect of annealing temperature on mechanical properties of the reference and CP-free specimens was investigated. One can see (Figure 2b) that the UTS of these specimens is almost unaffected by low-temperature annealing below 100 °C, while at higher temperatures, the strength decreases monotonically with temperature. Apparently, some ageing-related microstructural changes affecting the strength occur in the alloy at temperatures above 100 °C. Therefore, these changes are likely to take some part in the observed deterioration of UTS in the pre-exposed specimens subjected to air storage above 100 °C. However, the opposite trend, i.e., the increase in UTS, in the pre-exposed specimens stored in air in the temperature range below 80 °C, cannot be explained by any microstructural changes in the metallic matrix. Furthermore, the decrease in strength in the pre-exposed specimens, which is observed at higher temperatures, cannot be solely attributed to the microstructural transformations either, because it starts in the range between 80 and 100 °C and not between 100 and 120 °C as for the reference and CP-free specimens. Furthermore, the plateau between 120 and 150 °C featuring the dependence of UTS on the temperature of the pre-exposed specimens is absent on such dependencies for the specimens of other kinds. Thus, the other factors responsible for the decrease in UTS, which is attributed specifically to the specimens embrittled by pre-exposure, do exist. It is important to note that the microstructural changes, which likely affect the strength of the alloy at elevated temperatures, do not influence the ductility. It follows from Figure 2a that the EF of the reference and CP-free specimens scatters randomly around the nearly constant average value within the whole investigated temperature range. Therefore, the increasing EF of the pre-exposed specimens with post-exposure storage temperature should be entirely associated with the elimination of PESCC.

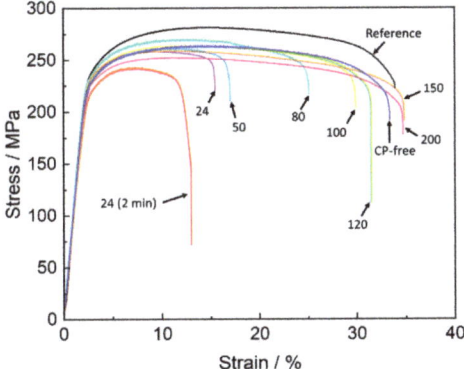

Figure 1. Stress–strain diagrams corresponding to the specimens of the ZK60 alloy which were SSRT tested in the reference and CP-free state as well as after pre-exposure to the corrosion solution followed by storage in air. Temperatures of air storage (in °C) are provided at the arrows. The duration of storage in air was 24 h, if not marked otherwise.

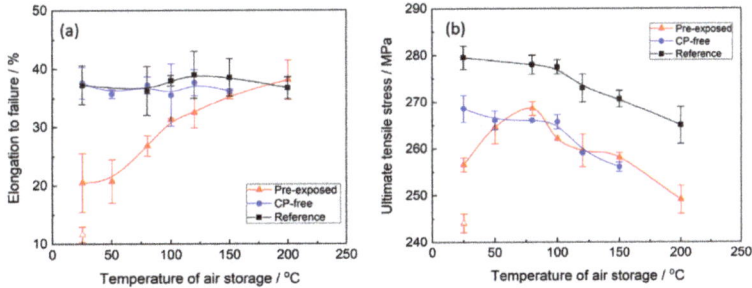

Figure 2. The effect of temperature of air storage on elongation to failure (**a**) and ultimate tensile strength (**b**) of the reference, CP-free, and pre-exposed specimens of the ZK60 alloy. The open and filled symbols indicate the pre-exposed specimens after air storage for 2 min and 24 h, respectively.

3.2. Fractographic and Side Surface Observations

The fracture surface of the specimens suffering from PESCC is characterised by the peripheral brittle-like "PESCC zone" surrounding the dimpled ductile region in the central part of the fracture surface. The similar appearance of the fracture surface was frequently reported to be characteristic of the pre-exposure effects in Mg alloys, including ZK60, used in the present study [7,16,23]. The quantitative fractographic analysis showed that, in general, the increase in the time and temperature of post-exposure storage in air results in the significant reduction in both the total area of the fracture surface and the area of the PESCC zone (as well as its areal fraction with respect to the total area of the fracture surface) (c.f., Figures 3 and 4a). It also follows from Figure 3f–h that when the temperature of air storage exceeds 80 °C, the PESCC zone becomes discontinuous and is represented by several small isolated "islands" with a brittle-like morphology at the edge of the fracture surface. After post-exposure storage at 150–200 °C, the PESCC zone is almost completely absent. However, some specimens occasionally exhibited a few tiny areas with the brittle topology (e.g., Figure 3g,h) even after storage at 150–200 °C. Despite the generally increasing apparent ductility of the fracture surface with the temperature of post-exposure storage, one can notice the anomalous increase in the fraction of the brittle relief on the fracture surface after storage at 120 °C (Figure 4a). As was shown above, the decrease in strength was supposedly associated with the microstructural changes occurring at the same temperature in the reference specimens.

The fractographic observations corroborate well with the results of mechanical testing showing the increasing ductility of the pre-exposed specimens due to the suppression of PESCC by post-exposure storage in air. The confident linear correlation with Pearson's $R^2 > 0.91$ is found between EF and the total area of their fracture surface as well as the area of the PESCC zone and its areal fraction (Figure 4b), thus confirming the link between the propagation of PESCC and the material's embrittlement unveiled by SSRT testing in ambient conditions.

The side surface observations in combination with the fractographic analysis show that the PESCC zone is formed due to the propagation of multiple cracks nucleating at the side surface along the whole gauge part of the specimens (c.f., Figures 5 and 6). These cracks coalesce during their propagation towards the specimen's centre as well as in the transverse direction. The coalescence of the cracks occurs in a stepwise manner by mutual axial shearing of the two halves of the fracture surface. The location of the cracks' nucleation sites on the peripheral part of the fracture surface corresponds to relatively flat regions separated from each other by the axial shearing steps. As an example, the shearing steps and the crack initiation sites between them are indicated by the inclined red and vertical white arrows, respectively, in Figures 7a and 8a. As is seen in the SEM side surface images in Figures 5 and 6, the number and size of the side surface cracks decrease considerably with the temperature of post-exposure storage. This visual observation is testified by the results of the quantitative fractographic analysis provided in Figure 9a. According to these results, the number of cracks' nucleation sites on the fracture surface increases considerably along with the storage temperature. Furthermore, as the temperature increases, the side surface cracks become shorter in the transverse direction and acquire a more ductile appearance, as is indicated by their less sharp and more round geometry (c.f., Figure 6). It follows from Figure 5a,b and Figure 9a that the increase in the time of post-exposure storage in air at 24 °C from 2 min to 24 h results in the remarkable growth in the number of side surface cracks and the cracks' nucleation sites on the fracture surface. Thus, the number of cracks decreases with temperature and increases with the time of air storage. It is worth noting that, similarly to the area of the PESCC zone, the number of the cracks' nucleation sites exhibits a sharp increase after air storage at 120 °C (Figure 9a), whereas just a few cracks can be found in the specimens subjected to holding in air at 150–200 °C.

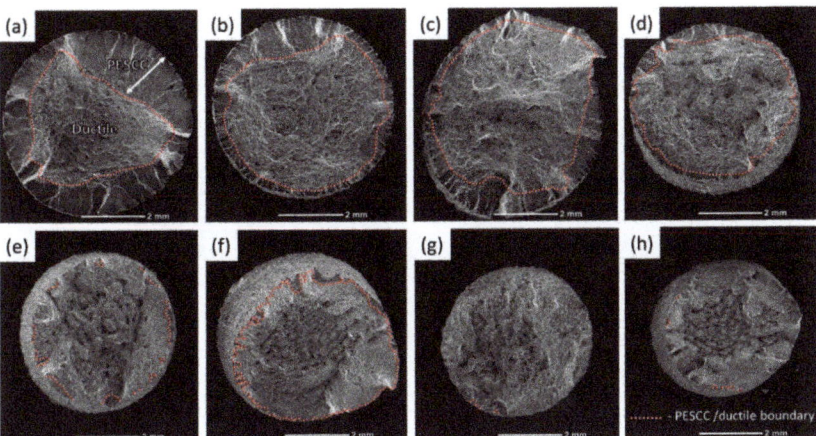

Figure 3. Whole views of the fracture surfaces of the specimens SSRT tested after pre-exposure followed by storage in air at 24 °C for 2 min (**a**) and by 24 h, at 24 °C (**b**), 50 °C (**c**), 80 °C (**d**), 100 °C (**e**), 120 °C (**f**), 150 °C (**g**), and 200 °C (**h**).

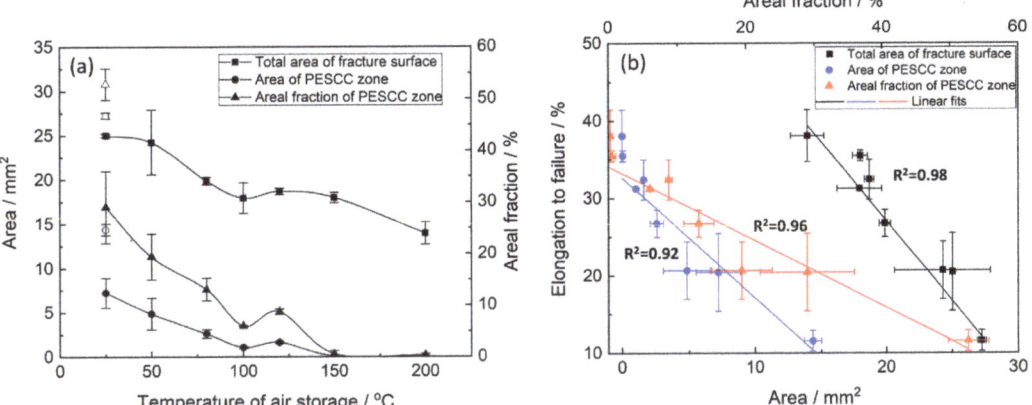

Figure 4. The effect of temperature of post-exposure storage on the areal fractographic characteristics of the ZK60 specimens tested after pre-exposure followed by air storage (**a**) and the correlation between these characteristics and the elongation to failure of the specimens (**b**). The open and filled symbols correspond to the pre-exposed samples subjected to air storage for 2 min and 24 h, respectively.

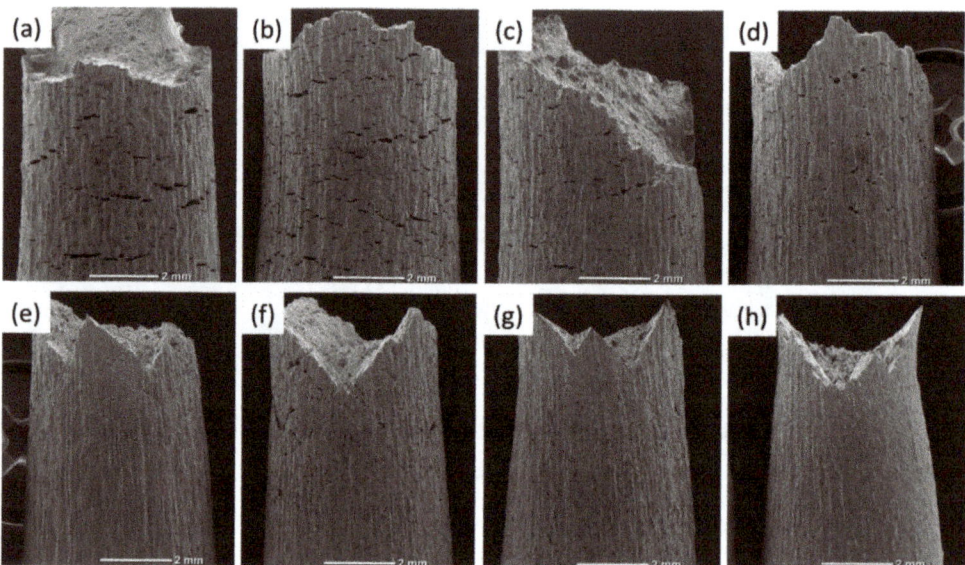

Figure 5. SEM images of side surfaces of the specimens SSRT tested after pre-exposure followed by air storage at 24 °C for 2 min (**a**), and by 24 h at 24 °C (**b**), 50 °C (**c**), 80 °C (**d**), 100 °C (**e**), 120 °C (**f**), 150 °C (**g**), and 200 °C (**h**).

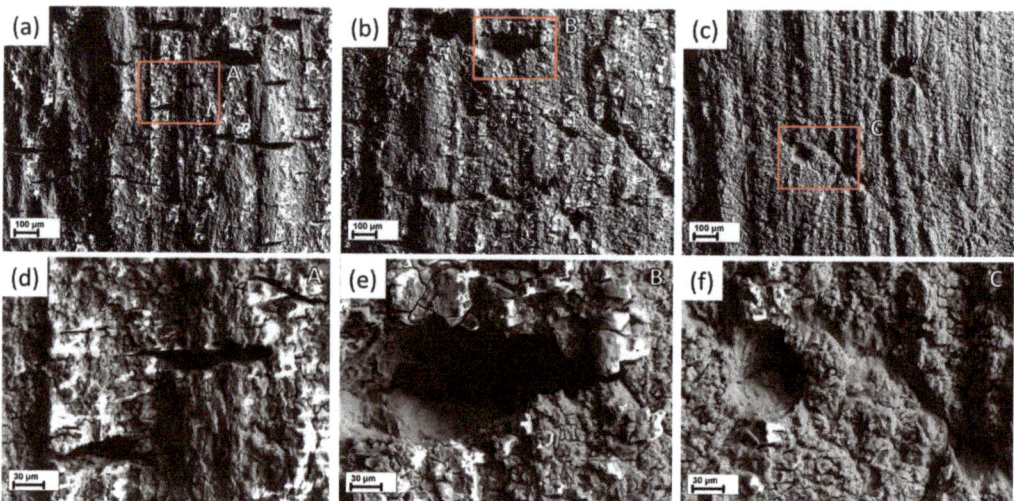

Figure 6. The appearance of the side surface cracks in the specimens SSRT tested after pre-exposure followed by 24 h of air storage at: 24 °C (**a,d**), 80 °C (**b,e**), and 100 °C (**c,f**). The magnified SEM images of the regions outlined by the frames A–C in (**a**–**c**) are represented in (**d**–**f**) to show the microscopic characteristics of side surface cracks.

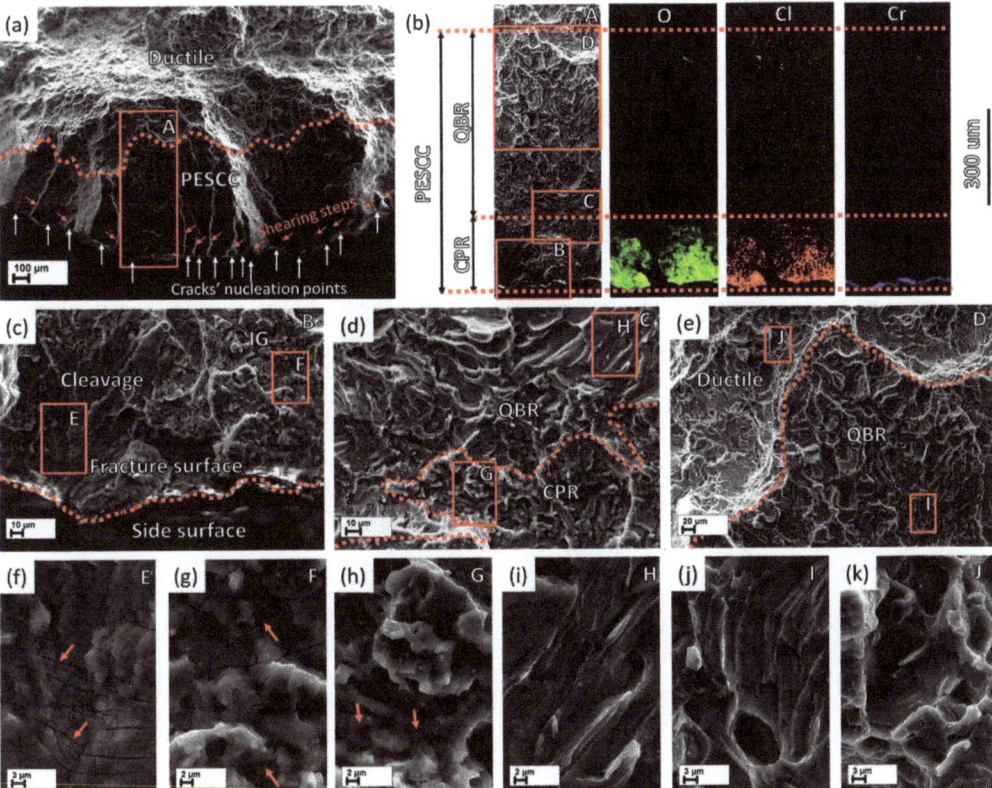

Figure 7. The fractographic features of the specimen SSRT tested after pre-exposure followed by 24 h of air storage at 24 °C: (**a**) the peripheral region of the fracture surface with PESCC zone delimited from the ductile region by the dotted curve. The vertical white and inclined red arrows indicate the nucleation points of the multiple cracks and the shearing ligaments between them, respectively; (**b**) the magnified SEM image with elemental EDS maps of the region outlined by the frame A in (**a**) showing the boundaries of the specific regions; (**c**) the magnified SEM image of the region outlined by the frame B in (**b**) showing the crack's nucleation part of the corrosion product region (CPR) characterised by cleavage and intergranular facets and corrosion products; (**d**) the magnified SEM image of the region outlined by the frame C in (**b**) showing the transition region between CPR and the quasi-brittle region (QBR); (**e**) the magnified SEM image of the region outlined by the frame D in (**b**) showing the transition region between QBR and the ductile region; (**f–k**) the magnified SEM images of the regions outlined by the frames E–J in (**c–e**) showing the characteristic morphologies of the specific regions.

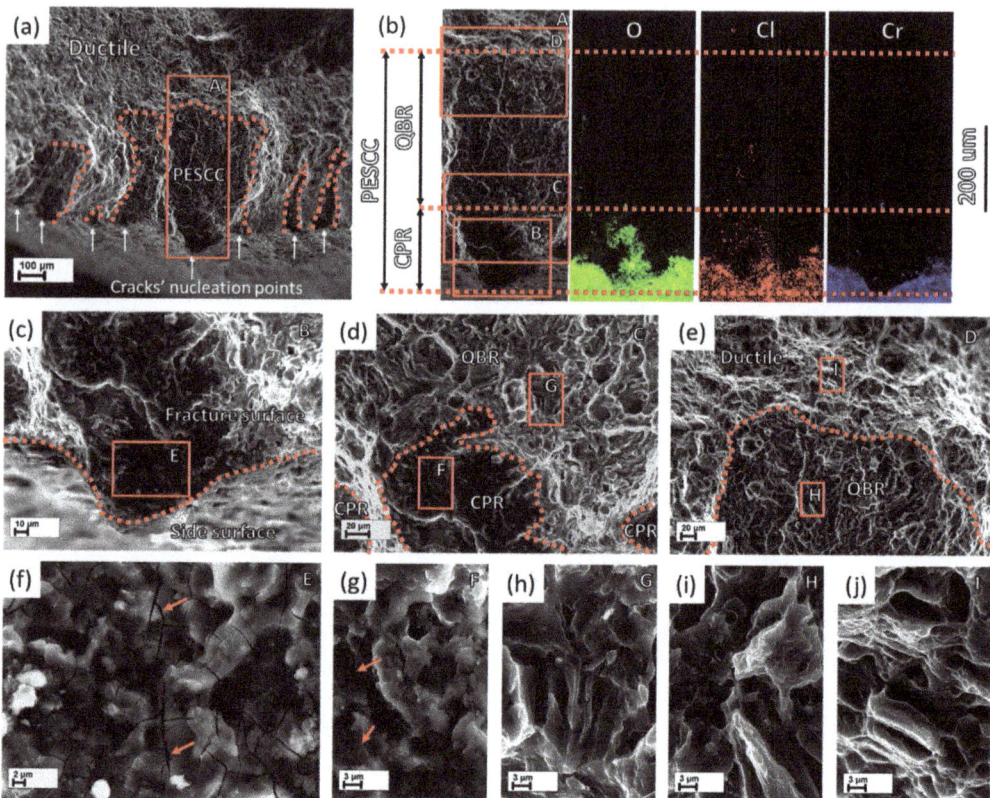

Figure 8. The fractographic features of the specimen SSRT tested after pre-exposure followed by 24 h of air storage at 80 °C: (**a**) the peripheral region of the fracture surface with isolated islands of PESCC zone delimited from the ductile region by the dotted curve. The vertical white arrows indicate the nucleation points of the multiple cracks; (**b**) the magnified SEM image with elemental EDX maps of the region outlined by the frame A in (**a**) showing the boundaries of the specific regions; (**c**) the magnified SEM image of the region outlined by the frame B in (**b**) showing the crack's nucleation part of CPR; (**d**) the magnified SEM image of the region outlined by the frame C in (**b**) showing the transition region between CPR and QBR; (**e**) the magnified SEM image of the region outlined by the frame D in (**b**) showing the transition region between QBR and ductile region; (**f**–**j**) the magnified SEM images of the regions outlined by the frames E–I in (**c**–**e**) showing the morphologies characteristic of the specific regions.

The PESCC zone in all pre-exposed specimens is composed of two distinct regions referred to as the corrosion product region (CPR) and the quasi-brittle region (QBR), which are seen one after the other from the edge of the specimen to the centre (Figures 7 and 8). The corrosion product region starts to develop immediately at the edge of the fracture surface and is covered by the cracked crust of corrosion products (Figure 7b,c,f–h and Figure 8b,c,f,g). It has been shown recently that the intergranular cracking assisted by the corrosion solution sealed in the side surface layer of corrosion products can be responsible for the formation of the corrosion product region on the fracture surface [6]. Thus, the specific "melted" relief, which is observed within this region under the crust of corrosion products (Figure 7g,h and Figure 8f,g), is probably composed of fine intergranular facets attacked by corrosion. In addition, the large transgranular cleavage facets exhibiting corrosion damage can also be occasionally distinguished within the corrosion product region on the fracture surface (Figure 7c,f). The thickness of the corrosion product layer on the fracture surface decreases with the distance from the edge of the fracture surface towards the following quasi-brittle region, as is firmly evidenced by the EDX analysis (Figures 7b and 8b). Additionally, the EDX elemental map obtained for the PESCC zone undeniably shows that this layer is strongly enriched with oxygen and chlorine. It should be stressed that the concentration of these elements decreases progressively along the crack path within the corrosion product region; however, it drops down abruptly to the negligible level at the boundary between the corrosion products and quasi-brittle regions. Thereby, the latter region is always completely free of traces of aggressive elements. Alternatively, the CPR–QBR boundary can be reliably distinguished by the morphological features in the SEM images with no aid from the EDX. The characteristic microcracks, such as those indicated by the arrows in Figures 7f–h and 8f,g, are found to be indispensable attributes of the crust of corrosion products. As such, they serve as independent markers of corrosion products on the fracture surface. The size and density of these microcracks are gradually reduced with the thickness of the corrosion product layer throughout the respective region (see Figures 7f–h and 8f,g). However, the cracks cease to appear in the quasi-brittle region beyond the CPR–QBR boundary (c.f., Figures 7i–k and 8h–j). Moreover, in contrast to the region that has been covered by corrosion products, the morphology of the quasi-brittle region is distinctively characterised by various tear ridges, small dimples, and fluted facets (Figure 7j,i and Figure 8h,i). These features indicate the appreciable contribution of plastic deformation to the crack growth mechanism.

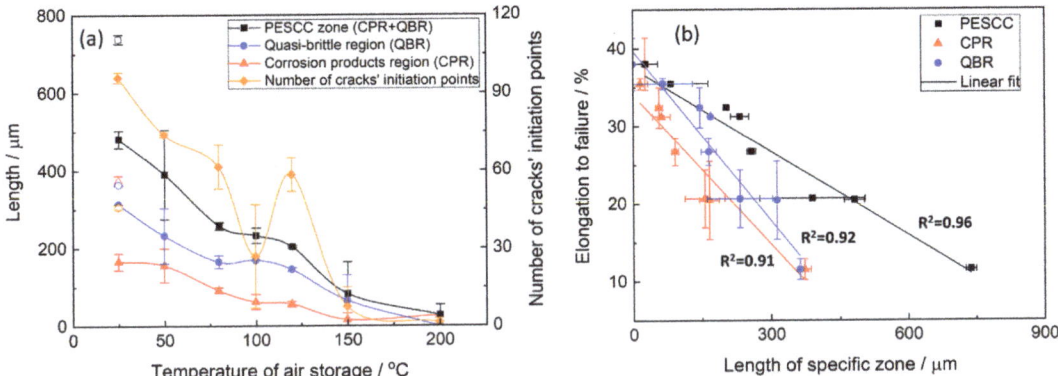

Figure 9. The effect of temperature of post-exposure storage on the linear fractographic characteristics of the ZK60 specimens tested after pre-exposure followed by air storage (**a**) and the correlation between these characteristics and the elongation to failure of the specimens (**b**). The open and filled symbols correspond to the pre-exposed samples subjected to air storage for 2 min and 24 h, respectively. CPR—the region covered by the crust of corrosion products, QBR—the quasi-brittle region free of corrosion products.

Notably, the CPR–QBR boundary recognised by the topological features on SEM images matches precisely with that distinguished by the EDX analysis (Figures 7b and 8b). Thus, both methods can be used to characterise and identify two morphologically distinct regions. Apparently, the secondary electron imaging is less laborious in comparison with the EDX mapping. That is why the former was prioritised for further measurements. The length of the corrosion product zone, l_{CPR}, as well as the length of the quasi-brittle region, l_{QBR}, and the PESCC zone, l_{PESCC}, were measured for each crack within the PESCC zone, as is schematically shown in Figures 7b and 8b. The average values of these characteristics were calculated for the specimens tested under specific experimental conditions. The effect of air storage temperature and time on the measured fractographic properties is illustrated in Figure 9a: the lengths of all specific zones, including the corrosion product region, decrease significantly with temperature and time. In particular, the corresponding reduction in l_{CPR} is also evidently illustrated by the EDX oxygen maps shown in Figure 10. The obtained results corroborate the conclusions made in the previous study that the corrosion products on the fracture surface are produced during the SSRT testing. Indeed, if the corrosion product region was formed during pre-exposure, the size of this region would not be affected by subsequent air storage. It is worth noting that reduction in l_{CPR} becomes notable at the temperatures of air storage higher than 50 °C, while the l_{QBR} considerably decreases after storage at a lower temperature. In the temperature range of 100–120 °C, the plateau followed by the further decay up to near complete vanishing at higher temperatures of air storage is seen on the l_{CPR}, l_{QBR}, and l_{PESCC} curves in Figure 9a. It is established that the lengths of all three characteristic regions, including the corrosion product region, correlate well with the elongation to failure of the specimens embrittled by pre-exposure to the aggressive environment (Figure 9b). This finding suggests the crucially active role of the corrosion solution sealed/stored within the side surface layer of corrosion products in the mechanism of PESCC.

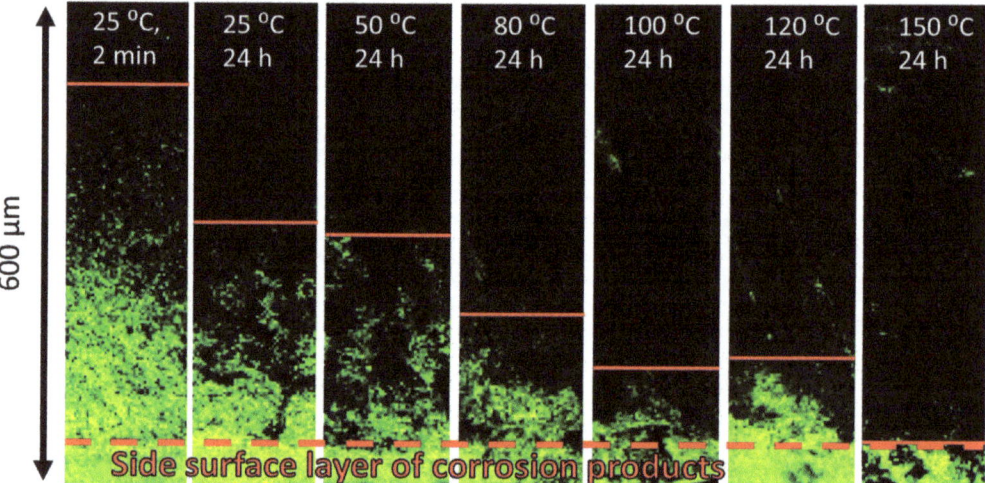

Figure 10. The distribution of oxygen on the peripheral part of the fracture surface for the specimens SSRT tested after pre-exposure followed by storage in air at different temperatures and durations. The red solid horizontal lines show the boundary between the CPR and QBR. The dashed line indicates the boundary between the side and fracture surfaces.

As has been mentioned, the corrosion products on the fracture surface can be associated with the corrosion solution, which is likely sealed within the corrosion product layer on the side surface. In conjunction with this suggestion, it is important to note that according to the results of EDX analysis, the corrosion product region on the fracture surface almost does not contain chromium (Figures 7b and 8b), which is, however, abundantly present in the side surface layer of corrosion products (Figure 11a). This feature of the corrosion product region also drastically differs the fracture surface of the tensile tested pre-exposed specimens from that of the specimens which have been SSRT tested right in the corrosive solution containing chromates. The EDX maps obtained from the fracture surface of the specimen SSRT tested in the 4% NaCl + 4% $K_2Cr_2O_7$ solution in the previous study [19] are provided in Figure 11b. It is clear that the fracture surface of this specimen demonstrates high concentrations of Cl, O, and Cr. This finding implies that the corrosion solution responsible for the formation of corrosion products on the fracture surface of the pre-exposed specimens is Cr-depleted and thus, its chemical composition can be different from that of the liquid initially used in the pre-exposure process.

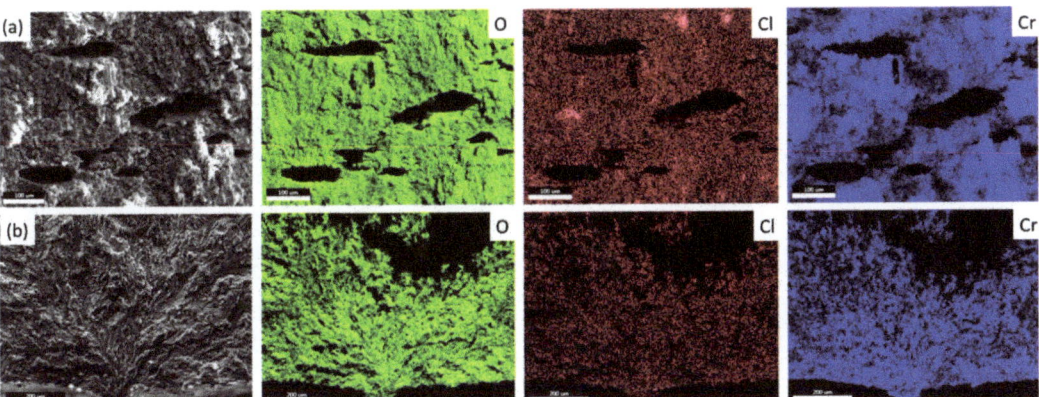

Figure 11. The SEM images and elemental maps obtained by EDX from the side surface of the specimen SSRT tested after pre-exposure in the 4% NaCl + 4% $K_2Cr_2O_7$ corrosion solution (**a**) and from the fracture surface of the specimen SSRT tested in the same corrosion solution (**b**).

3.3. Hydrogen Desorption Analysis

The thermal desorption analysis showed that the concentration of hydrogen extracted from the pre-exposed samples in the temperature range between 24 and 450 °C, $C_{H24-450}$, decreases notably with increasing temperature and time of air storage. The dependence of $C_{H24-450}$ on the temperature of air storage is non-linear and is characterised by the plateau within the interval between 24 and 80 °C, followed by the remarkable decay at higher temperatures (Figure 12a). Moreover, the anomalous increase in the hydrogen concentration is observed after air storage at 120 °C. The correlation between $C_{H24-450}$ and elongation to failure of the pre-exposed samples subjected to air storage at different temperatures is poor (Figure 13a), indicating that the relationship between $C_{H24-450}$ and the degree of PESCC is unlikely. It is generally accepted that diffusible hydrogen, i.e., chemically free hydrogen possessing high mobility within a metal, is responsible for HE in steels and other alloys [32]. It has been well established that this hydrogen completely escapes from metals at temperatures below 300 °C [33]. As was shown in our previous reports, the concentration of diffusible hydrogen in the matrix of the pre-exposed specimens of ZK60 and AZ31 alloys was negligible [20,23]. However, chemically free hydrogen can be, probably, accumulated within the layer of corrosion products. The extraction temperature of this hydrogen is also expected to be below 300 °C. The hydrogen extracted from the metals at temperatures above 300 °C is usually considered to be immobile, and hence, its role in hydrogen-assisted cracking is relatively insignificant. Thus, the concentrations of hydrogen extracted from the pre-exposed samples in the temperature intervals of 24–300, $C_{H24-300}$, and 300–450 °C, $C_{H300-450}$, were additionally assessed using the thermal desorption curves shown in Figure 12b. It was found that the behaviour of $C_{H300-450}$, depending on the temperature of air storage, is similar to that of $C_{H24-450}$ exhibiting the plateau at 24–80 °C, followed by the overall decay with the hump at 120 °C. In contrast, $C_{H24-300}$ notably decreases in the whole temperature range of air storage, including the 24–80 °C interval, where $C_{H24-450}$ and $C_{H300-450}$ remain almost unchanged. Furthermore, the concentration of hydrogen extracted below 300 °C during the thermal desorption analysis is negligible in the samples subjected to air storage at 150 and 200 °C. A strong correlation is found between $C_{H24-300}$ and EF of the pre-exposed specimens, which were exposed to air at different temperatures and times (Figure 13a), whereas the correlation between $C_{H300-450}$ and EF is even worse than that of $C_{H24-450}$. Moreover, $C_{H24-300}$ appreciably correlates with the size of specific zones on the fracture surface of the pre-exposed specimens, including the PESCC zone, corrosion product region and quasi-brittle region (Figure 13b).

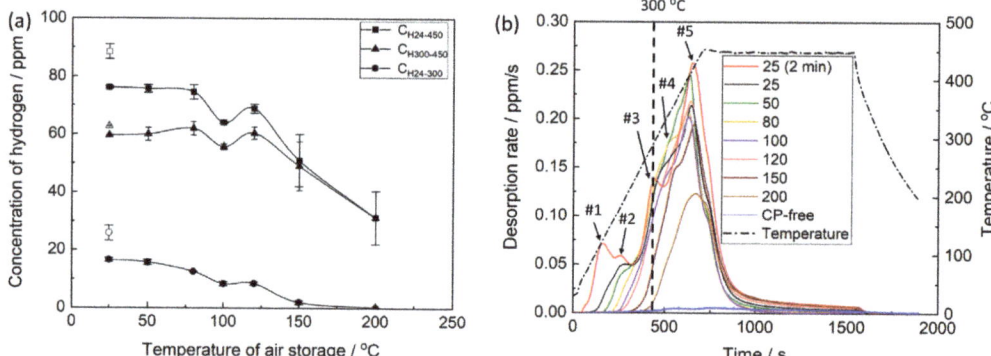

Figure 12. The effect of temperature and time of air storage on the concentration of hydrogen extracted in different temperature intervals (**a**) and on the thermal desorption spectra of hydrogen (**b**) for ZK60 subjected to pre-exposure followed by air storage at different temperatures. The open and filled symbols in (**a**) correspond to the pre-exposed samples subjected to air storage for 2 min and 24 h, respectively.

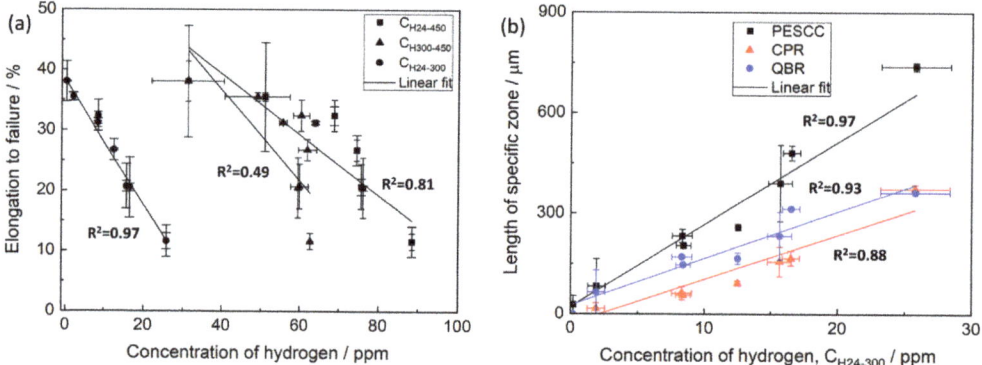

Figure 13. The correlation between the concentration of hydrogen extracted in different temperature intervals and the elongation to failure (**a**) and the lengths of specific zones on the fracture surface (**b**) of the specimens SSRT tested after pre-exposure followed by air storage at different temperatures and durations.

The thermal desorption spectra of hydrogen obtained from the specimens, which were subjected to pre-exposure, exhibit a few distinct superimposed peaks, as can be seen in Figure 12b. It was conclusively established in the previous studies [20,23,24] that all the observed peaks are associated with the corrosion products on the side surface layer because no considerable desorption of hydrogen from the matrix of the pre-exposed specimens occurs after chemical removal of those corrosion products. In support of this observation, the desorption spectrum for the specimen with the removed corrosion product layer is provided in Figure 12b (the plot is denoted as "CP-free") for reference. Thus, all peaks observed on the thermal desorption diagram in Figure 12b should be associated either with chemically bonded hydrogen evolving from the thermally decomposing components of corrosion products or with the chemically free hydrogen. The latter can be either trapped in the corrosion product layer or produced during the corrosion of the Mg matrix interacting with the corrosion solution sealed within that layer. All hydrogen other than that linked to chemical components of corrosion products can be considered to be potentially capable for facilitating crack growth and, thus, inducing embrittlement associated with PESCC. The large peaks #3–5 appearing above 300 °C likely correspond to such components of corrosion products as $Mg(OH)_2$ and/or MgH_2, which decompose in the temperature range of 280–450 °C [13,34,35]. It can be seen that at least peaks #4 and #5 are still present on the desorption spectra after air storage at 150 and 200 °C, though PESCC is fully eliminated at these temperatures. Thus, hydrogen related to the high-temperature peaks #4 and #5 is not likely responsible for PESCC. This conclusion corroborates the fact that the concentration of hydrogen extracted in the temperature intervals 25–450 and 300–450 °C poorly correlates with the propagation of PESCC (Figure 13a). The low-temperature peak indicated as #1 in Figure 12b is attributed to hydrogen, which freely evolves from the CP layer at room temperature. This is evidenced by the absence of this peak on the thermal desorption spectra corresponding to the pre-exposed specimens subjected to air storage at 24 °C for 24 h. Moreover, the bubbles of hydrogen gas emanating from the pre-exposed specimen can be seen by the naked eye when this specimen is submerged in an inert liquid such as CCl_4 [26]. This hydrogen can be associated with the molecular or atomic hydrogen liberated from the corrosion product layer as well as with hydrogen, which is produced in situ by the corrosion reaction [26]. Since the pre-exposed specimens do suffer from appreciable embrittlement after 24 h of air storage at 24 °C but do not exhibit peak #1 in the thermal desorption spectra, the hydrogen associated with this peak is not sufficient to be the sole reason for PESCC. Nevertheless, the contribution of this hydrogen to the observed embrittlement is possible because the reduction in peak #1 due to air storage at 24 °C is accompanied by the partial recovery of ductility. The nature of peaks #2 and #3 is still unknown. Presumably, they can be associated with the desorption of atomic or molecular hydrogen stored within the corrosion product layer or with the thermal decomposition of some hydrogen-containing components in this layer. As follows from Figure 12b, both peaks contribute considerably to the concentration of hydrogen extracted below 300 °C, which is found to be correlated with the extension of PESCC (Figure 13a), as well as with the size of the specific zones on the fracture surface (Figure 13b). Thus, hydrogen associated with peaks #1–3 can likely be involved in the PESCC phenomenon. However, the detailed analysis of origin and activation energies of hydrogen in different traps in the surface layer with corrosion products has yet to be carried out, e.g., methodologically similarly to that used in [36] to clarify the hydrogen distribution in the environmentally embrittled Al-Cu-Mg alloy.

3.4. Diffusible Hydrogen in the Matrix

The results of the present study show that PESCC of the alloy ZK60 can be completely eliminated by air storage of the specimens pre-exposed to the liquid corrosive environment. For the explanation of such a prominent effect of air storage, the contribution of a few factors, which can potentially affect the PESCC behaviour, is to be considered here.

Several authors suggested that the partial elimination of PESCC of Mg and its alloys caused by air storage in a desiccator at room temperature or by vacuum annealing at elevated temperatures can be attributed to the removal of diffusible hydrogen, which had been accumulated in the bulk metal during pre-exposure [5,7,10]. However, the concentration of diffusible hydrogen in the matrix of the pre-exposed specimens is negligible, as has been shown in the present and previous studies [16,19,20]. Additionally, let us reiterate that the ductility loss caused by PESCC can be completely eliminated by the removal of corrosion products from the side surface. Consequently, the absorption of diffusible hydrogen in the bulk of Mg being exposed to the corrosion solution is likely quite limited. In particular, it was suggested that the deep penetration of hydrogen into the metal is hindered by the surface layers of Mg hydroxide or hydride forming on the Mg surface during corrosion in aqueous solutions [20]. The diffusion rates of hydrogen in both these compounds are extremely low [9]. Thus, the PESCC phenomenon and its suppression by the storage in air can hardly be associated with the concentration of diffusible hydrogen in the matrix.

3.5. The Corrosion Solution in the Corrosion Product Layer

The present and other reports [6,16,20,23] have documented that the peripheral part of the fracture surface of ZK60 specimens, which were SSRT tested in air after pre-exposure, is abundantly covered by corrosion products. Several independent experimental observations convincingly demonstrated that these products are created during the SSRT testing in air, when none of the external aggressive environments interact with the surface of the specimens. To be more specific, it is strongly supported by the following observations: (i) no secondary cracks containing corrosion products inside are found in the pre-exposed specimens before mechanical testing [6], and (ii) the length and area of corrosion product region on the fracture surface are reduced with the increasing strain rate [6] and (iii) the temperature of air storage [present study]. If the corrosion product regions were formed during pre-exposure, they would be observed in the cross-section of the specimen right after pre-exposure, while their size would not be affected by the strain rate or conditions of air storage. Assuming that ambient air, in which the SSRT testing is carried out, cannot cause the formation of extensive corrosion products, we can conclude that there must be some corrosive medium present in the side surface layer of corrosion products or at the interface between this layer and the surface of the specimen during SSRT testing. The interaction of this corrosive medium with the crack tip is, therefore, decided to be responsible for the formation of the corrosion product region on the fracture surface. Moreover, the correlation between the length of this region and the degree of PESCC observed in the present study suggests that the corrosion solution contained within the corrosion product layer plays a crucial role in the mechanism of PESCC.

Although the nature of the corrosive medium being stored within the corrosion product layer is questionable and additional comprehensive investigation addressing this issue is required, some plausible explanations can be proposed. For example, it might be supposed that during pre-exposure, some amount of the original corrosion solution is sealed within the discontinuities, such as microcracks and voids, which are abundantly present within the corrosion product layer. Alternatively, the corrosive medium can probably be produced inside this layer after extraction of the specimen from the corrosion solution. It has been reported that the Mg hydroxide can react with CO_2 from ambient air, leading to the formation of magnesite $MgCO_3$ and water [37,38]. Probably, some components of corrosion products, such as $MgCl_2$, can be dissolved in this water, thus making the solution even more aggressive. In favour of this scenario, one may recall the fact that the corrosion product region is always Cr-free. The original corrosion solution contains $K_2Cr_2O_7$ producing the passive film on the Mg surface. The presence of this film is evidenced by the high concentration of Cr in the side surface layer of corrosion products in the pre-exposed specimens as well as on the fracture surface of the specimens tested in the corrosion solution. In contrast to $MgCl_2$, the chromates, which are formed on the Mg surface, are sparingly soluble in water. Thus, the lack of Cr in the newly formed corrosion solution is explainable. The reaction of the saline water with the crack surface would produce MgO, $Mg(OH)_2$, and $MgCl_2$, providing the high concentration of oxygen and chlorine on the elemental maps obtained by EDX from the corrosion product region on the fracture surface.

Regardless of its origin, the corrosion solution within the corrosion product layer should contain water to activate and maintain the electrochemical corrosion reaction producing corrosion products on the fracture surface. It can be suggested that air storage after pre-exposure is accompanied by the evaporation of this water from the corrosion product layer. The amount of evaporated water should grow with increasing temperature and time of air storage, causing consumption of the available corrosion solution. This can explain the reduction in corrosion product region on the fracture surface with time and temperature of air storage. As has been discussed above, the size of this region decreases progressively within the temperature range from 50 to 120 °C and almost vanishes after storage in air at 150–200 °C. The complete evaporation of water in this temperature range is well expected. The size of the corrosion product region is also reduced with the increasing time of air storage. However, unlike the effect of temperature, the longer time of air storage results in an increasing number of surface cracks. Probably, during longer post-exposure storage, more local volumes of hydroxide transform into magnesite with the release of water, thus creating a greater number of favourable sites for crack initiation. Nevertheless, the size of the corrosion product region decreases with the time of air storage because the total amount of the corrosion solution available at the cracks' initiation sites decreases due to the evaporation of water.

Despite the obvious interaction of the corrosive medium with the surface of the propagating cracks during mechanical testing of the pre-exposed specimens, the exact role of this interaction in the mechanism of the crack growth remains unclear. It was shown previously that the corrosion product region is mainly produced by intergranular cracking [6]. For example, the anodic dissolution along the grain boundaries, which are enriched with noble secondary phase particles, might be responsible for this kind of fracture mode. On the other hand, the anodic dissolution of Mg is always accompanied by the cathodic reaction of hydrogen evolution. The adsorption of this hydrogen at the crack tip or its absorption within a few atomic layers beneath the surface can cause intergranular cracking, as well as transgranular cleavage through the HE mechanism, referred to as adsorption-induced dislocation emission (AIDE) [21,39]. Furthermore, the corrosion solution being in contact with the juvenile metal surface at the crack tip can possibly act as the surface active liquid, inducing the embrittlement through Rehbinder's effect [40]. Thus, it is not necessary that the anodic dissolution accompanied by the formation of corrosion products on the fracture surface is the rate-controlling factor for the crack propagation.

3.6. Hydrogen in the Corrosion Product Layer

It is found that the corrosion product region is always followed by the quasi-brittle region, which exhibits no signs of any interaction with the corrosion solution. All features indicating the signs of corrosion on the fracture surface, including the elevated concentration of oxygen and chlorine as well as the microcracks attributed to the crust of corrosion products, cease to appear abruptly at the boundary between the corrosion product region and the quasi-brittle region. Thus, there should be a factor other than the direct contact of the crack tip with the corrosion solution, which drives the crack growth producing the quasi-brittle region. This factor should also be associated with the side surface corrosion product layer because the quasi-brittle region is absent on the fracture surfaces of the reference and corrosion product-free specimens. Furthermore, similarly to that responsible for the corrosion product region, this factor should be eliminated by storage in air, since the quasi-brittle region is found to be reduced under increasing time and temperature of air storage. Probably, the only factor which might act this way is hydrogen.

As was mentioned above, the evolution of hydrogen is an inalienable part of the corrosion process of Mg in aqueous solutions. Hydrogen can be produced during pre-exposure of the specimen as well as during the subsequent air storage if the corrosion solution sealed within the corrosion product layer interacts with the bare metal of the specimen. Indeed, it was shown recently that a large portion of hydrogen gas evolved from the pre-exposed specimen during its 24 h storage in CCl_4 [26]. The volume of this hydrogen summed with that of hydrogen extracted from the same specimen during subsequent gas analysis was twice as high as the volume of hydrogen extracted from the counterpart specimen subjected to the gas analysis right after pre-exposure. This was concluded to be evidence for the generation of hydrogen via the corrosion reaction, which occurs within the surface layer of the pre-exposed specimen stored in the inert environment. The part of hydrogen being produced both during pre-exposure and subsequent air storage can likely be accumulated inside the discontinuities of the side surface corrosion product layer or at the interface between this layer and the bare metal. Presumably, this hydrogen can stay within the corrosion product layer in the molecular and atomic forms. Along with time after extraction of the specimen from the corrosion solution, the weakly bonded part of this hydrogen desorbs from the corrosion product layer even at room temperature. This is witnessed by the low-temperature desorption peak in Figure 12b as well as by the naked-eye-visible hydrogen bubbles released from the specimen immersed in CCl_4 right after pre-exposure [26]. The remaining part of hydrogen sitting within the corrosion product layer is bonded more strongly and, therefore, desorbs only at higher temperatures and longer times. At least a part of hydrogen extracted from the specimen below 300 °C during the thermal desorption analysis can be associated with this strongly bonded hydrogen. It is established that the concentration of hydrogen extracted below 300 °C correlates well with both ductility and fractographic features of the pre-exposed specimens; these features include the lengths of the PESCC zone, corrosion product region, and quasi-brittle region. The observed correlation favours the suggestion that hydrogen being stored inside the corrosion product layer and extracted at relatively low temperatures can be involved in the mechanism of PESCC. Apparently, the post-exposure air storage results in the desorption of hydrogen from the corrosion product layer, thus suppressing PESCC. The correlation between the concentration of hydrogen and the size of the corrosion product region indicates that there is also a relationship between the amount of hydrogen and the corrosion solution stored within the corrosion product layer. Probably, both substances are contained together inside the same collectors because the corrosion solution can generate hydrogen.

As has been mentioned, it is unclear whether hydrogen is responsible for the intergranular crack growth producing the corrosion product region on the fracture surface or not. However, the formation of the quasi-brittle region is likely a hydrogen-assisted process. Various morphological signatures of ductile fracture, which are typical of the quasi-brittle region, indicate that plastic deformation should be involved to some extent into the mechanism of hydrogen-assisted cracking observed. Taking into account that adsorbed hydrogen, rather than absorbed one, plays the key role in this process, one can plausibly suggest that AIDE is the most probable mechanism of the crack growth producing the quasi-brittle region. This mechanism implies that the dislocation emission from the crack tip is facilitated due to the adsorption of hydrogen atoms, thus promoting the locally ductile crack growth [21,41]. The propagation of such crack produces fine slip markings, dimples, and tear ridges on the fracture surface, which represent the time-extended plastic processes rather than instant brittle failures. It was also suggested that AIDE is responsible for the fluted facets on the fracture surface of pure Mg failing due to SCC [39]. A similar fluted morphology is also commonly observed within the quasi-brittle region on the fracture surface of the ZK60 and AZ31 alloys SSRT tested in the corrosion solution [24,42] or after pre-exposure [6,20,23].

3.7. Microstructural Transformations

As has been shown above, the long post-exposure storage in air at temperatures above 100 °C affects the strength of the alloy ZK60 in the reference and corrosion product-free state. This degradation of strength is likely attributed to some ageing-related microstructural transformations, which are not completed during the preceding short vacuum annealing at 250 °C. It is important to note that there is an apparent interplay between these transformations in the microstructure and the PESCC phenomenon occurring in the alloy. This can be clearly traced by the anomalous behaviour of mechanical properties, fractographic features, and hydrogen concentration at around 120 °C on the graphs shown in Figures 2, 4a, 9a and 12a. Among all fractographic characteristics, the number of cracks is affected by these microstructural transformations to the greatest extent. Although the exact nature of the microstructural transformations and their role in PESCC were not investigated in the present study, a probable explanation for the observed results can be proposed. It is known that the mechanical properties of the alloy ZK60 can be influenced by ageing at temperatures above 100 °C [43,44] due to precipitation of some secondary phase particles such as Mg_2Zn_3 [45]. It has been reported that the precipitation of such particles leads to the increase in strength and hardness of Mg alloys in general and in ZK60 in particular, provided that the specimens are not overaged [46]. Otherwise, the coalescence of secondary phase particles results in the concomitant decrease in strength [43,44]. In particular, it was shown that ageing at 150 °C induced the reduction in strength and hardness of ZK60 if ageing lasted longer than 20 h [46]. Thus, the observed decrease in strength of ZK60 after 24 h of air storage above 100 °C is quite reasonable. Since most of the secondary phase particles in Mg alloys are noble with respect to the Mg matrix, they commonly provoke local anodic dissolution of surrounding metal submersed to an electrolyte. The corrosion pits produced at such anodic sites act as favourable nucleation points for the stress corrosion cracks [47]. Thus, assuming that the corrosion solution is likely to present within the layer of corrosion products and considering that it interacts with the bare metal during SSRT testing, it is reasonable to expect the greater number of cracks in the pre-exposed specimens to have a larger volume fraction of secondary phase particles in the microstructure. The greater number of active corrosion sites should also produce more hydrogen gas and corrosion products; this is supported by the anomalous increase in the concentration of hydrogen extracted in the temperature intervals of 24–300 and 300–450 °C, respectively.

3.8. The Scope of Further Research

Experimental results of the present study as well as the proposed discussion on the mechanisms of PESCC are specifically relevant to the ZK60 alloy. Since this alloy is representative of a broad class of wrought Mg alloys, one can likely expect the similar PESCC and SCC behaviour in other alloys of this class of materials. Nevertheless, the additional data pertinent to these phenomena occurring in Mg alloys of other chemical compositions is largely demanded to gain deeper understanding the generality of the observed phenomena on the one hand and to relate them to the chemistry of the alloy interacting with the aggressive environment on the other. In particular, the effect of the side surface layer of corrosion products in the mechanism of PESCC in different Mg alloys is to be elucidated. Furthermore, some other materials, e.g., structural steels and Ti alloys, experience appreciable embrittlement after pre-exposure to aggressive environments [48–51]. This embrittlement is generally accepted to be due to diffusible hydrogen, which, however, plays a less important role in PESCC of Mg alloys, as was shown above. Thus, the generality of the findings gained in the present study in regard to the driven mechanisms of PESCC is to be clarified also on a more global scale accounting for a wider range of metallic materials. The more specific experimental aspects related to the PESCC phenomenon in Mg alloys, such as the nature of the individual peaks on hydrogen thermal desorption spectra, the relationship between the microstructure and PESCC, as well as the origin and the role of the corrosion solution contained within the corrosion product layer, are of particular interest and will be considered in the forthcoming studies.

4. Conclusions

The results obtained in the present paper show that the pre-exposure stress corrosion cracking (PESCC) of the alloy ZK60 can be fully or partially inhibited by air storage after pre-exposure in corrosion solution. The increase in time and temperature of air storage promotes the suppression of PESCC. The complete elimination of PESCC of the alloy ZK60 pre-exposed to the 4% NaCl + 4% $K_2Cr_2O_7$ corrosive solution for 1.5 h can be achieved by 24 h of storage in air at 150–200 °C. The suppression of PESCC due to air storage is accompanied by the decrease in hydrogen concentration in the corrosion product layer and is manifested by the recovery of mechanical properties of the alloy as well as by the reduction in the brittle-like (PESCC) zone on the fracture surface, including both its peripheral region covered with corrosion products and the subsequent quasi-brittle region. It is suggested that the side surface layer of the corrosion product on the pre-exposed specimens serves as a "container" for the embrittling agents such as hydrogen and the corrosion solution, which interact with the bare metal during SSRT testing in air and are responsible for the formation of characteristic corrosion product region and quasi-brittle region on the fracture surface, respectively. The strong correlations between the length of the corrosion product region, the concentration of hydrogen extracted below 300 °C, and the extent of PESCC of the pre-exposed specimens suggest that both hydrogen and the corrosion solution stored within the side surface layer of corrosion products play vital roles in the mechanism of PESCC. The suppression of PESCC by air storage is due to the desorption of hydrogen and evaporation of the corrosion solution from the side surface layer of corrosion products, as is witnessed by the decrease in the hydrogen concentration as well as by the reduction in the size of characteristic regions on the fracture surface with the increasing temperature and time of the air storage.

Author Contributions: E.M.: Conceptualisation, methodology, investigation, formal analysis, writing—original draft preparation, review and editing. V.P.: Investigation, formal analysis. P.M.: Investigation, formal analysis. D.M.: Conceptualisation, resources, supervision, project administration. A.V.: Conceptualisation, writing—original draft preparation, review and editing, supervision, project administration, funding acquisition. All authors have read and agreed to the published version of the manuscript.

Funding: Financial support from the Russian Science Foundation through the grant-in-aid no. 18-19-00592 and 21-79-10378 is gratefully appreciated. The effect of temperature of air storage on the mechanical performance of the reference and corrosion product-free specimens was studied within project no. 21-79-10378. All other data relate to project no. 18-19-00592.

Institutional Review Board Statement: Not applicable.

Informed Consent Statement: Not applicable.

Data Availability Statement: The raw/processed data required to reproduce these findings cannot be shared at this time as the data also form part of an ongoing study. Specific requests can be directed to the corresponding author.

Conflicts of Interest: The authors declare no conflict of interest.

References

1. Makar, G.L.; Kruger, J.; Sieradzki, K. Stress Corrosion Cracking of Rapidly Solidified Magnesium-Aluminum Alloys. *Corros. Sci.* **1993**, *34*, 1311–1342. [CrossRef]
2. Wang, S.D.; Xu, D.K.; Wang, B.J.; Sheng, L.Y.; Han, E.H.; Dong, C. Effect of Solution Treatment on Stress Corrosion Cracking Behavior of an As-Forged Mg-Zn-Y-Zr Alloy. *Sci. Rep.* **2016**, *6*, 29471. [CrossRef] [PubMed]
3. Choudhary, L.; Singh Raman, R.K. Magnesium Alloys as Body Implants: Fracture Mechanism under Dynamic and Static Loadings in a Physiological Environment. *Acta Biomater.* **2012**, *8*, 916–923. [CrossRef] [PubMed]
4. Cai, C.; Song, R.; Wen, E.; Wang, Y.; Li, J. Effect of Microstructure Evolution on Tensile Fracture Behavior of Mg-2Zn-1Nd-0.6Zr Alloy for Biomedical Applications. *Mater. Des.* **2019**, *182*, 108038. [CrossRef]
5. Stampella, R.S.; Procter, R.P.M.; Ashworth, V. Environmentally-Induced Cracking of Magnesium. *Corros. Sci.* **1984**, *24*, 325–337. [CrossRef]
6. Merson, E.; Poluyanov, V.; Myagkikh, P.; Merson, D.; Vinogradov, A. Effect of Strain Rate and Corrosion Products on Pre-Exposure Stress Corrosion Cracking in the ZK60 Magnesium Alloy. *Mater. Sci. Eng. A* **2022**, *830*, 142304. [CrossRef]
7. Kappes, M.; Iannuzzi, M.; Carranza, R.M. Pre-Exposure Embrittlement and Stress Corrosion Cracking of Magnesium Alloy AZ31B in Chloride Solutions. *Corrosion* **2014**, *70*, 667–677. [CrossRef]
8. Song, R.G.; Blawert, C.; Dietzel, W.; Atrens, A. A Study on Stress Corrosion Cracking and Hydrogen Embrittlement of AZ31 Magnesium Alloy. *Mater. Sci. Eng. A* **2005**, *399*, 308–317. [CrossRef]
9. Dubey, D.; Kadali, K.; Panda, S.S.; Kumar, A.; Jain, J.; Mondal, K.; Singh, S.S. Comparative Study on the Stress Corrosion Cracking Susceptibility of AZ80 and AZ31 Magnesium Alloys. *Mater. Sci. Eng. A* **2020**, *792*, 139793. [CrossRef]
10. Song, Y.; Liu, Q.; Wang, H.; Zhu, X. Effect of Gd on Microstructure and Stress Corrosion Cracking of the AZ91-extruded Magnesium Alloy. *Mater. Corros.* **2021**, *2*, 1189–1200. [CrossRef]
11. Qi, F.; Zhang, X.; Wu, G.; Liu, W.; Wen, L.; Xie, H.; Xu, S.; Tong, X. Effect of Heat Treatment on the Stress Corrosion Cracking Behavior of Cast Mg-3Nd-3Gd-0.2Zn-0.5Zr Alloy in a 3.5 Wt% NaCl Salt Spray Environment. *Mater. Charact.* **2022**, *183*, 111630. [CrossRef]
12. Chen, L.; Blawert, C.; Yang, J.; Hou, R.; Wang, X.; Zheludkevich, M.L.; Li, W. The Stress Corrosion Cracking Behaviour of Biomedical Mg-1Zn Alloy in Synthetic or Natural Biological Media. *Corros. Sci.* **2020**, *175*, 108876. [CrossRef]
13. Kappes, M.; Iannuzzi, M.; Carranza, R.M. Hydrogen Embrittlement of Magnesium and Magnesium Alloys: A Review. *J. Electrochem. Soc.* **2013**, *160*, C168–C178. [CrossRef]

14. Chakrapani, D.G.; Pugh, E.N. Hydrogen Embrittlement in a Mg-Al Alloy. *Metall. Trans. A* **1976**, *7*, 173–178. [CrossRef]
15. Bobby Kannan, M.; Dietzel, W. Pitting-Induced Hydrogen Embrittlement of Magnesium-Aluminium Alloy. *Mater. Des.* **2012**, *42*, 321–326. [CrossRef]
16. Jafari, S.; Raman, R.K.S.; Davies, C.H.J. Stress Corrosion Cracking of an Extruded Magnesium Alloy (ZK21) in a Simulated Body Fluid. *Eng. Fract. Mech.* **2018**, *201*, 47–55. [CrossRef]
17. Prabhu, D.B.; Nampoothiri, J.; Elakkiya, V.; Narmadha, R.; Selvakumar, R.; Sivasubramanian, R.; Gopalakrishnan, P.; Ravi, K.R. Elucidating the Role of Microstructural Modification on Stress Corrosion Cracking of Biodegradable Mg–4Zn Alloy in Simulated Body Fluid. *Mater. Sci. Eng. C* **2020**, *106*, 110164. [CrossRef]
18. Chen, K.; Lu, Y.; Tang, H.; Gao, Y.; Zhao, F.; Gu, X.; Fan, Y. Effect of Strain on Degradation Behaviors of WE43, Fe and Zn Wires. *Acta Biomater.* **2020**, *113*, 627–645. [CrossRef]
19. Jiang, P.; Blawert, C.; Bohlen, J.; Zheludkevich, M.L. Corrosion Performance, Corrosion Fatigue Behavior and Mechanical Integrity of an Extruded Mg4Zn0.2Sn Alloy. *J. Mater. Sci. Technol.* **2020**, *59*, 107–116. [CrossRef]
20. Merson, E.; Poluyanov, V.; Myagkikh, P.; Merson, D.; Vinogradov, A. On the Role of Pre-Exposure Time and Corrosion Products in Stress-Corrosion Cracking of ZK60 and AZ31 Magnesium Alloys. *Mater. Sci. Eng. A* **2021**, *806*, 140876. [CrossRef]
21. Lynch, S.P. Hydrogen Embrittlement Phenomena and Mechanisms. *Corros. Rev.* **2012**, *30*, 63–133. [CrossRef]
22. Djukic, M.B.; Bakic, G.M.; Sijacki Zeravcic, V.; Sedmak, A.; Rajicic, B. The Synergistic Action and Interplay of Hydrogen Embrittlement Mechanisms in Steels and Iron: Localized Plasticity and Decohesion. *Eng. Fract. Mech.* **2019**, *216*, 106528. [CrossRef]
23. Merson, E.; Poluyanov, V.; Myagkikh, P.; Merson, D.; Vinogradov, A. Inhibiting Stress Corrosion Cracking by Removing Corrosion Products from the Mg-Zn-Zr Alloy Pre-Exposed to Corrosion Solutions. *Acta Mater.* **2021**, *205*, 116570. [CrossRef]
24. Merson, E.; Myagkikh, P.; Poluyanov, V.; Merson, D.; Vinogradov, A. On the Role of Hydrogen in Stress Corrosion Cracking of Magnesium and Its Alloys: Gas-Analysis Study. *Mater. Sci. Eng. A* **2019**, *748*, 337–346. [CrossRef]
25. Merson, E.D.; Poluyanov, V.A.; Myagkikh, P.N.; Merson, D.L.; Vinogradov, A.Y. Effect of Grain Size on Mechanical Properties and Hydrogen Occluding Capacity of Pure Magnesium and Alloy MA14 Subjected to Stress-Corrosion Cracking. *Lett. Mater.* **2020**, *10*, 94–99. [CrossRef]
26. Merson, E.; Poluyanov, V.; Myagkikh, P.; Merson, D.; Vinogradov, A. Evidence for the Presence of Corrosive Solution within Corrosion Products Film in Magnesium Alloy ZK60. *Lett. Mater.* **2022**, *12*, 76–80. [CrossRef]
27. Merson, E.D.; Myagkikh, P.N.; Klevtsov, G.V.; Merson, D.L.; Vinogradov, A. *Effect of Hydrogen Concentration and Strain Rate on Hydrogen Embrittlement of Ultra-Fine-Grained Low-Carbon Steel*; Springer International Publishing: Cham, Switzerland, 2021; Volume 143, ISBN 9783030669485.
28. Merson, E.D.; Krishtal, M.M.; Merson, D.L.; Eremichev, A.A.; Vinogradov, A. Effect of Strain Rate on Acoustic Emission during Hydrogen Assisted Cracking in High Carbon Steel. *Mater. Sci. Eng. A* **2012**, *550*, 408–417. [CrossRef]
29. Merson, E.; Vinogradov, A.; Merson, D.L. Application of Acoustic Emission Method for Investigation of Hydrogen Embrittlement Mechanism in the Low-Carbon Steel. *J. Alloys Compd.* **2015**, *645*, S460–S463. [CrossRef]
30. Linderov, M.; Brilevsky, A.; Merson, D.; Danyuk, A.; Vinogradov, A. On the Corrosion Fatigue of Magnesium Alloys Aimed at Biomedical Applications: New Insights from the Influence of Testing Frequency and Surface Modification of the Alloy ZK60. *Materials* **2022**, *15*, 567. [CrossRef]
31. Chen, J.; Tan, L.; Yang, K. Effect of Heat Treatment on Mechanical and Biodegradable Properties of an Extruded ZK60 Alloy. *Bioact. Mater.* **2017**, *2*, 19–26. [CrossRef]
32. Robertson, I.M.; Sofronis, P.; Nagao, A.; Martin, M.L.; Wang, S.; Gross, D.W.; Nygren, K.E. Hydrogen Embrittlement Understood. *Metall. Mater. Trans. A* **2015**, *46*, 2323–2341. [CrossRef]
33. Laureys, A.; Depover, T.; Petrov, R.; Verbeken, K. Influence of Sample Geometry and Microstructure on the Hydrogen Induced Cracking Characteristics under Uniaxial Load. *Mater. Sci. Eng. A* **2017**, *690*, 88–95. [CrossRef]
34. Chen, J.; Wang, J.; Han, E.; Dong, J.; Ke, W. States and Transport of Hydrogen in the Corrosion Process of an AZ91 Magnesium Alloy in Aqueous Solution. *Corros. Sci.* **2008**, *50*, 1292–1305. [CrossRef]
35. Evard, E.A.; Gabis, I.E.; Murzinova, M.A. Kinetics of Hydrogen Liberation from Stoichiometric and Nonstoichiometric Magnesium Hydride. *Mater. Sci.* **2007**, *43*, 620–633. [CrossRef]
36. Safyari, M.; Moshtaghi, M.; Kuramoto, S.; Hojo, T. Influence of Microstructure-Driven Hydrogen Distribution on Environmental Hydrogen Embrittlement of an Al–Cu–Mg Alloy. *Int. J. Hydrogen Energy* **2021**, *46*, 37502–37508. [CrossRef]
37. Jönsson, M.; Persson, D.; Thierry, D. Corrosion Product Formation during NaCl Induced Atmospheric Corrosion of Magnesium Alloy AZ91D. *Corros. Sci.* **2007**, *49*, 1540–1558. [CrossRef]
38. Jönsson, M.; Persson, D. The Influence of the Microstructure on the Atmospheric Corrosion Behaviour of Magnesium Alloys AZ91D and AM50. *Corros. Sci.* **2010**, *52*, 1077–1085. [CrossRef]
39. Lynch, S.P.; Trevena, P. Stress Corrosion Cracking and Liquid Metal Embrittlement in Pure Magnesium. *Corrosion* **1988**, *44*, 113–124. [CrossRef]
40. Malkin, A.I. Regularities and Mechanisms of the Rehbinder's Effect. *Colloid J.* **2012**, *74*, 223–238. [CrossRef]
41. Lynch, S.P. Environmentally Assisted Cracking: Overview of Evidence for an Adsorption-Induced Localised-Slip Process. *Acta Metall.* **1988**, *36*, 2639–2661. [CrossRef]

42. Merson, E.; Poluyanov, V.; Myagkikh, P.; Merson, D.; Vinogradov, A. Fractographic Features of Technically Pure Magnesium, AZ31 and ZK60 Alloys Subjected to Stress Corrosion Cracking. *Mater. Sci. Eng. A* **2020**, *772*, 138744. [CrossRef]
43. Wu, Y.J.; Zhang, Z.M.; Li, B.C. Effect of Aging on the Microstructures and Mechanical Properties of AZ80 and ZK60 Wrought Magnesium Alloys. *Sci. Sinter.* **2010**, *42*, 161–168. [CrossRef]
44. Li, Y.; Zhang, Z.M.; Xue, Y. Influence of Aging on Microstructure and Mechanical Properties of AZ80 and ZK60 Magnesium Alloys. *Trans. Nonferrous Met. Soc. China* **2011**, *21*, 739–744. [CrossRef]
45. Pan, F.S.; Wang, W.M.; Ma, Y.L.; Zuo, R.L.; Tang, A.T.; Zhang, J. Investigation on the Alloy Phases in As-Aged ZK60 Magnesium Alloy. *Mater. Sci. Forum* **2005**, *488–489*, 181–184. [CrossRef]
46. Chen, X.; Liu, L.; Pan, F. Strength Improvement in ZK60 Magnesium Alloy Induced by Pre-Deformation and Heat Treatment. *J. Wuhan Univ. Technol. Mater. Sci. Ed.* **2016**, *31*, 393–398. [CrossRef]
47. Winzer, N.; Atrens, A.; Song, G.; Ghali, E.; Dietzel, W.; Kainer, K.U.; Hort, N.; Blawert, C. A Critical Review of the Stress Corrosion Cracking (SCC) of Magnesium Alloys. *Adv. Eng. Mater.* **2005**, *7*, 659–693. [CrossRef]
48. Skilbred, E.S.; Palencsár, S.; Dugstad, A.; Johnsen, R. Hydrogen Uptake during Active CO2-H2S Corrosion of Carbon Steel Wires in Simulated Annulus Fluid. *Corros. Sci.* **2022**, *199*, 110172. [CrossRef]
49. Nagaoka, A.; Yokoyama, K.; Sakai, J. Evaluation of Hydrogen Absorption Behaviour during Acid Etching for Surface Modification of Commercial Pure Ti, Ti-6Al-4V and Ni-Ti Superelastic Alloys. *Corros. Sci.* **2010**, *52*, 1130–1138. [CrossRef]
50. Brass, A.M.; Chêne, J. Hydrogen Uptake in 316L Stainless Steel: Consequences on the Tensile Properties. *Corros. Sci.* **2006**, *48*, 3222–3242. [CrossRef]
51. Li, L.; Mahmoodian, M.; Li, C.Q.; Robert, D. Effect of Corrosion and Hydrogen Embrittlement on Microstructure and Mechanical Properties of Mild Steel. *Constr. Build. Mater.* **2018**, *170*, 78–90. [CrossRef]

Article

Effect of the Severe Plastic Deformation on the Corrosion Resistance of a Tantalum–Tungsten Alloy

Guoqiang Ma [1], Man Zhao [2], Song Xiang [3], Wanquan Zhu [2], Guilin Wu [1,4,*] and Xinping Mao [1,4]

[1] Beijing Advanced Innovation Center for Materials Genome Engineering, University of Science and Technology Beijing, Beijing 100083, China
[2] College of Materials Science and Engineering, Chongqing University, Chongqing 400044, China
[3] College of Materials and Metallurgy, Guizhou University, Guiyang 550025, China
[4] Yangjiang Branch, Guangdong Laboratory for Materials Science and Technology (Yangjiang Advanced Alloys Laboratory), Yangjiang 529500, China
* Correspondence: guilinwu@ustb.edu.cn; Tel.: +86-18513671811

Abstract: Tantalum and its alloys are regarded as equipment construction materials for processing aggressive acidic media due to their excellent properties. In this study, the influence of severe rolling (90%) on the dissolution rate of a cold-rolled Ta-4%W sheet in different directions was investigated during immersion testing and the corresponding mechanism was discussed. The results show that the dissolution rate of the cold-rolled sample is significantly lower than that of the undeformed sample. The corrosion resistance followed the sequence of "initial" < "90%-ND" < "90%-RD" < "90%-TD", while the strength is in positive correlation with the corrosion resistance. Severe rolling promotes grain subdivision accompanied by long geometrically necessary boundaries and short incidental dislocation boundaries on two scales in the cold-rolled sample. The volume elements enclosed by geometrically necessary boundaries form preferential crystallographic orientations. Such preferential crystallographic orientations can greatly weaken the electrochemical process caused by adjacent volume elements, resulting in greatly reduced corrosion rates in the severely deformed sample. The unexpected finding provides a new idea for tailoring the structures of tantalum alloys to improve both their strength and corrosion resistance.

Keywords: tantalum alloy; severe plastic deformation; grain subdivision; corrosion; crystallographic orientation

1. Introduction

Tantalum (Ta) is a typical rare transition metal. Ta and its alloys have received widespread attention in many fields such as nuclear engineering, the chemical industry, biomaterials, high temperature applications and the electronics industry. This is mainly because Ta and its alloys have excellent properties including high melting points, high corrosion resistance, good biocompatibility, mechanical and electrical properties [1–6]. Due to the excellent corrosion resistance, Ta alloy mill products, especially the binary Ta alloys containing tungsten (W), are used to manufacture processing equipment meant for harsh chemical environments, for example, bayonet heaters, tank liners, heat exchangers and various other components for sulfuric acid production [7]. While many previous studies have mainly focused on the corrosion behaviors of Ta and its alloys in hot, highly corrosive environments [8–11], limited attention has been paid to the relationship between thermomechanical processing and corrosion resistance.

Regardless of the final form of Ta alloy mill products, thermomechanical processing is an essential process. During the thermomechanical processing, the alloy undergoes different forms and/or levels of plastic deformation. Strain path and strain degree, etc., affect the microstructure and final properties of the alloy. According to earlier research results, the corrosion rate increases when metals and alloys undergo plastic deformation [12–14].

The theoretical basis for such results is that dislocations and other defects are generated during the deformation. These defects usually increase the active sites on the metal surface [14]. Plastic deformation increases the rate of anode dissolution but has little effect on the cathode process [13]. For example, Luo et al. [15] investigated a duplex stainless steel in a saturated $Ca(OH)_2$ solution containing 3.5 wt.% NaCl and noted that plastic deformation accelerated the corrosion rate due to the enhancement of metastable pitting susceptibility. Stefec et al. [16] adopted quantitative metallography to statistically analyze the corrosion pits and found the total area and number of pits increased with increasing deformation.

In recent years, a series of breakthroughs have been realized in tailoring nanostructured grains to achieve both high strength and good ductility by severe plastic deformation [17]. However, severe plastic deformation also introduces large proportions of metastable microstructures, grain boundaries and textures, which leads to a more complex corrosion behavior compared to conventional polycrystalline metals and alloys [18]. Wang et al. [19,20] obtained nanocrystalline ingot iron after 93% rolling reduction from conventional polycrystalline and found that the corrosion resistance of the obtained samples was greatly improved in both 1 M HCl and 0.05 M H_2SO_4 + 0.25 M Na_2SO_4 solutions. Lv et al. [21] found that pure iron had similar textures after a rolling reduction of 95.8% at room temperature and in liquid nitrogen conditions, but the grain refinement between them was different. Compared to the original annealed sample, the corrosion resistances of rolled samples were both improved, but the corrosion resistance of the sample rolled at room temperature was better than that of the sample rolled in liquid nitrogen.

The above results show that the presence of metastable microstructures, dislocations, and dislocation boundaries increase the corrosion rate, while the formations of textures and changes in the second phase may inhibit the corrosion rate during deformation. Recently, the potential to improve corrosion resistance by severe plastic deformation has been demonstrated in several separate studies for different metals and alloys. However, the interpretive reasons are different for such improvements [22–25]. The question still remains therefore how the change in corrosion resistance relates to the severe deformation structure for metals and alloys. In our previous studies on the Ta-4%W alloy [26–30], we have reported the microstructural evolution after cold rolling and the corrosion behavior on rolling plane in sulfuric acid. However, the effects of macroscopic orientation and crystallographic texture in different directions on the surface dissolution of cold-rolled sheets are still unknown and they are the key components of the present research.

In this study, corrosion rates in different macroscopic orientations of the cold-rolled sample were tested, and the corrosion morphology and microstructure were characterized. The study focuses on three aspects: (1) a corrosion degree comparison in different macroscopic orientations after severe rolling, (2) the microstructural source of the texture and (3) the relationships between texture and corrosion behavior. Our paper represents the first investigation to establish grain subdivision, dislocation substructure and texture evolution, as well as to establish a direct relationship to corrosion resistance and mechanical properties.

2. Materials and Methods

Ta-4%W ingots were obtained from electron beam melting under high-vacuum conditions. Ingots are forged and annealed in preparation for starting material, which has been described in our previous work [28,30]. The starting plates have a thickness of 10 mm, and then they are rolled at room temperature to a reduction in thickness of 90% (corresponding to a von Mises strain of 2.7). The rolled sheet has a thickness of 1 mm after rolling. In order to reveal the corrosion behaviors on different sections of the rolled sheet, i.e., sections perpendicular to the rolling direction (RD), the transverse direction (TD) and the normal direction (ND), the samples for immersion corrosion were cut from the rolled sheet with a size of 10 mm in length, 1 mm in width, and 0.3 mm in thickness. To obtain precision results, 9 samples were prepared for each tested section.

All the samples were grounded with final 5000 grit sandpaper, cleaned with alcohol and dried with a hair dryer before the corrosion tests. Then, the samples were immersed in a mixture solution of hydrofluoric acid, sulfuric acid and deionized water (1:50:49 by mass) from 7 days to 30 days. To explore the effect of severe plastic deformation on the corrosion rate of Ta-4%W, the initial undeformed sample was also immersed in the test solution and cut to the same size as the deformed sample to avoid possible size effects. The schematic illustration and experimental pictures of the samples' preparation for immersion testing is shown in Figure 1. The samples were weighed with a high precision electronic scale both before and after immersion testing in order to record the mass loss of the alloy, and then the total weight changes for each tested section were converted into the corrosion rate, which is calculated according to the following equation:

$$X \text{ (mm/a)} = 87600 \times /(W/DAT) \tag{1}$$

where W is the weight loss in grams, D is the sample density taken to be 16.6 g·cm^{-3}, A is the area of the sample in cm^2, and T is the time of exposure of the Ta-4%W sample in hours. After immersion, the micro-hardness of each test surfaces was also measured using a load of 500 g and a dwell time of 10 s. The corrosion morphologies of each sample immersed in the test solution after 30 days were characterized by a ZEISS Auriga scanning electron microscope (SEM) attached with an electron backscatter diffraction (EBSD) detector. Transmission electron microscope (TEM) foils were observed in a JOEL JEM 2100 TEM equipped with an online Kikuchi-line analysis system for crystallographic orientation determination [31].

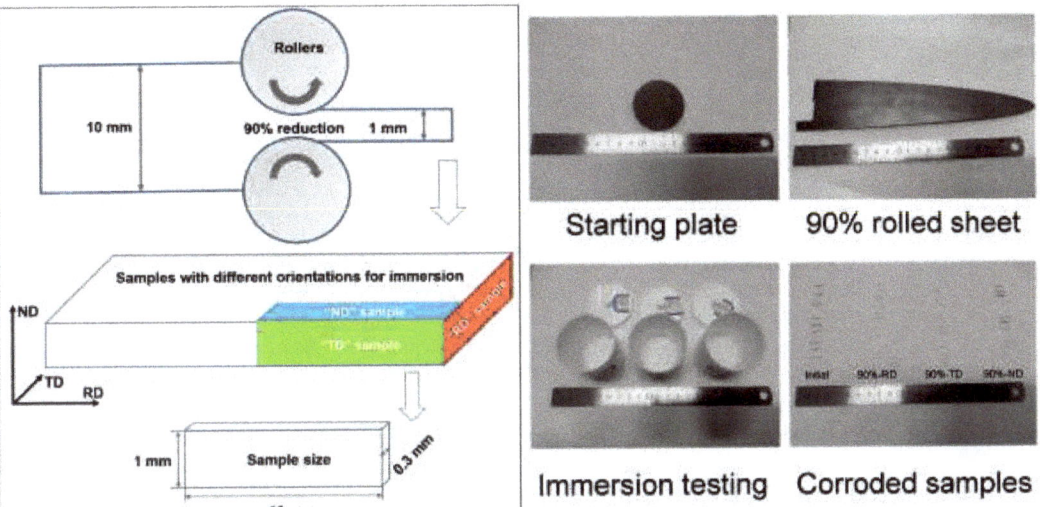

Figure 1. Schematic illustration and experimental pictures of samples for immersion testing: the "initial" sample was cut from the 10 mm-thick plates before cold rolling. The observed surfaces of the "90%-RD", "90%-TD" and "90%-ND" samples were perpendicular to the rolling direction, transverse direction and normal direction, respectively, of the 90% cold-rolled plate.

3. Results and Discussion

Figure 2 shows the corrosion rates of samples in the test solution from 7 days to 30 days and the micro-hardness of the corresponding test surfaces. As shown in Figure 2a, the corrosion rates of all the samples decrease with increasing corrosion time. This may be due to corrosion products gradually accumulating and covering the surfaces of the tested samples. Ta alloys show excellent passivation tendency due to a protective passive stable Ta$_2$O$_5$ oxide film. However, when Ta alloys are immersed in a mixture of hydrofluoric and sulfuric acid, several nanometers of Ta$_2$O$_5$ on the sample surface will dissolve and a corrosion product of fluoride complex (H_2TaF_7) will form. Such corrosion products act as a barrier to further corrosion [11], thus slowing down the corrosion process. Another surprising result is that the corrosion rates of the cold-rolled sample are significantly smaller than those of the initial sample regardless of the macroscopic orientations of the test samples. Among the cold-rolled samples in different orientations, the corrosion rate of the 90%-ND sample is higher than those of the 90%-RD sample and the 90%-TD sample. The microhardness results show that the 90%-TD sample has a maximum value of 261 HV, while the initial sample has a minimum value of 197 HV (Figure 2b). Generally, the corrosion resistance is in an inverse correlation with the strength. However, present results seem to support the conclusion that the harder the surface, the more resistant it is to corrosion. Similar results were also seen in BN-304SS by severe rolling technology [32].

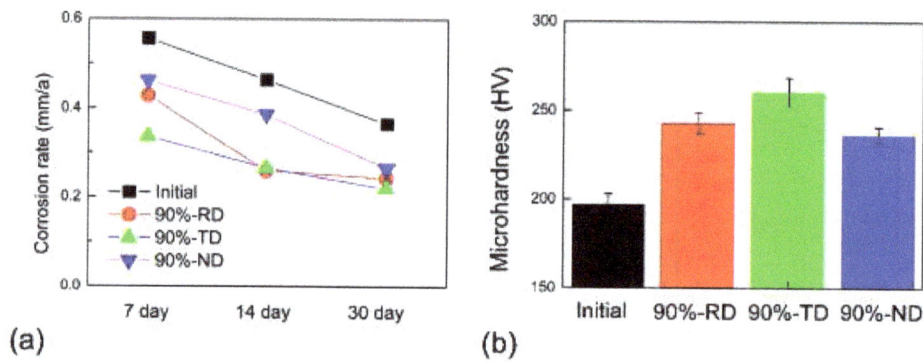

Figure 2. The corrosion rates of Ta-4%W in a 50 wt.% H_2SO_4 solution containing 1 wt.% fluoride ions (**a**) and the micro-hardness of the corresponding test surfaces (**b**).

Figure 3 shows SEM images of the morphologies for the Ta-4%W samples after the immersion test. The corroded surface is uneven between the adjacent grains, which means the shape of the grains is faintly visible for the initial sample (Figure 3a). After 90% cold rolling, the corroded surfaces of the samples with three macroscopic orientations become quite uniform. Based on the comparison of the corrosion morphology for each of the samples, the "90%-TD" sample can be assumed as relatively less corroded, followed by the "90%-RD" sample, then the "90%-ND" sample, while the initial sample can be assumed as more corroded.

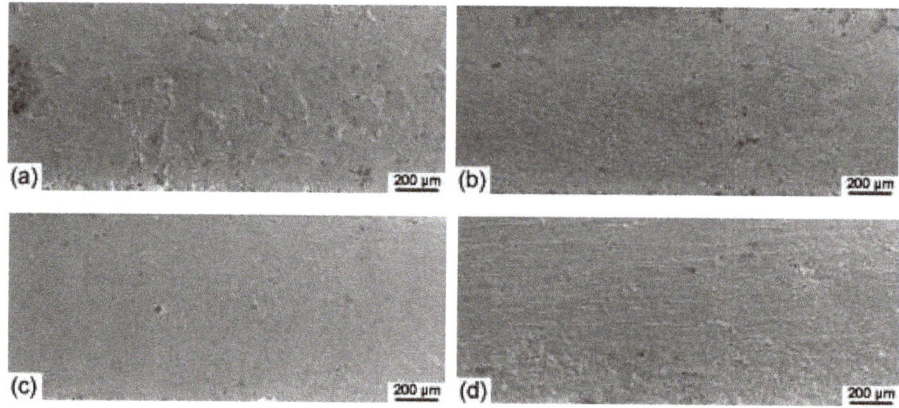

Figure 3. SEM images of the Ta-4%W corrosion samples after 30 days of immersion testing: (**a**) "initial" sample, (**b**) "90%-RD" sample, (**c**) "90%-TD" sample, and (**d**) "90%-ND" sample.

To further understand the difference in corrosion resistance, the corroded surfaces after 30 days of immersion testing were also characterized using EBSD. Due to a high density of 16.6 g/cm^3 for the present alloy, a high absorption ratio will happen when the electron beam falls into the pit areas. Thus, the surface quality of Ta alloys has a significant impact on the detection of EBSD signals, and the corroded sites or pit areas on corroded surfaces formed in immersion testing will not be resolved by EBSD [33]. Figure 4 shows the EBSD orientation maps and inverse pole figure (IPF) of grains detected in the initial sample. As shown in the orientation maps (Figure 4a), "blue" and "green" grains tend to have more severely corroded sites or pit areas (indicated by the black color), while "red" grains have fewer pit areas. The orientations of all the grains detected are shown in Figure 4b. It can be concluded that the {100} grains exhibit a significantly lower corrosion degree compared with {110} and {111} grains, which can also be supported by previous studies for Ta [30,34] and other body-centered cubic (bcc) metals such as Nb [35] and pure Fe [21,36].

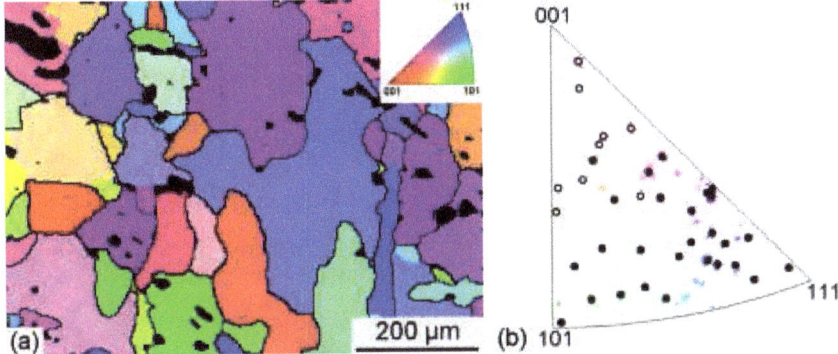

Figure 4. EBSD orientation maps (**a**) and IPF (**b**) of grains detected in the initial sample after 30 days of immersion testing. Note that the black regions in (**a**) are the preferentially corroded sites which cannot be resolved by EBSD. The grains without corroded sites are marked as hollow circles while the grains with corroded sites are marked as solid black circles in (**b**).

Figure 5 shows EBSD maps of the cold-rolled sample on corroded surfaces. Grain subdivision occurs in three directions and there are many dislocation substructures formed on the deformed sample. It is important that preferential corrosion still exists as the initial sample. The preferentially corroded sites, i.e., the black regions which cannot be detected by EBSD, are primarily located in the {111} grains.

Due to the resolution limit of the EBSD technique, finer-scale characterization of the deformation microstructure was studied using TEM (see Figure 6). Severe rolling promotes grain subdivision containing two types of dislocation boundaries. One is the long and extended planar boundaries with a larger scale, and the other is the highly curved morphology dislocation boundaries with a smaller scale (Figure 6a). The long and extended planar dislocation boundaries are known as geometrically necessary boundaries (GNBs), while the highly curved morphology dislocation boundaries are known as incidental dislocation boundaries (IDBs) [37]. The GNB misorientation angle can reach quite high values (average value of 11.9°) to accommodate lattice rotations and can lead to a strong texture evolution (preferential crystallographic orientations) [28]. The IDB misorientation angles (average value of 3.1°) are much smaller than those of GNB and cannot causes significant crystallographic orientation change across the boundary. A sketch of GNBs and IDBs has been shown in Figure 6b.

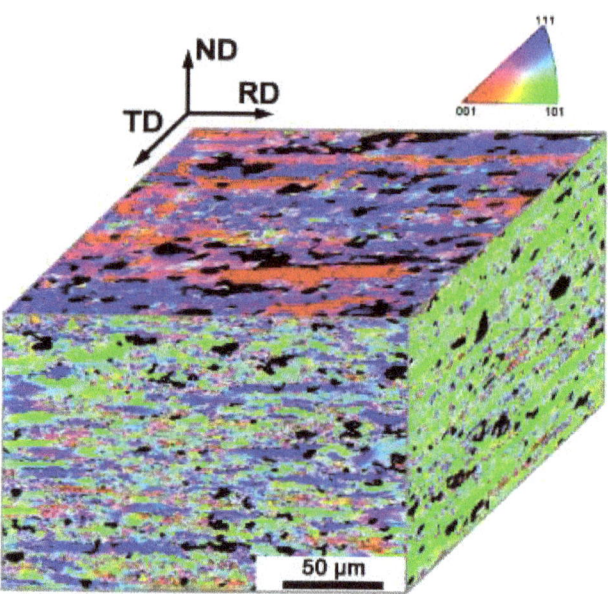

Figure 5. EBSD observation of preferentially corroded sites in the 90% cold-rolled samples after 30 days of immersion testing.

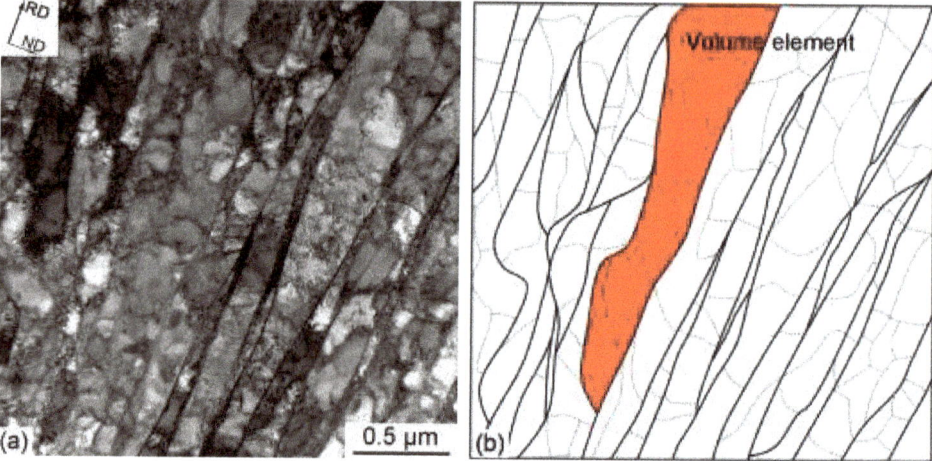

Figure 6. The microstructure of the 90% cold-rolled sample in the longitudinal plane: (**a**) TEM images of the lamellar boundary structure, (**b**) a sketch of the GNBs (solid black lines) and the IDBs (dashed grey lines) in (**a**).

Regarding the effect of plastic deformation on the corrosion behavior, a considerable investigation was carried out in a previous study but conflicting conclusions were reached. For example, experimental results by Stefec, Haraszit and Maric et al. [12,16,38] show that plastic deformation reduces the corrosion resistance of austenitic stainless steels as the strain increases (mostly from low to medium strain). Wang et al. [39] found that plastic deformation reduces the corrosion resistance of CrCoFeMnNi high entropy alloys even when the strain reaches 80%. However, experimental results by Lv and Wang et al. [32,40]

show that plastic deformation improves the corrosion resistance of austenitic stainless steels when the strain is high enough. The phenomenon of improved corrosion resistance due to plastic deformation is also found in BCC iron [19–21]. There are few controversial views on the explanation of plastic deformation reducing corrosion resistance, i.e., numerous dislocations and other crystallographic defects formed after plastic deformation promoting the electrochemical dynamics increase and dissolution rate. However, to the best of our knowledge, no unified view has been reached on the explanation of plastic deformation enhancing corrosion resistance. These mutually contradictory conclusions [19–22,24,25,32,40] mainly relate to (1) the conventional microstructure parameters such as dislocation density, grain size, twin boundaries, high angle boundaries, second phase, (2) the passivation film properties, and (3) even the valence electron configuration at an atomic scale.

It is worth noting that grain subdivision based on volume scales of dislocation boundaries classification and low energy dislocation structures has been observed in many metals and alloys. At certain strain levels (mostly at a von Mises strain of 0.8), the typical microstructural features of S-bands will form due to the emergence of new slip systems [41,42]. The S-band can also lead to the deformation texture being formed and the texture will continue to strengthen with increasing strain. A more efficient source of high-angle boundaries can be found if the overall texture evolution is considered, rather than just the evolution of individual directions. Therefore, grain subdivision, accompanied by a strong texture evolution, can lead to a significant increase in the fraction of high-angle boundaries during deformation [41]. Volume elements are defined as the regions enclosed by high-angle boundaries in TEM [31]. Since most high-angle boundaries belong to the GNBs, the volume elements enclosed by GNBs can be thus assumed to have the same electrochemical properties since they have the same crystallographic orientation and operate the same glide systems. One of the volume elements is illustrated by the color red in Figure 6b.

With the help of an online Kikuchi-line analysis system equipped on a JOEL JEM 2100 TEM [31], the crystallographic orientations of individual volume elements were measured and mapped to inverse pole figure (Figure 7). The crystallographic orientation with RD tends to be <110> in the unit triangle. The crystallographic orientation with TD ranges from <110> to <111> along the edge of the unit triangle. The crystallographic orientation with ND tends to be <111> primarily and tends to be <100> minorly. It is known that the <111> grains have the maximum Taylor factor value, while the <110> grains follow, and the <100> grains have the minimum Taylor factor value. Taylor factor is a parameter that reflects the difficulty of plastic deformation, representing the numbers of activated slip systems during deformation. Thus, the difference in micro-hardness between the macroscopic tests' surfaces is mainly due to different grain orientation distribution [13].

Compared to the uniform distribution of orientation in the initial undeformed sample (Figure 4b), the cold-rolled samples show significant preferential crystallographic orientations regardless of the sample direction on the rolled sheet. Previous studies have shown that the cathodic process preferred to occur in the <100> orientation volume element, while the anodic process mainly occurs in the <110> and <111> orientation volume element [30,35]. Such preferential crystallographic orientations in cold-rolled samples can greatly weaken the electrochemical process caused by the adjacent volume elements, resulting in greatly reduced dissolution rates for the Ta-4%W alloy. Among the cold-rolled samples, the "90%-ND" sample exhibited a worse corrosion resistance than that of the "90%-RD" and "90%-TD" samples. The possible reason for this result is the fact that the "90%-ND" sample has more <100> volume element than the "90%-RD" and "90%-TD" samples.

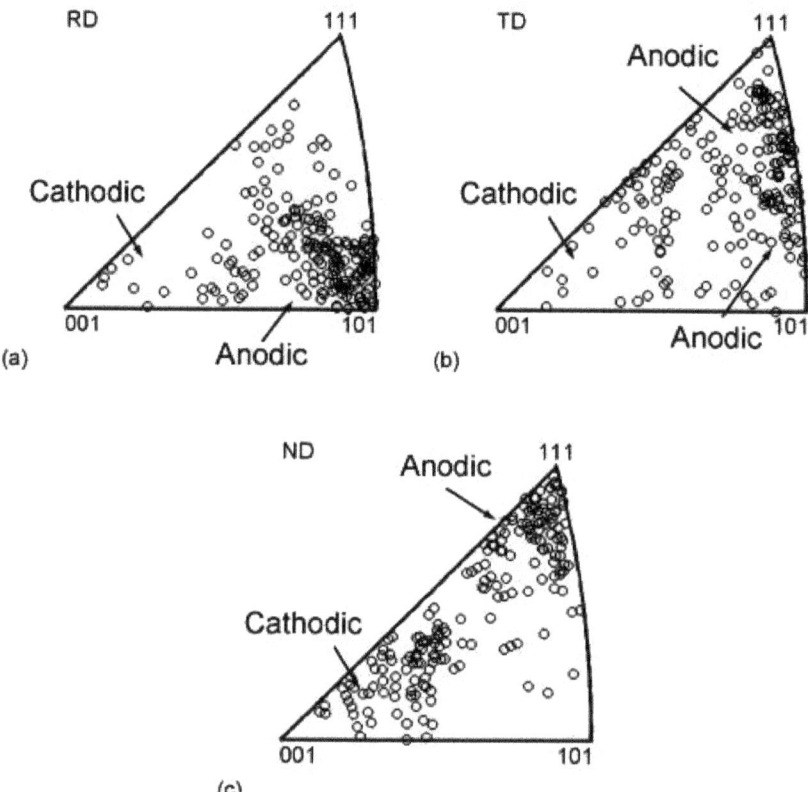

Figure 7. Inverse pole figure showing the (**a**) rolling direction, (**b**) transverse direction, and (**c**) normal directions of the subdivided grains in the 90% cold-rolled samples mapped by TEM.

4. Conclusions

The Ta-4%W alloy was cold rolled to achieve a reduction of 90% in sample thickness. The effect of severe plastic deformation on the corrosion behavior of the Ta-4%W alloy in a mixed solution of sulfuric acid and hydrofluoric acid was investigated. The results show that 90% cold rolling can significantly reduce the dissolution rate of the Ta-4%W alloy regardless of the macroscopic orientation of the cold-rolled samples. The corrosion resistance followed the sequence of "initial" < "90%-ND" < "90%-RD" < "90%-TD". During plastic deformation, two types of dislocation boundaries were formed. One type was the long and extended planar boundaries (GNBs) which subdivide the grains at a larger scale. The other type was the highly curved morphology dislocation boundaries (IDBs) formed between GNBs. Preferential crystallographic orientations were formed through volume elements enclosed by GNBs. The electrochemical process is different due to the difference in crystallographic orientations between the adjacent volume elements. However, the preferential crystallographic orientation might weaken this difference and thus greatly reduce the corrosion rate of the Ta-4%W alloy.

Author Contributions: Data curation, G.M. and M.Z.; formal analysis, G.M., M.Z., S.X. and W.Z.; investigation, G.M.; resources, X.M.; supervision, G.W. and X.M.; writing—original draft, G.M. and G.W; writing—review and editing, G.W. All authors have read and agreed to the published version of the manuscript.

Funding: This work was financially supported by the National Natural Science Foundation of China (NSFC, Nos. 52071038, 51974097).

Institutional Review Board Statement: Not applicable.

Informed Consent Statement: Not applicable.

Conflicts of Interest: The authors declare no conflict of interest.

References

1. Yan, M.; Wang, X.; Zhou, H.; Liu, J.; Zhang, S.; Lu, Y.; Hao, J. Microstructure, mechanical and tribological properties of graphite-like carbon coatings doped with tantalum. *Appl. Surf. Sci.* **2021**, *542*, 148404. [CrossRef]
2. McNamara, K.; Beloshapkin, S.; Hossain, K.M.; Dhoubhadel, M.S.; Tofail, S.A.M. Tantalum coating inhibits Ni-migration from titanium out-diffusion in NiTi shape memory biomedical alloy. *Appl. Surf. Sci.* **2021**, *535*, 147621. [CrossRef]
3. Novakowski, T.J.; Sundaram, A.; Tripathi, J.K.; Gonderman, S.; Hassanein, A. Deuterium desorption from ion-irradiated tantalum and effects on surface morphology. *J. Nucl. Mater.* **2018**, *504*, 1–7. [CrossRef]
4. Zhang, T.; Deng, H.W.; Xie, Z.M.; Liu, R.; Yang, J.F.; Liu, C.S.; Wang, X.P.; Fang, Q.F.; Xiong, Y. Recent progresses on designing and manufacturing of bulk refractory alloys with high performances based on controlling interfaces. *J. Mater. Sci. Technol.* **2020**, *52*, 29–62. [CrossRef]
5. Wang, S.; Wu, Z.H.; Xie, M.Y.; Si, D.H.; Li, L.Y.; Chen, C.; Zhang, Z.; Wu, Y.C. The effect of tungsten content on the rolling texture and microstructure of Ta-W alloys. *Mater. Charact.* **2020**, *159*, 110067. [CrossRef]
6. Liu, Y.; Liu, S.; Deng, C.; Fan, H.; Yuan, X.; Liu, Q. Inhomogeneous deformation of {111}<uvw> grain in cold rolled tantalum. *J. Mater. Sci. Technol.* **2018**, *34*, 2178–2182. [CrossRef]
7. Jakubowicz, J.; Adamek, G.; Sopata, M.; Koper, J.K.; Kachlicki, T.; Jarzebski, M. Microstructure and electrochemical properties of refractory nanocrystalline Tantalum-based alloys. *Int. J. Electrochem. Sci.* **2018**, *13*, 1956–1975. [CrossRef]
8. Gypen, L.A.; Brabers, M.; Deruyttere, A. Corrosion resistance of tantalum base alloys. Elimination of hydrogen embrittlement in tantalum by substitutional alloying. *Mater. Corros.* **1984**, *35*, 37–46. [CrossRef]
9. De Souza, K.A.; Robin, A. Influence of concentration and temperature on the corrosion behavior of titanium, titanium-20 and 40% tantalum alloys and tantalum in sulfuric acid solutions. *Mater. Chem. Phys.* **2007**, *103*, 351–360. [CrossRef]
10. Robin, A. Corrosion behavior of niobium, tantalum and their alloys in boiling sulfuric acid solutions. *Int. J. Refract. Metals Hard Mater.* **1997**, *15*, 317–323. [CrossRef]
11. Aimone, P.; Moser, K. Working with tantalum and tantalum alloys. In *CORROSION 2003*; NACE Int.: San Diego, CA, USA, 2003.
12. Maric, M.; Muránsky, O.; Karatchevtseva, I.; Ungár, T.; Hester, J.; Studer, A.; Scales, N.; Ribárik, G.; Primig, S.; Hill, M.R. The effect of cold-rolling on the microstructure and corrosion behaviour of 316L alloy in FLiNaK molten salt. *Corros. Sci.* **2018**, *142*, 133–144. [CrossRef]
13. Liu, Y.H.; Liu, S.F.; Zhu, J.L.; Deng, C.; Fan, H.Y.; Cao, L.F.; Liu, Q. Strain path dependence of microstructure and annealing behavior in high purity tantalum. *Mater. Sci. Eng. A* **2017**, *707*, 518–530. [CrossRef]
14. Greene, N.D.; Saltzman, G.A. Effect of plastic deformation on the corrosion of iron and steel. *Corrosion* **1964**, *20*, 293t–298t. [CrossRef]
15. Luo, H.; Wang, X.; Dong, C.; Xiao, K.; Li, X. Effect of cold deformation on the corrosion behaviour of UNS S31803 duplex stainless steel in simulated concrete pore solution. *Corros. Sci.* **2017**, *124*, 178–192. [CrossRef]
16. Štefec, R.; Franz, F. A study of the pitting corrosion of cold-worked stainless steel. *Corros. Sci.* **1978**, *18*, 161–168. [CrossRef]
17. Wu, X.; Yang, M.; Yuan, F.; Wu, G.; Wei, Y.; Huang, X.; Zhu, Y. Heterogeneous lamella structure unites ultrafine-grain strength with coarse-grain ductility. *Proc. Natl. Acad. Sci. USA* **2015**, *112*, 14501–14505. [CrossRef]
18. Rofagha, R.; Erb, U.; Ostrander, D.; Palumbo, G.; Aust, K.T. The effects of grain size and phosphorus on the corrosion of nanocrystalline Ni-P alloys. *Nanostruct. Mater.* **1993**, *2*, 1–10. [CrossRef]
19. Wang, S.G.; Shen, C.B.; Long, K.; Yang, H.Y.; Wang, F.H.; Zhang, Z.D. Preparation and electrochemical corrosion behavior of bulk nanocrystalline ingot iron in HCl acid solution. *J. Phys. Chem. B* **2005**, *107*, 2499–2503. [CrossRef]
20. Wang, S.G.; Shen, C.B.; Long, K.; Zhang, T.; Wang, F.H.; Zhang, Z.D. The Electrochemical Corrosion of Bulk Nanocrystalline Ingot Iron in Acidic SulfateSolution. *J. Phys. Chem. B* **2006**, *110*, 377–382. [CrossRef]
21. Lv, J.; Luo, H. The effects of cold rolling temperature on corrosion resistance of pure iron. *Appl. Surf. Sci.* **2014**, *317*, 125–130.
22. Abdulstaar, M.; Mhaede, M.; Wagner, L.; Wollmann, M. Corrosion behaviour of Al 1050 severely deformed by rotary swaging. *Mater. Des.* **2014**, *57*, 325–329. [CrossRef]
23. Wang, Z.; Zhu, F.; Zheng, K.; Jia, J.; Wei, Y.; Li, H.; Huang, L.; Zheng, Z. Effect of the thickness reduction on intergranular corrosion in an under–aged Al-Mg-Si-Cu alloy during cold–rolling. *Corros. Sci.* **2018**, *142*, 201–212. [CrossRef]
24. Xiang, S.; He, Y.; Shi, W.; Ji, X.; Tan, Y.; Liu, J.; Ballinger, R.G. Chloride-induced corrosion behavior of cold-drawn pearlitic steel wires. *Corros. Sci.* **2018**, *141*, 221–229. [CrossRef]
25. Zeng, Y.; Yang, F.; Chen, Z.; Guo, E.; Gao, M.; Wang, X.; Kang, H.; Wang, T. Enhancing mechanical properties and corrosion resistance of nickel-aluminum bronze via hot rolling process. *J. Mater. Sci. Technol.* **2021**, *61*, 186–196. [CrossRef]

26. Zhang, J.; Ma, G.Q.; Godfrey, A.; Shu, D.Y.; Chen, Q.; Wu, G.L. Orientation dependence of the deformation microstructure of Ta-4% W after cold-rolling. In Proceedings of the IOP Conference Series: Materials Science and Engineering, Risø, Denmark, 4–8 September 2017; Volume 219, p. 012051.
27. Ma, G.; Godfrey, A.; Chen, Q.; Wu, G.; Hughes, D.A.; Hansen, N. Microstructural evolution of Ta-4%W during cold rolling. In Proceedings of the IOP Conference Series: Materials Science and Engineering, Risø, Denmark, 2–6 September 2019; Volume 580, p. 012041.
28. Ma, G.; Hughes, D.A.; Godfrey, A.W.; Chen, Q.; Hansen, N.; Wu, G. Microstructure and strength of a tantalum-tungsten alloy after cold rolling from small to large strains. *J. Mater. Sci. Technol.* **2021**, *83*, 34–48. [CrossRef]
29. Ma, G.; He, Q.; Luo, X.; Wu, G.; Chen, Q. Effect of recrystallization annealing on corrosion behavior of Ta-4%W alloy. *Materials* **2018**, *12*, 117. [CrossRef] [PubMed]
30. Ma, G.; Wu, G.; Shi, W.; Xiang, S.; Chen, Q.; Mao, X. Effect of cold rolling on the corrosion behavior of Ta-4W alloy in sulphuric acid. *Corros. Sci.* **2020**, *176*, 108924. [CrossRef]
31. Liu, Q. A Simple and Rapid Method for Determining Orientations and Misorientations of Crystalline Specimens in TEM. *Ultramicroscopy* **1995**, *60*, 81–89. [CrossRef]
32. Wang, S.G.; Sun, M.; Liu, S.Y.; Liu, X.; Xu, Y.H.; Gong, C.B.; Long, K.; Zhang, Z.D. Synchronous optimization of strengths, ductility and corrosion resistances of bulk nanocrystalline 304 stainless steel. *J. Mater. Sci. Technol.* **2020**, *37*, 161–172. [CrossRef]
33. Fan, H.Y.; Liu, S.F.; Guo, Y.; Deng, C.; Liu, Q. Quantifying the effects of surface quality on texture measurements of tantalum. *Appl. Surf. Sci.* **2015**, *339*, 15–21. [CrossRef]
34. Xu, Y.; Zhang, Z.; Mao, S.; Ma, Y. Corrosion behavior of different tantalum crystal faces in NH4Br–ethanol solution and DFT calculation. *Appl. Surf. Sci.* **2013**, *280*, 247–255. [CrossRef]
35. Wang, W.; Alfantazi, A. Correlation between grain orientation and surface dissolution of niobium. *Appl. Surf. Sci.* **2015**, *335*, 23–226. [CrossRef]
36. Schreiber, A.; Schultze, J.W.; Lohrengel, M.M.; Kármán, F.; Kálmán, E. Grain dependent electrochemical investigations on pure iron in acetate buffer pH 6.0. *Electrochim. Acta* **2003**, *51*, 2625–2630. [CrossRef]
37. Kuhlmann-Wilsdorf, D.; Hansen, N. Geometrically necessary, incidental and subgrain boundaries. *Scripta Metall Mater.* **1997**, *25*, 1557–1562. [CrossRef]
38. Haraszti, F.; Kovacs, T. Plastic deformation effect of the corrosion resistance in case of austenitic stainless steel. In Proceedings of the IOP Conference Series: Materials Science and Engineering, Miskolc, Hungary, 3–7 October 2016; IOP Publishing Ltd.: Bristol, UK, 2017; Volume 175, p. 012048.
39. Wang, Y.; Jin, J.; Zhang, M.; Liu, F.; Wang, X.; Gong, P.; Tang, X. Influence of plastic deformation on the corrosion behavior of CrCoFeMnNi high entropy alloy. *J. Alloys Compd.* **2022**, *891*, 161822. [CrossRef]
40. Lv, J.; Luo, H.; Liang, T.; Guo, W. The effects of grain refinement and deformation on corrosion resistance of passive film formed on the surface of 304 stainless steels. *Mater. Res. Bull.* **2015**, *70*, 896–907.
41. Hughes, D.A.; Hansen, N. High angle boundaries formed by grain subdivision mechanisms. *Acta Mater.* **1997**, *45*, 3871–3886. [CrossRef]
42. Hughes, D.A.; Hansen, N. The microstructural origin of work hardening stages. *Acta Mater.* **2018**, *148*, 374–383. [CrossRef]

Article

Time-Dependent Seismic Fragility of Typical Concrete Girder Bridges under Chloride-Induced Corrosion

Xiaoxiao Liu [1,*,†], Wenbin Zhang [2,†], Peng Sun [3] and Ming Liu [4,*]

1. School of Civil Engineering and Architecture, Xi'an University of Technology, Xi'an 710048, China
2. Infrastructure Department, Northwestern Polytechnical University, Xi'an 710129, China; zhangwb@nwpu.edu.cn
3. School of Computer Science, Northwestern Polytechnical University, Xi'an 710129, China; sunpeng@nwpu.edu.cn
4. Center for Advancing Materials Performance from the Nanoscale (CAMP-Nano), State Key Laboratory for Mechanical Behavior of Materials, Xi'an Jiaotong University, Xi'an 710049, China
* Correspondence: xxliu@xaut.edu.cn (X.L.); liuming0313@xjtu.edu.cn (M.L.)
† These authors contributed equally to the work.

Abstract: Recent studies highlighted the importance of the combined effects of prestress loss and corrosion deterioration for concrete girder bridge structures when the effect of damage on the performance level is estimated. The multi-deterioration mechanisms connected with chloride erosion include the cross-sectional area loss of longitudinal bars and stirrups, the reduction in the ductility, the decrease in the strength of steels and the strength loss of concrete in RC columns. For the corroded RC columns and corroded elastomeric bridge bearings, analytical models of the material degradation phenomena were employed for performing the probabilistic seismic performance analysis, which could obtain the system seismic fragility of aging bridges by considering the failure functionality of multiple correlated components (e.g., columns, bearings). The combined effects of prestress loss and cracking were also considered when developing time-dependent system seismic fragility functions. Here, a typical multi-span reinforced concrete girder bridge was used as a case study for studying the time-variant seismic performance. The results revealed the importance of the joint effects of the multi-deterioration mechanisms when modeling the time-dependent seismic fragility of aging bridge systems, as well as the significance of considering the combined effects of prestress loss and cracking.

Keywords: corrosion; multi-deterioration mechanics; prestress and cracking; system seismic fragility; RC concrete girder bridges

Citation: Liu, X.; Zhang, W.; Sun, P.; Liu, M. Time-Dependent Seismic Fragility of Typical Concrete Girder Bridges under Chloride-Induced Corrosion. *Materials* **2022**, *15*, 5020. https://doi.org/10.3390/ma15145020

Academic Editor: Francesco Fabbrocino

Received: 2 June 2022
Accepted: 17 July 2022
Published: 19 July 2022

Publisher's Note: MDPI stays neutral with regard to jurisdictional claims in published maps and institutional affiliations.

Copyright: © 2022 by the authors. Licensee MDPI, Basel, Switzerland. This article is an open access article distributed under the terms and conditions of the Creative Commons Attribution (CC BY) license (https:// creativecommons.org/licenses/by/ 4.0/).

1. Introduction

Reinforced concrete (RC) highway bridges experience prolonged exposure to a chloride environment such that these structures suffer from various forms of lifetime degradation, which result in some changes in the material properties. Corrosion in reinforcing steels initiated by chloride ions penetration into the concrete will cause the loss of cross-sectional area of the steels and the cover cracking of RC columns. Some efforts were made to investigate the potential significance of the deterioration of seismic performance across a large number of aged bridges. Choe et al. [1,2] researched the capacity reduction of RC columns in a corroded single-bent bridge due to corrosion and developed a time-variant seismic fragility model of the RC columns; the results revealed the potential importance of the crucial material properties degradation and corrosion parameters. Li et al. [3] performed a series of experiments to evaluate the seismic fragility of corroded RC bridges. Alipur et al. [4] discussed the changes in seismic vulnerability of a group of detailed bridge models, and the effect of the corrosion process on the nonlinear time-dependent parameters was considered during the life-cycle cost. Recent work by Ghosh and Padgett [5] examined the joint effects of corrosion deterioration of the column and steel bearing on the seismic fragility of bridge systems. Li et al. [6] proposed a time-dependent fragility analysis method for bridges under the action of chloride ion corrosion and scour by determining the probability distribution types and statistical characteristics of various environmental and corrosion parameters. To quantify the effect of steel corrosion on seismic fragility estimates, Cui et al. [7] developed an improved time-dependent seismic fragility framework by taking into account the increase in the corrosion rate after concrete cracking and the reduction in the seismic capacity of RC bridge substructures during the service life. Meanwhile, the authors proposed an analytical method based on a backpropagation artificial neural network to provide probabilistic capacity estimates of deteriorating RC substructures. Li et al. [8] proposed a time-dependent seismic fragility assessment framework by considering the correlation of random parameters of aging highway bridges under non-uniform chloride-induced corrosion attacks. Abbasi et al. [9] developed and compared the overall time-dependent system and individual component fragility curves of older and newly designed multi-frame reinforced concrete bridges in California. Dey and Sil [10] proposed performing seismic fragility analysis of corrosion-affected bridges located in the coastal region of India by considering pitting corrosion as a realistic corrosion degradation mechanism. Fu et al. [11] developed a life-cycle fragility assessment method to establish the time-dependent seismic fragility curves of an illustrative cable-stayed bridge at the component and system levels. Fan et al. [12] established a fragility analysis framework for RC bridge structures subjected to the multi-hazard effect of vessel collisions and corrosion. Meanwhile, the modeling and analysis of the time-dependent seismic fragility due to the corrosion effects of RC bridges has received significant attention [13–18].

However, it should be noted that the above contributions neglect additional mechanisms, such as stirrup corrosion, cover defects and damaged confined concrete. Specifically, the corrosion of stirrups leads to the strain reduction of confined concrete (or the reduction in confinement levels of core concrete) and the reduction of the shear strength in the concrete components (i.e., the bottom and top of bridge columns [19]). Furthermore, the corrosion of steel reinforcements and prestressing strands of the concrete bridges leads to significant prestress losses and cracking, which induces excessive deflections during their service life [20,21]. Meanwhile, expansive forces that originate from the formation of the oxidation product may cause cracking of the concrete cover [22], and the strength of the concrete material is reduced at specific intervals during the life cycle. The deterioration mechanisms of different components of RC bridges depend on their physical conditions and environmental exposure scenarios. For multi-span continuous (MSC) bridges and multi-span simply supported (MSSS) girder bridges, the different forms of degradation due to corrosion can be found in the individual components, and the multiple types of deterioration among different components (i.e., column, bearing) of RC bridges should be considered as dependent. Meanwhile, the corrosion degradation of RC components of typical bridges mainly gives rise to the cross-sectional area loss of embedded steels, including the longitudinal bars and hoops. This type of degradation can also lead to stress reduction of unconfined concrete and the ductility decrease of confined concrete, along with cracking of the cover. These multi-deterioration mechanisms are common for bridge components, such as the decks and columns of RC bridges. In addition, the diameter loss of steel bearing anchor bolts and the strength of elastomeric bearing dowel bars can cause the deterioration of the bearing system. Some studies [5,23] revealed that the corrosion of the reinforcement steel in RC columns and the deterioration of the bridge bearings could be found in MSC steel girder bridges and MSSS concrete girder bridges in Central and Southeastern United States, which are seismically vulnerable regions. According to China's Ministry of Transportation, the aging and deterioration phenomenon of MSC concrete girder bridges are commonly found in seismic zones, such as northern China, because deicing salts are extensively used on highway bridges [24]. The existing contributions [6–21] also indicated that RC bridges subjected to the mechanisms of corrosion deterioration might be more vulnerable to earthquake hazards. Moreover, since previous studies [25,26] demonstrated that chloride from deicing salts can cause more degradation of bridges than that in a marine environment, the impacts of corrosion degradation due to chloride-laden deicing salts should be researched in detail.

Although the above works focused on the evaluation of the corrosion effect on the seismic vulnerability of RC bridge components/systems, the joint impacts of the multi-deterioration mechanisms among multiple bridge components, which have a significant impact on the seismic performance, were rarely considered in recent studies. Meanwhile, the corrosion of critical components, including the corrosion of RC columns and the aging of elastomeric bridge bearings, as well as the combined effects of prestress loss and cracking, also have a negative effect on seismic performance. Therefore, bridge columns are assumed to be subjected to salt mist formed from chloride-laden water [27], and the elastomeric bridge bearings were considered to be affected by chloride ions and thermal oxidation [28–30] in this study. For this purpose, the joint effects of multi-deterioration mechanisms on system seismic fragility for RC bridges were researched by considering the degradation correlations between multiple components. Then, dynamic behaviors of bridge systems under multiple related deterioration mechanisms could be calculated to support the development of time-variant seismic fragility curves. Subsequently, a probabilistic approach was proposed to evaluate the time-dependent seismic fragility, and the effects of multi-deterioration mechanisms on seismic vulnerability could be researched in detail by including the joint impact of columns and bearing corrosion deterioration. The multi-deterioration mechanisms, which are connected with chloride erosion, include the cross-sectional area loss of longitudinal bars and stirrups, the reduction in the ductility, the decrease in the strength of steels and the strength loss of concrete in RC columns because the cover cracking exists and the failure of the bearings system of bridges will appear due to the corrosion of bearing pads and dowel bars. The impact of multi-deterioration were applied to a typical MSC concrete girder bridge in the system fragility assessment framework of aged bridges. Meanwhile, the combined effects of the prestress loss and cracking on the seismic fragility were found in detail. For the corroded RC columns and corroded elastomeric bridge bearing, analytical models of material degradation phenomena were introduced into the probabilistic seismic performance evaluation, which obtained the system fragility of bridges by considering the failure functionality of multiple correlated critical components (e.g., columns, bearing). A three-dimensional nonlinear dynamic model of the MSC bridges, which considered its aging components as multi-deterioration mechanisms, was performed to evaluate the changes in the structural capacity and was also employed to obtain the seismic response of critical components and systems. Subsequently, time-dependent seismic fragility curves were analyzed to generate a comparison of the seismic performance at different time points during the service life when the parameter properties' variation, ground motions and deterioration parameters were considered simultaneously. The results showed that the multi-deterioration of the multiple correlated components had a joint effect on the seismic performance of the aged bridges subjected to seismic hazards, as well as the combined effects of prestress loss and cracking.

2. MSC Concrete Girder Bridge Geometry

To verify the proposed methodology for developing time-dependent fragility curves and offer insights on the effects of the multi-deterioration mechanism of multiple related components on seismic response and vulnerability, a sample MSC concrete girder bridge was considered as a case study in this research. Nearly 67.8% of all bridges in Central and Southeastern United States and approximately 74% in China are of this bridge type [24]. The MSC concrete girder bridge type was used for the case study since an overwhelming majority of 74% of bridges found in seismic zones are of this type and these bridges are seismically vulnerable due to inadequate detailing of the components. Previous studies on classes of bridges by Nielson [31] indicated that the vulnerability of multiple components of this type of bridge is necessary when considering aging and deterioration. This vulnerability phenomenon can be attributed to inadequately seat widths, bolted elastomeric pad bearings and insufficient transverse reinforcement, which inhibits the shear resistance and ductile capacity in RC columns. Consequently, higher seismic demands on the bridge system due to the above reasons can be generated during earthquake events.

The typical MSC concrete girder bridge identified by Nielson [31] is introduced in this study because of the similar bridge configuration in China [24]. As shown in Figure 1, this bridge has three spans with lengths of 11.9, 22.3 and 11.9 m, which give an overall bridge length of 46.1 m. The decks of width 15.01 m were constructed of eight AASHTO concrete bridge girders. Among these girders, type I was for the end span and type III was used for the middle span. A concrete parapet between the girders and deck, which can reduce the effects of live load, was used to concatenate the MSC bridge [31]. Two pile bent abutments and two three-column bents support a three-deck. The circular columns have a cross-sectional area of 641.2 mm^2. This type of column can be reinforced with 12-#29 longitudinal bars and #13 transverse stirrups spaced at 305 mm. The fixed and expanded bearings for this bridge class are elastomeric pads with two steel dowels, as characterized by Neilson [31].

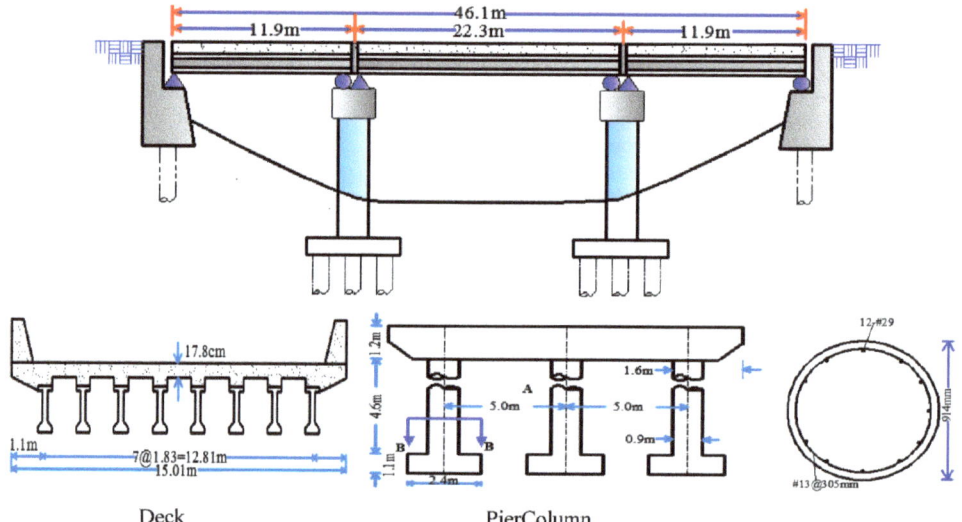

Figure 1. Case study multi-span continuous (MSC) concrete girder bridge.

3. Multi-Degradation Mechanisms of RC Columns Due to Corrosion

All equations used in Section 2 can be found in Table 1.

Table 1. Equation list for Section 2.

Equation Number	Equation Expression
Equation (1)	$\frac{\partial C(x,t)}{\partial t} = \frac{\partial}{\partial x} D \frac{\partial C(x,t)}{\partial x}$
Equation (2)	$C(x,t) = C_S \left[1 - erf\left(\frac{x}{2\sqrt{Dt}}\right) \right]$
Equation (3)	$T_I = \frac{x^2}{4D} \left[erf^{-1}\left(\frac{C_S - C_C}{C_S}\right) \right]^{-2}$
Equation (4)	$R(t) = \int_{T_I}^{t} r_{corr} dt$
Equation (5)	$\Phi(t) = \Phi(0) - 2R(t)$
Equation (6)	$\phi(t) = \phi(0) - 2R(t)$
Equation (7)	$\Phi(t) = \Phi_i - R(t)$

Table 1. Cont.

Equation Number	Equation Expression
Equation (8)	$\phi(t) = \phi_i - R(t)$
Equation (9)	$\eta = 2R(t)/\Phi(0);\ \eta \in [0,1]$
Equation (10)	$\lambda = 2R(t)/\phi(0);\ \lambda \in [0,1]$
Equation (11)	$A(t) = \begin{cases} [1-\eta_S(t)/2]A_S(0), & t \leq T_I \\ [1-\eta_S(t)]A_S(0), & T_I \leq t \leq T_I + \Phi_i/r_{corr} \\ 0, & t \geq T_I + \Phi_i/r_{corr} \end{cases}$
Equation (12)	$a(t) = \begin{cases} [1-\lambda_k(t)/2]a_s(0), & t \leq T_I \\ [1-\lambda_k(t)]a_s(0), & T_I \leq t \leq T_I + \phi_i/r_{corr} \\ 0, & t \geq T_I + \phi_i/r_{corr} \end{cases}$
Equation (13)	$\eta_S(t) = \eta(2-\eta)$
Equation (14)	$\varepsilon_{su}(t) = \begin{cases} \varepsilon_{su}(0) & 0 \leq \eta_S \leq 0.016 \\ 0.1521\eta_S^{-0.4583}\varepsilon_{su}(0), & 0.016 < \eta_S \leq 1 \end{cases}$
Equation (15)	$f_{sy}(t) = (1-1.077\eta_S(t))f_{sy}(0)$
Equation (16)	$f_{yh}(t) = (1-1.077\lambda_k(t))f_{yh}(0)$
Equation (17)	$f_{pc}(t) = \frac{f_{pc}(0)}{1+K\varepsilon_t(t)/\varepsilon_c(0)}$
Equation (18)	$\varepsilon_t(t) = \frac{n_{bars}w}{r_i}$
Equation (19)	$w = k_w(\Delta A_s - \Delta A_{s0})$
Equation (20)	$\eta_S(0) = 1 - \left[1 - \frac{\alpha}{\phi(0)}\left(7.53 + 9.32\frac{x}{\phi(0)}\right)10^{-3}\right]^2$
Equation (21)	$\varepsilon_{cu}(t) = 0.004 + 0.9\rho_s(t)\frac{f_{yh0}}{300}$

3.1. The Diffusion Process and Corrosion Initiation Time

Chloride-induced corrosion is considered to be one of the major causes of the aging of reinforced concrete bridges. Corrosion of reinforcing steel in an RC column is initiated by the ingress of chloride ions through the concrete cover to the bar's surface. The main source of chloride ions is considered to be the deicing salts used in winter. The diffusion process of chloride ions that result in corrosion can be described by Fick's second law [25]. The formulation can be seen as Equation (1) in Table 2, where C is the chloride ion concentration, D is the chloride diffusion coefficient of concrete, x is the depth of the concrete cover and t is the time step in years. We assumed that the initial condition (initial chloride ion content), boundary condition (surface chloride content) and material property (chloride diffusion coefficient) were equal to zero, the mean-invariant C_s and the mean-invariant D, respectively. The closed solution of the chloride ion content is shown as Equation (2) [32], where $erf(\cdot)$ is the Gaussian error function. The corrosion parameters involved in the diffusion process were assumed to be random variables whose distribution type, mean u and coefficient of variation COV are listed in Table 2 [5,19].

Table 2. Probability distribution and the random parameters involved in the diffusion process.

Random Variable	Unit	Distribution Type	Mean	COV
Concrete cover, x	mm	Lognormal	40.00	0.20
Diffusion coefficient, D	cm^2/year	Lognormal	1.29	0.10
Surface chloride concentration, C_s	wt%/cem	Lognormal	0.10	0.10
Critical chloride concentration, C_C	wt%/cem	Lognormal	0.040	0.10
Corrosion rate, r_{corr}	mm/year	Lognormal	0.127	0.3

The diffusion process of chloride ions can be described in probabilistic terms by using the Monte Carlo method. It was assumed that the depth of concrete cover x was 40 mm and the service life t was 50 years; the resulting probability density functions (PDFs) of chloride ion concentrations that were obtained using a Monte Carlo simulation with a sample size of 50,000 are shown in Figure 2. With reference to Fick's second law of diffusion and Equation (2), the corrosion initiation time T_I was evaluated using Equation (3) [5], where T_I is the corrosion initiation time in years, C_S is the chloride concentration of concrete surface and C_C is the critical chloride concentration that can dissolve the protective passive film around the reinforcement steels. Therefore, a Monte Carlo simulation was performed with a sample size of 50,000 for random variables to evaluate the probability distribute of the corrosion initiation time T_I, as shown in Figure 3. A lognormal distribution with a mean $\mu = 1.0374$ years and standard deviation $\sigma = 0.2829$ was found to be a good estimation of the simulated distribution for the corrosion initiation time T_I. This distribution had an influence on the probabilistic modeling of the steel corrosion in the bridge RC columns.

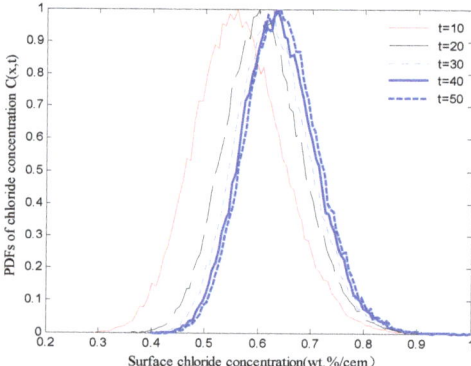

Figure 2. PDFs of the chloride concentration $C(x,t)$.

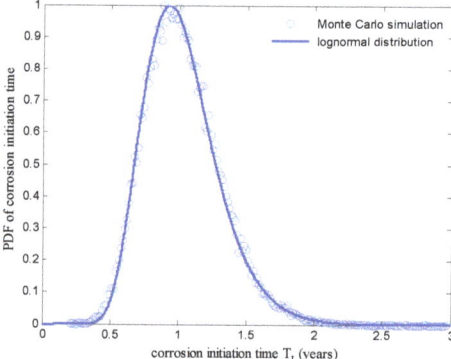

Figure 3. PDFs of the corrosion initiation time T_1.

3.2. Reduction in the Cross-Sectional Area of Longitudinal Bars and Stirrups

The corrosion of rebars in the existing RC column was initiated by reducing the cross-section. However, some recent studies [33,34] showed that the corrosion of stirrups, which causes a reduction in confinement behavior, is more severe than that of the longitudinal bars in an RC column. Furthermore, the corrosion can degrade the shear resistance of the column by potentially shifting the failure modes from a ductile failure mode to a shear failure mode. Therefore, the effect of both the loss of the cross-sectional area and the ductile degradation of stirrups on the seismic behavior of bridge systems should be considered in this study. Corrosion from deicing salts leads to a reduction in the effective area of both the longitudinal and transverse rebars. First, the corrosion penetration R is introduced as Equation (4) [19], where r_{corr} is the corrosion rate and represents the implicit function of the time-dependent area due to corrosion, while $t - T_I$ is the corrosion propagation time. The time-variant loss of longitudinal reinforcement and stirrup diameter in the BLG model can be evaluated using Equation (5) and Equation (6), respectively [19]. Combining Equation (4), Equation (5) and Equation (6), $\Phi(t)$ and $\phi(t)$ can be expressed as Equation (7) and Equation (8), where $\Phi(0)$ and $\phi(0)$ are the initial diameters of the longitudinal bar and stirrup, respectively; Φ_i and ϕ_i are the diameters of a corroding longitudinal bar and stirrup at time $t = i$, respectively; $\Phi(t)$ and $\phi(t)$ are the diameters of a longitudinal bar and stirrup at the end of $t - T_I$ years. The corrosion penetration index of rebars η is introduced as Equations (9) and (10) [35].

Then, the time-dependent area of a corroding longitudinal bar can be expressed as Equation (11), where $A_S(0)$ is the initial area of longitudinal reinforcement and $\eta_S(t)$ is the loss rate of a longitudinal bar. The time-dependent area of a corroding stirrup can be expressed in Equation (12), where $a_s(0)$ is the initial area of a stirrup and $\lambda_k(t)$ is the loss rate of a stirrup. It is acknowledged that the corrosion rate is considered a constant on average and it is assumed to be lognormally distributed, as listed in Table 1. Moreover, the diameters of the longitudinal rebar and stirrup are assumed to be lognormally distributed, whose mean u and coefficient of variation COV are listed in Table 3. On the basis of the above assumptions, the reduction in the diameter of the rebar due to corrosion could be evaluated for a lifetime of 50 years ($\Phi(0)$ = 28.58 mm and $\phi(0)$ = 12.70 mm). The time-variant mean of the longitudinal bar and the stirrups, as well as the corresponding standard deviations, are shown in Figure 4. The scattering region around the mean represents the stochastic time-dependent loss of both the longitudinal bars' and the stirrups' cross-sectional areas because of the effect of uncertainties in degradation parameters. This figure illustrates that the effects of stirrup corrosion were more significant than that of longitudinal reinforcement corrosion because the smaller diameter of a transverse bar led to higher levels of corrosion of the stirrup. Thus, the area loss of steel corrosion in RC columns could be modeled as the reduction in the cross-sectional area of both longitudinal reinforcing and stirrups in the BLG model with the fiber section.

Table 3. Probability distribution of random parameters for bridge modeling.

Uncertainty Parameter	Units	Distribution Type	Distribution Parameters A [1]	Distribution Parameters B [1]
Concrete compressive strength	Mpa	Lognormal	30	5/30
Reinforcing steel yield strength	Mpa	Lognormal	300	30/300
Reinforcing steel diameter	mm	Lognormal	28.58	0.10
Stirrup yield strength	Mpa	Lognormal	235	30/235
Stirrup diameter	mm	Lognormal	12.70	0.10
Elastomeric bearing shear modulus	Mpa	Uniform	0.66	2.07
Steel dowel lateral strength	Mpa	Uniform	200.96	381.70
Dowel bar tension strength	Mpa	Uniform	522.09	845.00

[1] Mean and standard deviation of the lognormal distribution, lower bound and upper bound of the uniform distribution.

3.3. Reduction in Strength and Ductility of Corroded Longitudinal Bars and Stirrups

The corrosion of stirrups is more serious than that of longitudinal reinforcement in RC columns and it leads to a reduction in confinement behavior. The corrosion can degrade the shear-resistant capacity of RC columns by potentially changing the ductile failure to brittle failure or even shear failure [34]. According to some studies [36–38], the reduction in both the strength and the ductility in longitudinal reinforcement can be considered an explicit function of the cross-sectional loss. The cross-sectional loss rate of the longitudinal bar is considered a function of the corrosion penetration index η in Equation (13) [19]. Therefore, the time-dependent ultimate strain of the longitudinal reinforcement in Steel 01 material can be expressed as Equation (14) [19], where $\varepsilon_{su}(0)$ is the initial nominal value of the ultimate strain of the longitudinal rebar. The time-variant ultimate strain is shown in Figure 5.

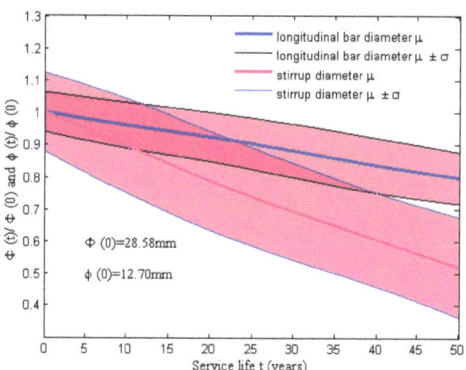

Figure 4. Time-evolving longitudinal bar and stirrup diameters for $\Phi(0) = 28.58$ mm and $\phi(0) = 12.70$ mm.

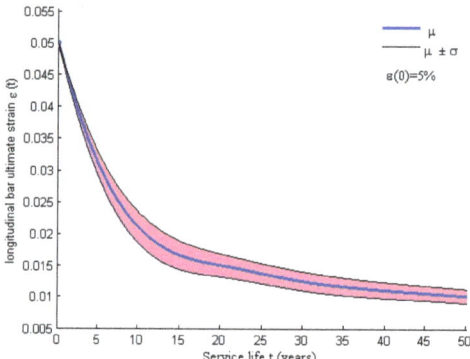

Figure 5. Ultimate strain of a longitudinal bar over time.

Furthermore, the residual yield strength of the longitudinal rebar depends on the cross-sectional loss rate $\eta_S(t)$, which is expressed as Equation (15) [39], where $f_{sy}(0)$ is the initial nominal yield strength of the longitudinal bar. Similarly, the time-dependent yielding strength of the hooping in the BLG model depends on the cross-sectional loss rate $\lambda_k(t)$. The yielding strength is expressed as Equation (16) [39], where $f_{yh}(0)$ is the initial yielding strength of the stirrups. The yielding strengths of both the longitudinal bar and stirrup were assumed to be lognormally distributed, whose mean u and the coefficient of variation COV are listed in Table 1. The time-dependent yielding strengths $f_{sy}(t)$ and $f_{yh}(t)$ can be respectively evaluated using the cross-sectional loss rate η_S and λ_k, which depend on the reduction in the diameter in both the corroding longitudinal rebar and hooping.

3.4. Reduction in Strength and Ductility in Corroded Concrete

For the corrosion initiation phase, the confinement behavior of the concrete can be enhanced due to the expansion of the corrosion products. However, the propagation of longitudinal cracks and the cracking of concrete cover inversely result in a reduction in the confinement with the accumulation of rust products. According to the unconfined and confined stress–strain rules (i.e., BLG model, Figure 6), the residual strength of the unconfined concrete can be evaluated using Equation (17) [40], where K is a coefficient related to the rebar roughness and the diameter (a value $K = 0.1$ is used for medium-diameter ribbed rebars), $\varepsilon_c(0)$ is the initial strain of the unconfined concrete at the peak compressive stress $f_{pc}(0)$ and $\varepsilon_t(t)$ is the average tensile strain in the cracked concrete at time t. The average tensile strain can be evaluated using Equation (18) [40], where n_{bars} is the number of longitudinal rebars, r_i is the width of the cross-sectional area in a pristine state and w is the crack width for each longitudinal rebar. The relationship between the crack width and cross-sectional loss of the longitudinal steel bar can be expressed as Equation (19) [41], where $k_w = 0.0575$ mm^{-1}, ΔA_s is the cross-sectional loss of a longitudinal steel bar for cracking propagation and ΔA_{s0} is the cross-sectional loss of the longitudinal reinforcement for cracking initiation. Then, one obtains the following expressions: $\Delta A_s = \eta_S(t) A_S(0)$ and $\Delta A_{s0} = \eta_S(0) A_S(0)$. The cross-sectional loss rate of the longitudinal rebar for the cracking initiation $\eta_S(0)$ can be evaluated using Equation (20) [40], where $\phi(0)$ is the initial diameter of the longitudinal rebar, x is the depth of the concrete cover and α is the pit concentration factor. For uniform corrosion, one has $\alpha = 2$, and for the localized corrosion, one obtains $4 < \alpha < 8$. The initial compressive strength of unconfined concrete is assumed to be lognormally distributed, whose mean u and the coefficient of variation COV are listed in Table 1. The residual strength $f_{pc}(t)$ can be evaluated using the cross-sectional losses ΔA_s and ΔA_{s0}, which are dependent on the swedged steel bars, as shown in Figure 7.

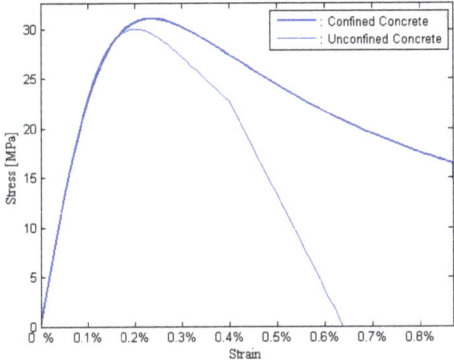

Figure 6. Unconfined and confined stress–strain law obtained with the BLG model.

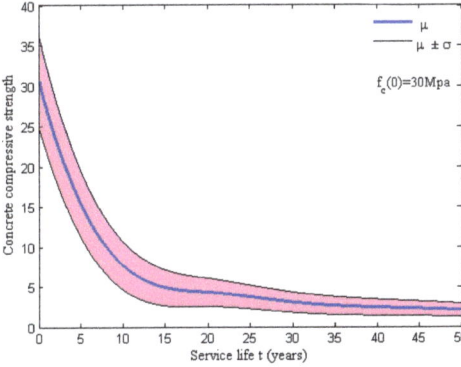

Figure 7. Time variance of the compression strength of unconfined concrete.

Since a failure criterion for confined concrete is not provided in the BLG model, the ultimate compressive strain proposed by Scott et al. [42] was assigned to the concrete core fibers. The time-dependent ultimate compressive strain of the confined concrete related to the first stirrup facture can be estimated with the failure criterion. Furthermore, the strain is expressed as Equation (21), where $\rho_s(t)$ is the volume–stirrup ratio at time t and f_{yh0} is the initial yield strength of a transverse stirrup. The ultimate compressive strain of the confined concrete $\varepsilon_{cu}(t)$ can be calculated using the stirrup ratio $\rho_s(t)$, which has a direct dependency on the percentage loss of the total swedged reinforcement in the cross-section of an RC column, as shown in Figure 8. The result indicated that the ultimate compressive strain of the confined concrete is also time-dependent because the progressive reduction of stirrup area was closer to the concrete surface, which can be more susceptible to corrosion. Therefore, the material degradation of the RC column was properly simulated by using the BLG model and Steel 01.

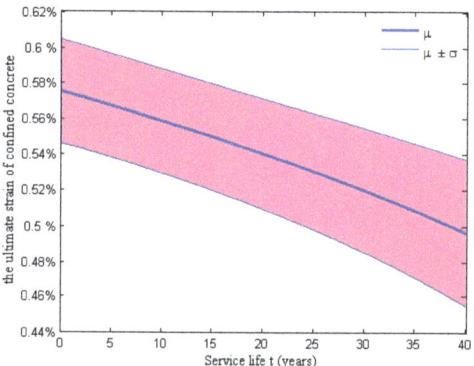

Figure 8. Time-dependence of the ultimate compressive strain of confined concrete.

4. Degradation of Elastomeric Bridge Bearings Due to Corrosion and Thermal Oxidation

All equations for Section 3 can be found in Table 4.

Elastomeric bridge bearings are widely used in concrete girders, where they allow for force transmission from the superstructure to the substructure. These types of bearing systems consist of an elastomeric nature rubber (NR) pad and steel dowels, but they often form "walk-out" bearings during seismic events due to the effects of aging and deterioration [43]. Corrosion of these assemblies, caused by chloride-laden water from deicing salts, may potentially result in a larger deformation of the bearing systems under seismic loading. For the steel dowels of the elastomeric bridge bearings, corrosion deterioration can lead to a reduction in both the cross-section and the shear strength. Moreover, the elastomeric NR pads suffer an increase in both the shear modulus and shear stiffness due to corrosion and thermal oxidation [28–30].

Additionally, each member of this type of bearing system makes a contribution to the transfer of forces. For example, an elastomeric NR pad transfers a lateral load by developing a frictional force, while steel dowels provide resistance via the action of the beam. The assemblies of the elastomeric bridge bearings consist of the fixed and expansion types, which depend on the dimensions of the slot in the NR pad [31], as shown in Figure 9. The sliding behavior of the elastomeric NR pad and the yield of steel dowels in pristine bridge bearings can be modeled by using the proper composite action.

As above-mentioned, aging and thermal oxidation may lead to an increase in the shear modulus in the elastomeric NR pads, where these deterioration mechanisms can be modeled with a changed value of the constitutive parameter in Steel 01 materials. Normally, the value of the shear modulus in the elastomeric bridge bearings recommended by AASHTO [44] is assumed to be a mean constant in the analytical modeling of bridges. However, a battery of accelerated exposure tests proposed by Itoh et al. [28] illustrated that the shear modulus of the NR pad is not a constant and can be changed over time due to thermal oxidation. The derivation process of the time-dependent shear modulus for an NR pad can be expressed as follows.

Table 4. Equation list for Section 3.

Equation Number	Equation Expression	Equation Number	Equation Expression
Equation (22)	$B_s/B_0 = k_s\sqrt{t_r} + 1$	Equation (32)	$\frac{G(t)}{G_0} = 1 + \tau\Delta G_s$
Equation (23)	$k_s = a_1\zeta^3 + a_2\zeta^2 + a_3\zeta + a_4$	Equation (33)	$\tau = \begin{cases} ((m-h)/h)^2 & (0 \leq m \leq h) \\ 0 & (h \leq m \leq l-h) \\ ((m-(l-h))/h)^2 & (l-h \leq m \leq l) \end{cases}$
Equation (24)	$B_s = G_s(\lambda_1^2 + \lambda_2^2 + \lambda_3^2 - 3)$ $B_0 = G_0(\lambda_1^2 + \lambda_2^2 + \lambda_3^2 - 3)$	Equation (34)	$G(t) = G_0(1 + \tau\Delta G_s)$
Equation (25)	$G_s/G_0 = B_s/B_0 = k_s\sqrt{t_r} + 1$	Equation (35)	$t_r/t = e^{\frac{E}{R}(\frac{1}{T_r} - \frac{1}{T})}$
Equation (26)	$\Delta G_s = (G_s - G_0)/G_0 = G_s/G_0 - 1$	Equation (36)	$\Delta G_s(t) = k_s\sqrt{t}$
Equation (27)	$h = \alpha\exp(\beta/T^*)$	Equation (37)	$k(t) = G(t)A_p/S_p$
Equation (28)	$G(t)/G_0 = B(t)/B_0 = 1 + \Delta G_s \quad (m = 0 \text{ or } l)$	Equation (38)	$F_y(t) = f_y A_d(t)/\sqrt{3}$
Equation (29)	$G(t)/G_0 = 1 \quad (h_0 \leq m \leq l - h)$	Equation (39)	$F_u(t) = f_u A_d(t)/\sqrt{3}$
Equation (30)	$dG(t)/dm = 0$ $(m = h \text{ or } l - h)$		
Equation (31)	$\frac{G(t)}{G_0} = v_1 m^2 + v_2 m + v_3$		

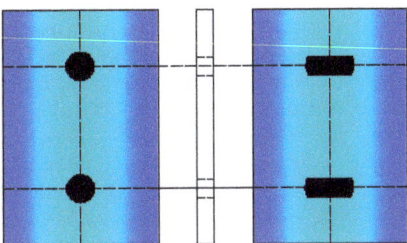

Figure 9. Fixed and expansion elastomeric bearing types, depending on the dimensions of the slot.

First, an aging model of NR bearings proposed by Itoh and Gu [29,30] showed the relation between the variation of strain energy and the aging time under accelerated exposure test conditions (i.e., the simulated thermal oxidation). This aging model can be provided by Equation (22), where B_s/B_0 represents the relative strain energy versus its initial state at the NR block surface, k_s is the strain-dependent coefficient for the strain energy and t_r is the aging time at the test temperature in hours. Here, 60 °C was taken as the reference temperature of the NR, and the empirical formula for k_s was calculated using Equation (23), where $a_1 = 0.54$, $a_2 = -4.19$, $a_3 = 8.16$ and $a_4 = 9.59$ at 60 °C, and ζ is the nominal strain.

Second, a relationship between the strain energy and the shear modulus can be expressed using a one-parameter neo-Hoolean material model [45], as shown in Equation (24), where B is the strain energy; G is the shear modulus; and λ_1, λ_2 and λ_3 are the stretches due to the uniaxial tension. If this material is incompressible, then $\lambda_1^2\lambda_2^2\lambda_3^2 = 1$. Combining Equations (22) and (23), a relative change in the shear modulus of the NR bearing due to thermal oxidation can be calculated using Equation (25), where G_s/G_0 is the relative shear modulus versus its initial state for the NR bearing at time $t = t_r$. Thus, the normalized shear modulus variation ΔG_s can be obtained using Equation (26). Finally, the shear modulus variation in the outer region from the NR bearing's surface to the critical depth h should be expressed by using an equation. The relationship between the critical depth and the temperature can be obtained using Equation (27) [29], where T^* is the absolute temperature and α and β are the coefficients determined by the thermal oxidation test. It is assumed that the shear modulus variation $G(t)/G_0$ might be a function of the position m. The boundary conditions can be derived using Equations (28) to (30), where $G(t)$ and G_0 are the shear modulus at time t and the initial state, respectively; ΔG_s is the normalized shear modulus variation; and l is the width of the NR bearing. The normalized shear modulus $G(t)/G_0$ can be also assumed to be a square relation of the position m, and the function is given as Equation (31). Combining the boundary conditions, the normalized shear modulus $G(t)/G_0$ can be written as Equations (32) and (33). The time-dependent shear modulus in a bearing NR pad can be calculated using Equation (34), where $G(t)$ is the shear modulus at real time t; τ is coefficient correlated with the position m, the critical depth h and the width l of the bridge NR bearing. Since thermal oxidation is commonly assumed to be a first-order chemical reaction for NR materials, the relationship between the deterioration time under the service condition and the time in the accelerated exposure tests can be expressed as Equation (35) [29], where t_r is the test time, t is the aging time, T_r is the absolute temperature in the thermal oxidation test, T is the absolute temperature under the service condition, R is the gaseous constant (8.314 J/mol K) and E is the activation energy of the rubber (94,900 J/mol). Consequently, the normalized shear modulus variation ΔG_s can be written as Equation (36).

The initial shear modulus of the elastomeric bearing pad is assumed to be uniformly distributed, whose upper bound a and lower bound b are listed in Table 1, and the time-dependent shear modulus $G(t)$ is plotted in Figure 10. However, the time-variant shear stiffness of the NR pad is considered a critical element in the analytical modeling of elastomeric bearing pads, and the aging model of the shear stiffness can be expressed as Equation (37), where $G(t)$ is the shear modulus of the NR pad at the real time t, A_p is the area of the elastomeric bearing NR pad and S_p is the thickness of the bearing pad. For the assemblies of the fixed and the expansion steel dowels, corrosion deterioration may lead to a loss of the cross-sectional area of this component type. Then, this deterioration effect is modeled by considering the variation of parameters in hysteretic material, which includes a reduction in the yield strength and ultimate lateral strength. The time-dependent yield strength and the ultimate lateral strength can be calculated using Equations (38) and (39) [46], where f_y and f_u are the tensile strength and the ultimate shear strength of the steel dowels, respectively, and $A_d(t)$ is the time-dependent cross-sectional area of the steel dowels. The initial area of the steel dowels and dowel tensile strength, as well as the lateral strength of the bearing dowel, are assumed to be uniformly distributed and whose upper bound a and lower bound b are listed in Table 1, respectively. Thus, the yield strength $F_y(t)$ and the ultimate lateral strength $F_u(t)$ can be evaluated using the cross-sectional loss of dowel bars $A_d(t)$, as shown in Figure 11. The results indicated that a reduction in the dowels' area has a significant impact on the strength over time due to corrosion.

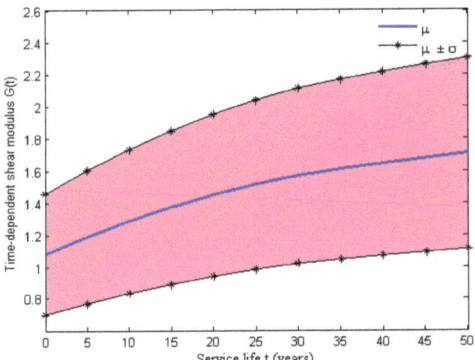

Figure 10. Time evolution of the shear modulus in a bearing NR pad.

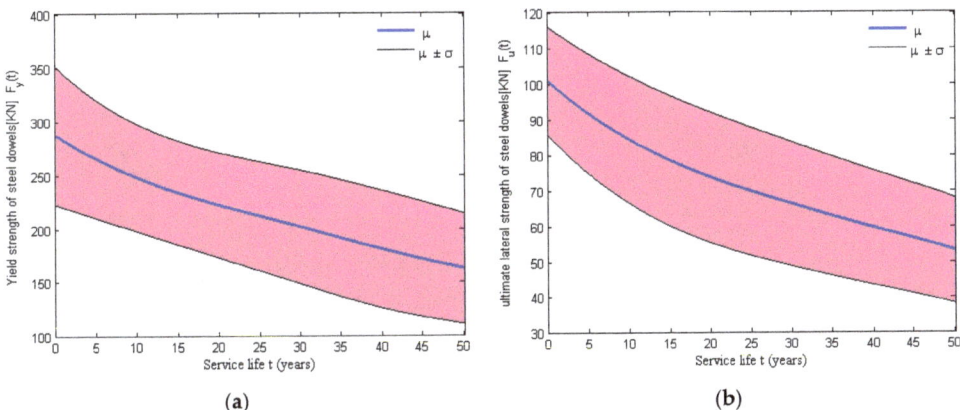

Figure 11. Time evolution of the (**a**) yield strength and (**b**) ultimate lateral strength for swedged steel dowels.

5. Impact of Corrosion on the Seismic Response of Bridge Components

All equations for Section 4 can be found in Table 5.

To analyze the impact of corrosion multi-deterioration mechanisms on the seismic fragility of the MSC concrete girder bridge, the seismic responses of the MSC bridge components, which considers the effect of the time-dependent aging, should be presented using dynamic simulations. For the geometry of this bridge type considered in a pristine state, the first two fundamental modes along longitudinal and transverse directions were 0.49 s and 0.37 s, respectively. Moreover, the foundations for the MSC bridge were assumed to be on medium hard soil represented by site class II (shear wave velocity between 500 m/s and 250 m/s), and the target response spectrum of 0.53 s was calculated by considering the site soil conditions. Then, twenty real earthquake records, whose response spectra had the greatest similarity to the target response spectrum, were selected to capture the uncertainty of ground motions. Thus, the dynamic responses of the bridge were illustrated through nonlinear time history analysis with the above twenty seismic inputs. The impact of corrosion multi-deterioration mechanisms on the evolving dynamic response of the bridge is presented using probabilistic seismic demand models (PSDMs) of aging components (Equation (42)). The influence of multi-deterioration mechanisms of a single component and the joint effects of both column and bearing on the seismic demands are compared in the following sections.

Table 5. Equation list for Sections 4 and 5.

Equation Number	Equation Expression	Equation Number	Equation Expression
Equation (40)	$u_\theta = \frac{\theta_m}{\theta_y}$	Equation (49)	$T_p = f_p A_p(\eta)$
Equation (41)	$P_f(t) = P[D(t) \geq C(t)|IM]$	Equation (50)	$\rho = 4\pi \left(R_0^2 - R_\rho^2\right)/A_p$
Equation (42)	$\ln[\hat{D}_i(t)] = \ln[a_i(t)] + b_i(t)\ln(IM)$	Equation (51)	$\begin{cases} w_c = \frac{4\pi(R_t-R_0)}{A^*+B^*} - H^* \\ A^* = (1-v_c)(R_0/R_c)^{\sqrt{a}} \\ B^* = (1+v_c)(R_c/R_0)^{\sqrt{a}} \\ H^* = 2\pi R_c f_t / E_c \end{cases}$
Equation (43)	$\beta_{Dm,i}(t) = \sqrt{\frac{\sum\left(\ln(d_{m,i}(t)) - \ln\left(a_i(t)IM_m^{b_i(t)}\right)\right)^2}{N-2}}$		
Equation (44)	$P[D_i(t) \geq d_i(t)|IM] = \Phi\left(\frac{\ln(IM)-\lambda_i(t)}{\zeta_i(t)}\right)$		
Equation (45)	$P_i(DS|IM) = \Phi\left(\frac{\ln(IM)-\gamma_i(t)}{\zeta_i(t)}\right)$		
Equation (46)	$\gamma_i(t) = \frac{\ln \hat{C}_i(t) - \ln a(t)}{b(t)}$		
Equation (47)	$\zeta_i(t) = \frac{\sqrt{\beta_{Dm,i}^2(t)+\beta_{C,i}^2(t)}}{b(t)}$		
Equation (48)	$P_{sys}(DS|IM) = \Phi\left(\frac{\ln(IM)-\ln[\gamma_{sys}(t)]}{\zeta_{sys}(t)}\right)$		

5.1. The Impact of Deterioration on the Seismic Demand of RC Columns

As elaborated on earlier, the aging of RC columns is modeled via a relationship between stress and strain, which includes multiple deterioration mechanisms. There is a great shift in the demands of the aging RC columns due to the loss of both the steel area (i.e., longitudinal bars, stirrups) and concrete cover when the bridge experiences seismic loading. Then, the seismic demand placed on the column can be expressed using Equation (40), where u_θ is the curvature ductility demand ratio, θ_m is the maximum curvature demand of a column under seismic loading and θ_y is the yield curvature in the column. The peak curvature ductility demands at 0 years and 50 years are shown in Figure 12. It can be seen that if only column multi-degradation was considered, it had a great impact on the column demands, which were significantly increased with aging from 0 to 50 years. Additionally, when the joint effects of both the column and bearing multi-deterioration were taken into account, the column demands at a specific time can be higher than when the multi-deterioration of an individual column component is considered. It is worth noting that the only bearing multi-deterioration had a minimal impact on the column demands. The results indicated the significance of considering the multi-deterioration mechanisms of the multiple components.

5.2. The Impact of Deterioration on the Seismic Demand of Bridge Bearings

Similarly, Figure 13 shows the influences of the multi-deterioration of the multiple components and a single component on the seismic response of the fixed bearing along the longitudinal direction. When only the multi-degradation of the fixed bearing was considered, there was a significant influence on the bearing deformation, which constantly increased from 0 to 50 years. If only column multi-degradation was considered, there was a small impact on the bearing deformation. Both the joint considerations of the column and bearing degradations did not have a higher effect on the bearing demand along the longitudinal direction than when only bearing deterioration is considered. However, Figure 14 shows the dependency between the multi-deterioration mechanisms of the column and expansion bearings in the transverse directions.

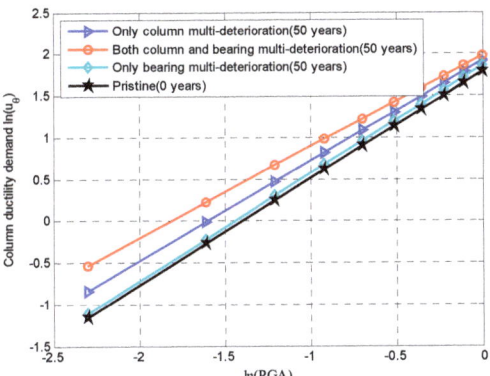

Figure 12. Mean value of the demand placed on an RC column through PSDM.

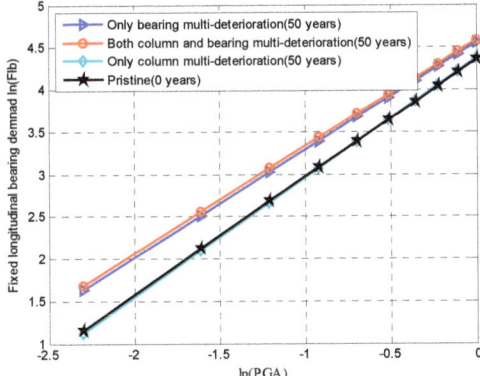

Figure 13. Mean values of the demand placed on a fixed bearing in the longitudinal direction using a PSDM.

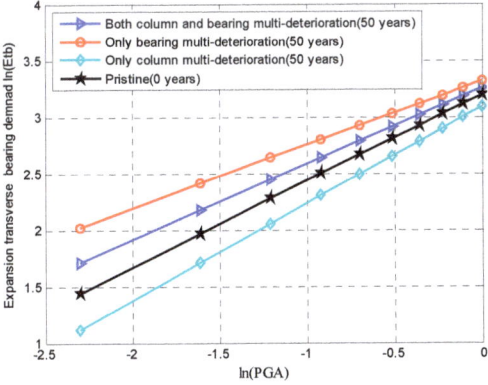

Figure 14. Mean values of the demand placed on an expansion bearing in the transverse direction using a PSDM.

If only column multi-deterioration is considered, the demands on the expansion transverse bearing at 50 years were lower than the demands on the pristine expansion bearing due to the domination of the corroded RC column. When the multi-deterioration of the expansion transverse bearing was considered, the demands on the expansion bearing at 50 years were higher than that of the pristine demands. The reason for this was that the increase in strength reduction of steel dowels was higher than the increase in the shear stiffness of the NR pad induced by thermal oxidation. The multi-deteriorations of both the column and the expansion transverse bearing had a greater impact on the bearing deformation than when only column deterioration was considered. However, the deterioration mechanisms of multiple components had a lower influence on the bearing deformation than when only bearing degradation was taken into account. The reason for this was that the individual consideration of the corrosion degradation of the expansion bearing system had a greater role on the seismic responses after 50 years of exposure to deicing salts.

6. Impact of Corrosion on the Seismic Fragility of an Aged Bridge System

All equations for Section 5 can be found in Table 5.

The impact of multi-deterioration mechanisms on the seismic performance of aged bridge components can be evaluated by developing time-dependent bridge fragility curves at a system level. Moreover, the increase in the probability of a damaged state exceedance with continuous degradation of the bridge along its service life can be quantified using such fragility curves. A methodology for the development of time-dependent fragility curves is presented in the following section using a typical case study of an aged MSC concrete girder bridge. The uncertainties in bridge modeling attributes, ground motion and deterioration parameters were considered in probabilistic terms for this case.

6.1. Time-Dependent Probabilistic Seismic Demand Models (PSDMs)

Time-dependent fragility curves represent the probability of damage exceedance of structures under earthquake excitation at different points in time throughout the service life. Such time-evolving fragility curves show the impact of multi-deterioration mechanisms on the seismic performance of aging structures at the system level. The generalized time-dependent seismic fragility function can be expressed as Equation (41), where $P_f(t)$ is the probability of damage state exceedance of a specific MSC bridge at an aging time t. $D(t)$ and $C(t)$ are the seismic demand and the capacity of an aged bridge at time t, respectively. IM is the intensity of the ground motion. Before developing the time-dependent seismic fragility curves, the relationship between the time-varying seismic demand and capacity should be established by using probabilistic seismic demand models (PSDMs). Such PSDMs for bridge critical components are constructed using nonlinear time history analysis to capture the impact of multi-deterioration mechanisms on the dynamic responses. Thus, to consider the uncertainties in the ground motion and structural parameters, a total of 20 real ground motions from PEER were used in the analysis and an equal number of random bridge samples in pristine states were generated through Latin hypercube sampling. The peak ground acceleration (PGA) of the 20 random samples was modulated ranging from 0.095 g to 1.05 g. In addition, to generate 20 random aging bridge samples, the uncertainties in the corrosion parameters that affect the multi-deterioration mechanisms of the column and bearing components were also taken into account in the finite element modeling.

To develop probabilistic seismic demand models for bridge components at different points in time (e.g., 0, 25 and 50 years), the 20 random bridge samples for each time instant at a specific intensity level of the ground motion were performed using nonlinear time history analysis. Then, probabilistic seismic demand models, which reflect the relationship between the peak demands of aged bridge components and ground motion intensity, were developed using linear regression. The demands considered in this research included RC columns and fixed and expansion elastomeric bearings. Consequently, the time-dependent probabilistic seismic demand models can be expressed as Equation (42) [31], where $\hat{D}_i(t)$ is the time-dependent median value of the seismic demand for bridge component i, $a_i(t)$ and $b_i(t)$ are the linear regression parameters, and IM is the intensity of ground motion.

A logarithmic seismic demand for the specific component $\ln D_i(t)$ was assumed to be normally distributed, and then $\ln D_i(t)$ and $\beta_{Dm,i}(t)$ were the median value and dispersion, respectively. $\beta_{Dm,i}(t)$ can be expressed as Equation (43), where $d_m(t)$ is the mth peak seismic demand for bridge component i and IM_m is the mth PGA. Then, the time-evolving PSDMs can also be expressed by using Equation (44), where $\lambda_i(t)$ is the natural logarithm of the seismic demand related to the median value of PGA and $\lambda_i(t) = (\ln d_i(t) - \ln a_i(t))/b_i(t)$; $\zeta_i(t)$ is the lognormal standard deviation and $\zeta_i(t) = \beta_{Dm,i}(t)/b_i(t)$.

6.2. Time-Dependent Seismic Fragility of an Aged Bridge System

In addition to probabilistic seismic demand models, the structural capacities of different bridge components should also be estimated during time-dependent seismic fragility analysis. The limit state capacities of different components considered in this study are lognormal and presented in Table 6 [31]. When the seismic demands and the limit state capacities for different bridge components are assumed to be lognormally distributed, the time-dependent seismic fragility at the component level can be obtained using Equation (45), where $\gamma_i(t)$ and $\zeta_i(t)$ are median values (in units of g PGA) and logarithmic standard deviations of the ith component fragility, respectively. $\gamma_i(t)$ and $\zeta_i(t)$ can be expressed as Equations (46) and (47), where $\hat{C}_i(t)$ and $\beta_{C,i}(t)$ are the median and dispersion of the ith component capacity, respectively, and $\beta_{Dm,i}(t)$ is the dispersion of the ith component demand.

Table 6. Capacity limit state for different bridge components for an MSC concrete girder bridge.

Bridge Component	Slight		Moderate		Extensive		Complete	
	Med.	Disp.	Med.	Disp.	Med.	Disp.	Med.	Disp.
RC columns	1.29	0.59	2.10	0.51	3.52	0.64	5.24	0.65
Fixed bearing—longitudinal	28.9	0.60	104.2	0.55	136.1	0.59	186.6	0.65
Fixed bearing—transverse	28.8	0.79	90.9	0.68	142.2	0.73	195.0	0.66
Expansion bearing—longitudinal	28.9	0.60	104.2	0.55	136.1	0.59	186.6	0.65
Expansion bearing—transverse	28.8	0.79	90.9	0.68	142.2	0.73	195.0	0.66

The assessment of bridge system vulnerability is performed by assuming the bridge as a series system, as presented by Nielson [47,48]. The demands of the bridge components under seismic loading are considered dependent and then the correlation coefficient between the peak responses can be estimated by constructing a joint probability density function (JPDF) for component demands. The generalized formula for the aged bridge system fragility can be derived using joint probabilistic seismic demand models (JPSDMs) [47,48]. The JPSDMs can be written as Equation (48), where $\gamma_{sys}(t)$ and $\zeta_{sys}(t)$ are the median values (in units of g PGA) and logarithmic standard deviations of the system fragility at different points in time, respectively. Solutions to Equation (48) can be directly calculated by using Monte Carlo analysis.

6.3. Time-Dependent Seismic Fragilities of Aged Bridge System

The joint impacts of the multi-deterioration of multiple components on system vulnerabilities are quantified by developing time-dependent fragility curves of the overall bridge. Figure 15 shows the aging bridge system fragility curves at three different points in time (e.g., 0, 25 and 50 years) for moderate and complete states. It can be clearly seen that the seismic fragility curves at the system level increased steadily with age from 0 to 50 years. However, the seismic fragility of the individual components tended to show a reduction in vulnerability when the bridge component continued to be corroded due to the individual consideration of multi-deterioration for a single component. Furthermore, the seismic fragility of the bridge expansion bearing along the transverse direction can decrease at the initial time (e.g., 0, 15 and 25 years) due to the added stiffness of the bearing NR pads induced by thermal oxidation. The results indicated the importance of considering the joint effects of multi-deterioration mechanisms for multiple bridge components. Furthermore, it was found that the multi-deterioration of the aging MSC concrete girder bridge had a potentially negative influence on the overall seismic fragility at the system level.

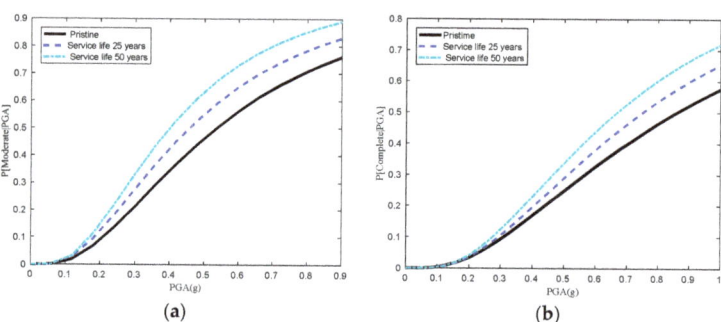

Figure 15. Time-dependent system seismic fragility curves for an MSC concrete girder bridge for (**a**) a moderate damage state and (**b**) a complete damage state under multi-deterioration of multiple components.

6.4. Time-Dependent Seismic Fragilities Considering Prestress Losses and Cracking

A reference model for the corrosion-induced presetress loss was developed [20,21] by considering the effects of concrete cracking. Corrosion-induced prestress loss can be modeled as the difference between the effective prestress in an uncorroded strand and that in the corroded strand [49]. Here, it was suggested that the strain compatibility and force equilibrium equations in concrete bridge-girders could also be used to evaluate the effective prestress in the corroded strand. If the pre-stressing force in the pre-tensioned concrete bridge-girders is released, then the strand pre-stress would transfer to the concrete via the bonding stress at the stand-concrete interface. Then, the effective prestress in the corroded strand can be calculated using Equation (49) [49], where T_p represents the tension force of the corroded strand and $A_p(\eta)$ denotes the residual cross-sectional area of the corroded strand.

During the corrosion process, RC components of bridges also suffer from the prestress and the expansive pressure. When the tensile stress induced by the expansive pressure exceeds the concrete tensile strength, the concrete is considered to be cracking. The concrete cover can contain a cracked inner region and an uncracked outer region. Here, the outer wires can be considered to be first corroded when the strand suffers from corrosion. The corrosion loss of a strand can be expressed as Equation (50) [50], where R_0 and R_ρ are the radiuses of wire before and after corrosion, respectively, and A_p is the strand cross-sectional area. Meanwhile, it is assumed that the smeared cracks in the cracked region are distributed uniformly, and then a reduction factor can be used to reflect the residual tangential stiffness in the cracked concrete. Consequently, by combining stress equilibrium equations with the strain compatibilities, the crack width on the concrete surface can be written as Equation (51) [50], where R_t is the radius of wire with corrosion products; $R_c = R_0 + C$, where C is the concrete cover; $\nu_c = \sqrt{\nu_1 \nu_2}$, where ν_1 and ν_2 are the Poisson ratios of concrete in the radial and tangential directions, respectively; a is a reduction factor; f_t is the concrete tensile strength under the biaxial stress state; and E_c is the elastic modulus of concrete.

Subsequently, the combined effects of the prestress losses and cracking on the seismic fragilities of bridge systems can be quantified by using Equation (48), which also includes the effects of the multi-deterioration of multiple components. Figure 16 shows the aging bridge system fragility curves at three different points in time (e.g., 0, 25 and 50 years) for moderate and complete states. It can be derived from Figures 15 and 16 that the system fragility when considering the combined effects of prestress losses and cracking was more conservative than that without considering these effects. The results indicated that the combined effects of the prestress losses and cracking should not be neglected when performing the seismic fragility of aging bridge systems.

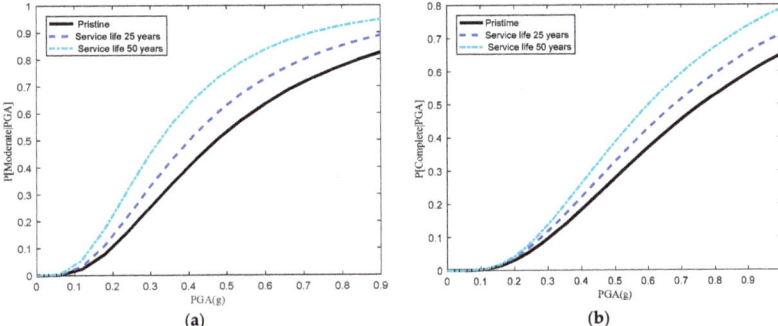

Figure 16. Time-dependent system seismic fragility curves for an MSC concrete girder bridge for (a) a moderate damage state and (b) a complete damage state subjected to prestress losses and cracking.

Finally, it is stressed that other time-variant approaches, such as the empirical experiment and trial-and-error [51–53], for the corrosion process of aging bridge systems are very expensive and very unrealistic when it comes to obtaining a series of multi-deterioration mechanisms among multiple components of aging bridge systems. Moreover, the empirical experiment and trial-and-error method are very resource-consuming when dealing with the time-dependent system fragility of such complex aging bridges when undergoing the earthquake excitation test. Therefore, when we would like to efficiently develop the overall time-variant deterioration process of aging bridge systems, the finite element modeling combined with the theoretical modeling of the corrosion process is the best choice for performing the time-dependent overall seismic fragility curves of aging bridge systems. The applicability of different approaches can be summarized in Table 7.

Table 7. Applicability of different approaches.

Different Approaches	Applicability
Theoretical modeling of the corrosion process combined with finite element modeling	1. Deals with multi-deterioration mechanisms among multiple components (e.g., RC columns, bearing systems); 2. Deals with the deterioration process of prestress loss and cracking; 3. Develops the time-dependent overall seismic fragility of aging bridge systems; 4. Updates the system fragility assessment of a bridge with real-time monitoring data.
Empirical experiment and trial-and-error method [51–53]	1. It is very suitable for a single component's corrosion process; 2. It is time-consuming for time-dependent overall fragility of aging bridge systems and the earthquake excitation test; 3. It is very expensive and very unrealistic to simultaneously obtain a series of multi-deterioration mechanisms, prestress loss and cracking among multiple components (e.g., RC columns, bearing systems) of aging bridge systems.

7. Conclusions

This paper provides a probabilistic method for identifying the time-dependent fragility of an aging bridge system under earthquake events by considering the impacts of multi-deterioration of multiple bridge components, as well as the combined effects of the prestress losses and cracking. A typical MSC concrete girder bridge is used to evaluate the seismic performance by considering the uncertainty models in structural material and corrosion parameters. Chlorine corrosion from deicing salts, which are widely utilized across northern China, can be considered the cause of the degradation and aging of bridge systems. Here, the multi-deterioration mechanisms include the corrosion deterioration of an RC column due to loss of both the steel area (i.e., longitudinal bars, stirrups) and concrete covers, and the deterioration of elastomeric bearing assemblies due to the reduction in both the shear stiffness of bearing NR pads and steel dowel area. The multi-deterioration mechanisms of RC components of the bridge systems affect the lateral force resistance and seismic responses. In addition, the models for corrosion-induced prestress loss and cracking are introduced to develop time-dependent seismic fragilities of bridge systems. The overall seismic demands found using PSDMs showed a steady increase as the growing service life when joint effects of multi-deterioration of bridge multiple components (e.g., RC columns and fixed bearings) were considered. However, there was a decrease in the demand (e.g., bearing deformation) on other multiple components (e.g., RC columns and expansion transverse bearings).

Time-variant PSDMs and the components' capacities were constructed for calculating the seismic fragility of the MSC concrete bridges, which considered the variability in attributes of bridges, ground motion and degraded parameters. The seismic fragility curves at the system level demonstrated a steady increase over time due to corrosion aging. However, individual components, such as expansion bearings in the transverse direction, revealed a decreased vulnerability at the initial time due to the added bearing pad shear stiffness. Subsequently, the dominant effect of the loss of steel dowel areas led to an increase in vulnerability. Moreover, the system seismic fragility considering the combined effects of prestress losses and cracking was more conservative. That meant the effects of both prestress losses and cracking should not be ignored when investigating the seismic fragility of the aging bridge systems. All these system fragility curves presented a more authentic estimation of the seismic vulnerability of aging bridges, and they offered a more accurate basis for life-cycle cost analysis.

Author Contributions: Data curation, W.Z.; Funding acquisition, X.L.; Investigation, P.S. and M.L.; Resources, P.S.; Visualization, W.Z.; Writing—original draft, X.L.; Writing—review & editing, M.L. All authors have read and agreed to the published version of the manuscript.

Funding: This research was funded by National Natural Science Foundation of China, grant numbers 51801149 and 11802224.

Institutional Review Board Statement: Not applicable.

Informed Consent Statement: Not applicable.

Data Availability Statement: No data were used to support this study.

Conflicts of Interest: The authors declare no conflict of interest.

References

1. Choe, D.-E.; Gardoni, P.; Rosowsky, D.; Haukaas, T. Seismic fragility estimates for reinforced concrete bridges subject to corrosion. *Struct. Saf.* **2009**, *31*, 275–283. [CrossRef]
2. Choe, D.-E.; Gardoni, P.; Rosowsky, D.; Haukaas, T. Probabilistic capacity models and seismic fragility estimates for RC columns subject to corrosion. *Reliab. Eng. Syst. Saf.* **2008**, *93*, 383–393. [CrossRef]
3. Li, J.; Gong, J.; Wang, L. Seismic behavior of corrosion-damaged reinforced concrete columns strengthened using combined carbon fiber-reinforced polymer and steel jacket. *Constr. Build. Mater.* **2009**, *23*, 2653–2663. [CrossRef]
4. Alipour, A.; Shafei, B.; Shinozuka, M. Performance evaluation of deteriorating highway bridges located in high seismic areas. *J. Bridge Eng.* **2011**, *16*, 597–611. [CrossRef]
5. Liu, M.; Cheng, X.; Li, X.; Yue, P.; Li, J. Corrosion behavior and durability of low-alloy steel rebars in marine environment. *J. Mater. Eng. Perform.* **2016**, *25*, 4967–4979. [CrossRef]
6. Li, T.; Lin, J.; Liu, J. Analysis of time-dependent seismic fragility of the offshore bridge under the action of scour and chloride ion corrosion. *Structures* **2020**, *28*, 1785–1801. [CrossRef]

7. Cui, F.; Li, H.; Dong, X.; Wang, B.; Li, J.; Xue, H.; Qi, M. Improved time-dependent seismic fragility estimates for deteriorating RC bridge substructures exposed to chloride attack. *Adv. Struct. Eng.* **2021**, *24*, 437–452. [CrossRef]
8. Li, H.; Li, L.; Zhou, G.; Xu, L. Time-dependent seismic fragility assessment for aging highway bridges subject to non-uniform chloride-induced corrosion. *J. Earthq. Eng.* **2020**, *26*, 3523–3553. [CrossRef]
9. Abbasi, M.; Moustafa, M.A. Time-dependent seismic fragilities of older and newly designed multi-frame reinforced concrete box-girder bridges in California. *Earthq. Spectra* **2019**, *35*, 233–266. [CrossRef]
10. Dey, A.; Sil, A. Advanced corrosion-rate model for comprehensive seismic fragility assessment of chloride affected RC bridges located in the coastal region of india. *Structures* **2021**, *34*, 947–963. [CrossRef]
11. Fu, P.; Li, X.; Xu, L.; Xin, L. Life-cycle seismic damage identification and components damage sequences prediction for cable-stayed bridge based on fragility analyses. *Bull. Earthq. Eng.* **2021**, *19*, 6669–6692. [CrossRef]
12. Fan, W.; Sun, Y.; Yang, C.; Sun, W.; He, Y. Assessing the response and fragility of concrete bridges under multi-hazard effect of vessel impact and corrosion. *Eng. Struct.* **2020**, *225*, 111279. [CrossRef]
13. Reshvanlou, B.A.; Nasserasadi, K.; Ahmadi, J. Modified time-dependent model for flexural capacity assessment of corroded RC elements. *KSCE J. Civ. Eng.* **2021**, *25*, 3897–3910. [CrossRef]
14. Lu, X.; Wei, K.; He, H.; Qin, S. Life-cycle seismic fragility of a cable-stayed bridge considering chloride-induced corrosion. *Earthq. Eng. Resil.* **2022**, *1*, 60–72. [CrossRef]
15. Hu, S.; Wang, Z.; Guo, Y.; Xiao, G. Life-cycle seismic fragility assessment of existing RC bridges subject to chloride-induced corrosion in marine environment. *Adv. Civ. Eng.* **2021**, *2021*, 9640521. [CrossRef]
16. Xu, J.G.; Wu, G.; Feng, D.C.; Cotsovos, D.M.; Lu, Y. Seismic fragility analysis of shear-critical concrete columns considering corrosion induced deterioration effects. *Soil Dyn. Earthq. Eng.* **2020**, *134*, 106165. [CrossRef]
17. Li, H.; Li, L.; Zhou, G.; Xu, L. Effects of various modeling uncertainty parameters on the seismic response and seismic fragility estimates of the aging highway bridges. *Bull. Earthq. Eng.* **2020**, *18*, 6337–6373. [CrossRef]
18. Mortagi, M.; Ghosh, J. Concurrent modelling of carbonation and chloride-induced deterioration and uncertainty treatment in aging bridge fragility assessment. *Struct. Infrastruct. Eng.* **2020**, *18*, 197–218. [CrossRef]
19. Biondini, F.; Camnasio, E.; Palermo, A. Life-cycle performance of concrete bridges exposed to corrosion and seismic hazard. In Proceedings of the Structures Congress 2012, Chicago, IL, USA, 29–31 March 2012; American Society of Civil Engineers (ASCE): Reston, VA, USA, 2012; pp. 1906–1918.
20. Limongelli, M.P.; Siegert, D.; Merliot, E.; Waeytens, J.; Bourquin, F.; Vidal, R.; Le Corvec, V.; Gueguen, I.; Cottineau, L.M. Damage detection in a post tensioned concrete beam—Experimental investigation. *Eng. Struct.* **2016**, *128*, 15–25. [CrossRef]
21. Bonopera, M.; Liao, W.C.; Perceka, W. Experimental–theoretical investigation of the short-term vibration response of uncracked prestressed concrete members under long-age conditions. *Structures* **2022**, *35*, 260–273. [CrossRef]
22. Coronelli, D.; Gambarova, P. Structural assessment of corroded reinforced concrete beams: Modeling guidelines. *J. Struct. Eng.* **2004**, *130*, 1214–1224. [CrossRef]
23. Ghosh, J.; Padgett, J.E. Impact of multiple component deterioration and exposure conditions on seismic vulnerability of concrete bridges. *Earthq. Struct.* **2012**, *3*, 649–673. [CrossRef]
24. Liu, M. Effect of uniform corrosion on mechanical behavior of E690 high-strength steel lattice corrugated panel in marine environment: A finite element analysis. *Mater. Res. Express* **2021**, *8*, 66510. [CrossRef]
25. Liu, M.; Cheng, X.; Li, X.; Jin, Z.; Liu, H. Corrosion behavior of Cr modified HRB400 steel rebar in simulated concrete pore solution. *Constr. Build. Mater.* **2015**, *93*, 884–890. [CrossRef]
26. Liu, X.X.; Liu, M. Reliability model and probability analysis method for structural pitting corrosion under mechanical loading. *Anti Corros. Methods Mater.* **2019**, *66*, 529–536. [CrossRef]
27. Liu, M.; Li, J. In-situ Raman characterization of initial corrosion behavior of copper in neutral 3.5% (wt.) NaCl solution. *Materials* **2019**, *12*, 2164. [CrossRef]
28. Itoh, Y.; Gu, H.; Satoh, K.; Kutsuna, Y. Experimental investigation on ageing behaviors of rubbers used for bridge bearings. *Doboku Gakkai Ronbunshuu A* **2006**, *62*, 176–190. [CrossRef]
29. Itoh, Y.; Gu, H.S. Prediction of aging characteristics in natural rubber bearings used in bridges. *J. Bridge Eng.* **2009**, *14*, 122–128. [CrossRef]
30. Gu, H.S.; Itoh, Y. Ageing behaviour of natural rubber and high damping rubber materials used in bridge rubber bearings. *Adv. Struct. Eng.* **2010**, *13*, 1105–1113. [CrossRef]
31. Nielson, B.G. Analytical Fragility Curves for Highway Bridges in Moderate Seismic Zones. Ph.D. Thesis, Georgia Institute of Technology, Ann Arbor, MI, USA, 2005.
32. Petcherdchoo, A. Time dependent models of apparent diffusion coefficient and surface chloride for chloride transport in fly ash concrete. *Constr. Build. Mater.* **2013**, *38*, 497–507. [CrossRef]
33. Liu, M.; Cheng, X.; Li, X.; Hu, J.; Pan, Y.; Jin, Z. Indoor accelerated corrosion test and marine field test of corrosion-resistant low-alloy steel rebars. *Case Stud. Constr. Mater.* **2016**, *5*, 87–99. [CrossRef]
34. Zhang, G.; Li, B. The corrosion of stirrups and its effect on the seismic fragility of a corroded reinforced concrete (RC) column. *Risk Anal. IX* **2014**, *47*, 331.
35. Biondini, F.; Bontempi, F.; Frangopol, D.M.; Malerba, P.G. Cellular automata approach to durability analysis of concrete structures in aggressive environments. *J. Struct. Eng.* **2004**, *130*, 1724–1737. [CrossRef]

36. Apostolopoulos, C.; Papadakis, V. Consequences of steel corrosion on the ductility properties of reinforcement bar. *Constr. Build. Mater.* **2008**, *22*, 2316–2324. [CrossRef]
37. Almusallam, A.A. Effect of degree of corrosion on the properties of reinforcing steel bars. *Constr. Build. Mater.* **2001**, *15*, 361–368. [CrossRef]
38. Kobayashi, K. The seismic behavior of RC member suffering from chloride-induced corrosion. In Proceedings of the Second Fib Congress, Naples, Italy, 5–8 June 2006; pp. 5–8.
39. Niu, D. *Durability and Life Forecast of Reinforced Concrete Structure*; China Sciences Press: Beijing, China, 2003.
40. Vecchio, F.J.; Collins, M.P. The modified compression-field theory for reinforced concrete elements subjected to shear. *ACI J. Proc.* **1986**, *83*, 219–231.
41. Vidal, T.; Castel, A.; Francois, R. Analyzing crack width to predict corrosion in reinforced concrete. *Cem. Concr. Res.* **2004**, *34*, 165–174. [CrossRef]
42. Scott, B.; Park, R.; Priestley, M. Stress-strain behavior of concrete confined by overlapping hoops at low and high strain rates. *ACI J. Proc.* **1982**, *79*, 13–27.
43. Imbsen, R. *Increased Seismic Resistance of Highway Bridges Using Improved Bearing Design Concepts*; Pittsburgh University: Pittsburgh, PA, USA, 1984.
44. Aashto, L. *Bridge Design Specifications*; American Association of State Highway and Transportation Officials: Washington, DC, USA, 1998.
45. Mase, G.T.; Mase, G.E. *Continuum Mechanics for Engineers*; CRC Press: Boca Raton, FL, USA, 2010.
46. Hwang, H.; Liu, J.B.; Chiu, Y.-H. *Seismic Fragility Analysis of Highway Bridges*; The Unversity of Memphis: Memphis, TN, USA, 2001.
47. Nielson, B.G.; DesRoches, R. Seismic fragility methodology for highway bridges using a component level approach. *Earthq. Eng. Struct. Dyn.* **2007**, *36*, 823–839. [CrossRef]
48. Nielson, B.G.; DesRoches, R. Seismic performance assessment of simply supported and continuous multispan concrete girder highway bridges. *J. Bridge Eng.* **2007**, *12*, 611–620. [CrossRef]
49. Dai, L.; Bian, H.; Wang, L.; Potier-Ferry, M.; Zhang, J. Prestress loss diagnostics in pretensioned concrete structures with corrosive cracking. *J. Struct. Eng.* **2020**, *146*, 4020013. [CrossRef]
50. Wang, L.; Dai, L.; Bian, H.; Ma, Y.; Zhang, J. Concrete cracking prediction under combined prestress and strand corrosion. *Struct. Infrastruct. Eng.* **2019**, *15*, 285–295. [CrossRef]
51. Nie, J.; Braverman, J.; Hofmayer, C.; Choun, Y.S.; Kim, M.K.; Choi, I.K. *Fragility Analysis Methodology for Degraded Structures and Passive Components in Nuclear Power Plants I Llustrated Using a Condensate Storage Tank (No. KAERI/TR–4068/2010)*; Korea Atomic Energy Research Institute: Daejeon, Korea, 2010.
52. Ayazian, R.; Abdolhosseini, M.; Firouzi, A.; Li, C.Q. Reliability-based optimization of external wrapping of CFRP on reinforced concrete columns considering decayed diffusion. *Eng. Fail. Anal.* **2021**, *128*, 105592. [CrossRef]
53. Sun, H.; Burton, H.V.; Huang, H. Machine learning applications for building structural design and performance assessment: State-of-the-art review. *J. Build. Eng.* **2021**, *33*, 101816. [CrossRef]

MDPI
St. Alban-Anlage 66
4052 Basel
Switzerland
Tel. +41 61 683 77 34
Fax +41 61 302 89 18
www.mdpi.com

Materials Editorial Office
E-mail: materials@mdpi.com
www.mdpi.com/journal/materials